DATE DUE

THE ECOLOGICAL RIFT

THE ECOLOGICAL RIFT
Capitalism's War on the Earth

by John Bellamy Foster, Brett Clark,
and Richard York

MONTHLY REVIEW PRESS
New York

Library of Congress Cataloging-in-Publication Data
Foster, John Bellamy.
The ecological rift : capitalism's war on the earth / by John Bellamy Foster, Brett Clark, and Richard York.
p. cm.
ISBN 978-1-58367-218-1 (pbk.) — ISBN 978-1-58367-219-8 (cloth)
1. Capitalism. 2. Environmental degradation. 3. Climate change. I. Clark, Brett. II. York, Richard, 1971- III. Title.
HB501.F658 2010
333.7—dc22

2010035831

Monthly Review Press
146 West 29th Street, Suite 6W
New York, NY 10001
www.monthlyreview.org

5 4 3 2 1

Contents

PART THREE

Dialectical Ecology / 213

PART FOUR

Ways Out / 375

Preface

The *ecological rift* referred to in the title of this book is the rift between humanity and nature. The world is really one indivisible whole. The rift that threatens today to tear apart and destroy that whole is a product of artificial divisions within humanity, alienating us from the material-natural conditions of our existence and from succeeding generations. Our argument, in brief, is that a deep chasm has opened up in the metabolic relation between human beings and nature—a metabolism that is the basis of life itself. The source of this unparalleled crisis is the capitalist society in which we live.

Ironically, most analyses of the environmental problem today are concerned less with saving the planet or life or humanity than saving capitalism—the system at the root of our environmental problems. As Derrick Jensen and Aric McBay cogently write in *What We Leave Behind*, we live in a culture in which there is an "inversion of what is real and not real," where "dying oceans and dioxin in every mother's breast milk" are considered less real than "industrial capitalism." Hence, we are constantly led to believe that "*the end of the world* is less to be feared than *the end of industrial capitalism*. . . . When most people in this culture ask, 'How can we stop global warming?' that's not really what they are asking. They're asking, 'How can we stop global warming without significantly changing this lifestyle [or deathstyle, as some call it] that is causing global warming in the first place?' The answer is that you can't. It's a stupid, absurd, and insane question."

Jensen and McBay go on to state: "Industrial capitalism can never be sustainable. It has always destroyed the land upon which it depends for raw materials, and it always will. Until there is no land (or water, or air) for it to exploit. Or until, and this is obviously the far better option, there is no industrial capitalism."[1]

We cannot say this any better. But we can offer an analysis in this book that helps us better understand the nature of this destruction—how it came about and why it is so difficult to change—and that envisions a path (if barely perceived as yet) away from the system that is killing the planet.

We write this book principally as professional environmental sociologists. More than most academic disciplines environmental sociology has emerged in direct response to a crisis: the crisis of the earth. The discipline is now polarized between two principal approaches. One is the attempt to bend nature even more to our will, to make it conform to the necessities of our production. (This general view is even more dominant and more dangerously expressed within environmental economics.) The other involves an analysis that examines the social drivers of ecological degradation, illuminating the contradictions of the social order. This approach is a call to change human society fundamentally in the direction of ecological sustainability and social equality. The former approach is known as *ecological modernization*. The latter consists of various *radical ecologies* that challenge the *treadmill of capitalist accumulation*, with the object of generating a new relation to the earth. Our work attempts to push forward the second, more critical view.

The chapters in this book emerged mostly from previously published pieces written during the last decade or so. Each of these writings has been revised for this book, some of them extensively. The introduction, chapter 7, and chapter 16 appear here for the first time, while chapter 6 and chapter 9 are effectively new, having been transformed unrecognizably from their earlier published versions. Given that most of these chapters had their origin in separate pieces, there is a certain amount of repetition that is unavoidable in the book as a whole, particularly around the central concept of the metabolic rift. We have made an effort to lessen such repetition and those instances that remain should be viewed as variations within a consistant, overriding theme. At all times our analysis returns to the fundamental, material problem of the ecological or metabolic rift arising from the treadmill of capitalist accumulation.

We have divided the book into four distinct parts. "Part One: Capitalism and Unsustainable Development" deals primarily with the conflict between capitalist economics and environmental sustainability. "Part Two: Ecological Paradoxes" examines various paradoxes (beyond the "Paradox of Wealth" in chapter 1) associated with economic growth, technological change, and nature under the capitalist system. "Part Three: Dialectical Ecology" addresses the complex, interrelatedness of society and the environment and the critical-theoretical tools for analyzing this. "Part Four: Ways Out" focuses on the problem of ecological reform/ecological revolution. Readers looking for fully worked out solutions to the environmental crisis will not find them here. But they will, we trust, find the kind of revolutionary hope and realistic-visionary perspective that we need and must build upon to generate a just and sustainable future.

Our indebtedness to the many individuals who have helped and inspired us along the way in the process of developing the analysis of this book is deep and wide-ranging. We would like to thank the many environmental sociologists we have interacted with in the course of our research and writing, and especially the members of the Environment and Technology Section of the American Sociological Association. Among these we would especially like to acknowledge Eugene Rosa, Tom Dietz, Riley Dunlap, Andrew Jorgenson, Ariel Salleh, Kenneth Gould, Marcia Hill Gossard, and David Pellow. We wish to express our debt as well to departed comrades: Fred Buttel, Stephen Bunker, and Allan Schnaiberg. We also thank Alf Hornborg, an anthropologist in the Human Ecology Division at Lund University, for his support and keen insight. Among journal editors we are particularly grateful for the encouragement of John Jermier at *Organization & Environment*, Karen Lucas at *Theory and Society*, Servaas Storm at *Development and Change*, Linda Kalof at *Human Ecology Review*, and Victor Wallis at *Socialism and Democracy*. We would also like to thank Bertell Ollman, Anthony Smith, and Marcello Musto for their support and insights as book editors in relation to two of the chapters reprinted in revised form below.

Among Marxist ecologists, who have helped forge the way for this kind of analysis, we point to the pathbreaking work of Paul Burkett, Elmar Altvater, Ted Benton, James O'Connor, and Peter Dickens. István Mészáros has provided direct and indirect inspiration for our discussion of the social metabolic order of capital. Stephen Jay Gould has been a source of personal and intellectual inspiration for all that we have done here. We have also benefited from the support of Richard Levins and Richard Lewontin, and from their dialectical approach to biology.

At *Monthly Review*, we would like to thank John Mage, John Simon, Michael Yates, Fred Magdoff, Claude Misukiewicz, Susie Day, and Yoshie Furuhashi, together with Martin Paddio, Scott Borchert, and Spencer Sunshine at Monthly Review Press. It is only fair to say that this book and many of the ideas contained within it would never have seen the light of day without their constant encouragement, support, and, not infrequently, hard work on our behalf.

Much of what we have sought to accomplish here has been deeply affected by our work with our students and colleagues in environmental sociology at (or formerly connected with) the University of Oregon. Here we would like to acknowledge in particular Jason Moore, Diana Gildea, Rebecca Clausen, Stefano Longo, Philip Mancus, Shannon Elizabeth Bell, Christina Ergas, Thembisa Waetjen, Paul Prew, Carlos Castro, Mark Hudson, Hannah Holleman, R. Jamil Jonna, Lora Vess, Laura Earles, Tony Silvaggio, Kari Norgaard, Eric Edwards, Dan Wilson, Jason Schreiner, Ryan Wishart, and Cade Jameson. Val Burris and Joseph Fracchia, at the University of Oregon, have continually shared with us their enormous knowledge of Marxism and critical social thought.

To write about ecology obviously has meaning only if one is connected to the earth. This is invariably a shared relationship that we experience with the people closest to us. In our own cases, we share our love of nature with our partners, Carrie Ann Naumoff, Kris Shields, and Theresa Koford—each of whom, in their own life endeavors, are as much involved in issues of sustainability and justice as we are. From them, we learn constantly. We therefore dedicate *The Ecological Rift*, and the hope for humanity and the earth that it represents, to them and to all in the world like them.

August 4, 2010
Eugene, Oregon
Raleigh, North Carolina

INTRODUCTION

A Rift in Earth and Time

> The nature that preceded human history . . . no longer exists anywhere (except perhaps on a few Australian coral islands of recent origin).
>
> —KARL MARX[1]

> Planet Earth, creation, the world in which civilization developed, the world with climate patterns that we know and stable shorelines, is in imminent peril. The urgency of the situation crystallized only in the past few years. We now have clear evidence of the crisis. . . . The startling conclusion is that continued exploitation of all fossil fuels on Earth threatens not only the other millions of species on the planet but also the survival of humanity itself—and the timetable is shorter than we thought.
>
> —JAMES HANSEN[2]

The term *Anthropocene* was coined a decade ago by the Nobel Prize–winning atmospheric chemist Paul Crutzen to mark the coming to an end, around the time of the late-eighteenth-century Industrial Revolution, of the Holocene epoch in planetary history. *Holocene* literally means "New Whole." It stands for the stable, interglacial geological epoch, dating back 10,000 to 12,000 years, in which civilization arose. Anthropocene, in contrast, means "New Human." It represents a new geological epoch in which humanity has become the main driver of rapid changes in the earth system.[3] At the same time it highlights that a potentially fatal ecological rift has arisen between human beings and the earth, emanating from the conflicts and contradictions of the modern capitalist society. The planet is now dominated by a technologically potent but alienated humanity—alienated from both nature and itself; and hence ultimately destructive of everything around it. At issue is not just the sustainability of human society, but the diversity of life on Earth.

It is common today to see this ecological rift simply in terms of climate change, which given the dangers it poses and the intractable problems for capitalism it presents has grabbed all the headlines. But recently scientists—in a project led by Johan Rockström at the Stockholm Resilience Centre, and including Crutzen and the leading U.S. climatologist, James Hansen—have developed an analysis of nine "planetary boundaries" that are crucial to maintaining an earth-system environment in which humanity can exist safely. Climate change is only one of these, and the others are ocean acidification, stratospheric ozone depletion, the nitrogen and the phosphorus cycles, global freshwater use, change in land use, biodiversity loss, atmospheric aerosol loading, and chemical pollution. For the last two, atmospheric aerosol loading and chemical pollution, there are not yet adequate physical measures, but for the other seven processes clear boundaries have been designated. Three of the boundaries—those for climate change, ocean acidification, and stratospheric ozone depletion—can be regarded as tipping points, which at a certain level lead to vast qualitative changes in the earth system that would threaten to destabilize the planet, causing it to depart from the "boundaries for a healthy planet." The boundaries for the other four processes—the nitrogen and phosphorus cycles, freshwater use, change in land use, and biodiversity loss—are better viewed as signifying the onset of irreversible environmental degradation.

Three processes have already crossed their planetary boundaries: climate change, the nitrogen cycle, and biodiversity loss. Each of these can therefore be seen, in our terminology, as constituting an extreme "rift" in the planetary system. Stratospheric ozone depletion was an emerging rift in the 1990s, but is now stabilizing, even subsiding. Ocean acidification, the phosphorus cycle, global freshwater use, and land system change are all rapidly emerging global rifts, though not yet extreme. Our knowledge of these rifts can be refined, and more planetary rifts may perhaps be discovered in the future. Nevertheless, the analysis of planetary boundaries and rifts, as they present themselves today, helps us understand the full scale of the ecological crisis now confronting humanity. The simple point is that the planet is being assaulted on many fronts as the result of human-generated changes in the global environment.[4]

In the planetary boundaries model developed by Rockström and his associates, each ecological process has a preindustrial value (that is, the level reached before the advent of industrial capitalism), a proposed boundary, and a current status. In the case of climate change the preindustrial value was 280 parts per million (ppm) of carbon dioxide concentration in the atmosphere; its proposed boundary is 350 ppm (necessary if tipping points such as a catastrophic rise in sea level are to be avoided); and its current status is 390 ppm. Biodiversity loss is measured by the rate of extinction (number of species lost per million species per year). The preindustrial annual rate, referred to as the "natural" or "background" rate of species loss, was 0.1–1 per million; the proposed boundary is 10 per million; whereas the current rate is greater than 100 per million (100–1,000 times the preindustrial background rate). With respect to the nitrogen cycle, the boundary is concerned with the amount of nitrogen removed from the atmosphere for human use in millions of tons per year. Before the rise of industrial capitalism (more specifically before the discovery of the Haber-Bosch process early in the twentieth century), the amount of nitrogen removed from the atmosphere was 0 tons. The proposed boundary, to avoid irreversible degradation of the earth system, is 35 million tons per year. The current status is 121 million tons per year.

In each of these extreme rifts, the stability of the earth system as we know it is being endangered. We are at red alert status. If business as usual continues, the world is headed within the next few decades for major tipping points along with irreversible environmental degradation, threatening much of humanity. Biodiversity loss at current and projected rates could result in the loss of upward of a third of all living species this century. The pumping of more and more nitrogen into the biosphere is resulting in the creation of dead zones in lakes and ocean regions (a phenomenon also affected by phosphorus). Each one of these rifts by itself constitutes a global ecological crisis. These ruptures reveal that the limits of the earth system are not determined by the sheer physical scale of the economy but by the particular rifts in natural processes that are generated.[5]

The emerging rifts in the other ecological processes, which have not yet overshot their boundaries, are scarcely less threatening. For the phosphorus cycle (categorized as part of a single planetary boundary together with the nitrogen cycle), the preindustrial quantity flowing into the oceans per year was approximately 1 million tons; the proposed boundary is 11 million tons (based on the assumption that ocean anoxic events begin at ten times the background rate); and its current status is already 8.5 to 9.5 million tons. In regard to ocean acidification, the value refers to a global mean saturation state of aragonite (a form of calcium carbonate) in surface seawater. A decline in the number indicates an increase in the acidity of the ocean. The preindustrial value was 3.44 (surface ocean aragonite saturation state); the proposed boundary—after which there would be a massive die-down of shell-forming organisms—is 2.75; and the current status is 2.90. In the case of freshwater use, the preindustrial annual consumption of freshwater in km^3 (cubic kilometers) was 415; the estimated boundary is 4,000 km^3 (marking a threshold beyond which the irreversible degradation and collapse of terrestrial and aquatic ecosystems is likely); and the current rate of consumption is 2,600 km^3. For change in land use, the parameters are set by the percentage of global ice-free land surface converted to cropland. In preindustrial times, this percentage was very low. The proposed boundary is 15 percent (after which there is the danger of triggering catastrophic effects on ecosystems), and the current status is 11.7 percent. In each of these emerging rifts, we are faced with an orange alert status, in which we are rapidly moving toward extreme conditions, whereby we will pass the planetary boundaries, undermining the earth system that supports the conditions of life.

No measure for chemical pollution has yet been determined, but proposals include measuring the effects of persistent organic pollutants (otherwise known as POPs), plastics, endocrine disruptors, heavy metals, and nuclear waste on ecosystems and the earth system in general. Likewise, no measure has yet been determined for atmospheric aerosol loading (the overall particulate concentration in the atmosphere on a regional basis), which can disrupt monsoon systems, lead to health problems, and interact with climate change and freshwater boundaries.

Stratospheric ozone depletion is the one previously emerging rift that was brought under control (as far as anthropogenic drivers were concerned) in the 1990s, reducing what was a rapidly growing threat to life on the planet due to an increase in ultraviolet radiation from the sun. The preindustrial value of ozone concentration was 290 (Dobson Units—the measurement of atmospheric ozone columnar density, where 1 Dobson Unit is defined as 0.01 millimeters thick under standard pressure and temperature); the proposed planetary boundary is a concentration of 276 (after which life on the planet would experience devastating losses); and the current status is 283. Between 60°S and 60°N latitude, the decline in stratospheric ozone concentrations has been halted. Nevertheless, it will take decades for the Antarctic ozone hole to disappear, and Arctic ozone loss will likely persist for decades. Life on the planet had a close call.[6]

The mapping out of planetary boundaries in this way gives us a better sense of the real threat to the earth system. Although in recent years the environmental threat has come to be seen by many as simply a question of climate change, protecting the planet requires that we attend to all of these planetary boundaries, and others not yet determined. The essential problem is the unavoidable fact that an expanding economic system is placing additional burdens on a fixed earth system to the point of planetary overload. It has been estimated that in the early 1960s humanity used half of the planet's biocapacity in a year. Today this has risen to an overshoot of 30 percent beyond the earth's regenerative capacity. Business-as-usual projections point to a state in which the ecological footprint of humanity will be equivalent to the regenerative capacity of two planets by the mid-2030s.[7]

Rockström and his associates concluded their article in *Nature* by stating: "The evidence so far suggests that, as along as the [planetary boundary] thresholds are not crossed, humanity has the freedom to pursue long-term social and economic development." Although this is undoubtedly true, what is obviously not addressed in this conclusion—but is clearly the point of their whole analysis—is that these thresholds have in some cases already been crossed and in other cases will soon be crossed with the continuation of business as usual. Moreover, this can be attributed in each and every case to a primary cause: the current pattern of global socioeconomic development, that is, the capitalist mode of production and its expansionary tendencies. The whole problem can be called "the global ecological rift," referring to the overall break in the human relation to nature arising from an alienated system of capital accumulation without end.

All of this suggests that the use of the term Anthropocene to describe a new geological epoch, displacing the Holocene, is both a description of a new burden falling on humanity and a recognition of an immense crisis—a potential terminal event in geological evolution that could destroy the world as we know it. On the one hand, there has been a great acceleration of the human impact on the planetary system since the Industrial Revolution, and particularly since 1945—to the point that biogeochemical cycles, the atmosphere, the ocean, and the earth system as a whole, can no longer be seen as largely impervious to the human economy. On the other hand, the current course on which the world is headed could be described not so much as the appearance of a stable new geological epoch (the Anthropocene), as an end-Holocene, or more ominously, end-Quarternary, terminal event, which is a way of referring to the mass extinctions that often separate geological eras. Planetary boundaries and tipping points, leading to the irreversible degradation of the conditions of life on

Earth, may soon be reached, science tells us, with a continuation of today's business as usual. The Anthropocene may be the shortest flicker in geological time, soon snuffed out.

The Ecological Crisis of Social Science

If natural science is now posing such serious questions about the continuation of life as we know it, what is the role of social science? Should it not be helping us to understand how humanity, by radically changing its system of social and economic production, which today is clearly the chief cause of the problem, might respond to this massive threat? Unfortunately, the role of social science in this respect is paradoxical. Tragically, the more pressing the environmental problem has become and the more urgent the call for ecological revolution has been articulated, the more quiescent social scientists seem to have become on the topic, searching for a kind of remediation of the problem, in which real change will not be required. Although thirty years ago it was common to find challenges to the capitalist exploitation of the environment emanating from social scientists who were then on the environmentalist fringe, today the main thrust of environmental social science has shifted to ecological modernization—a managerial approach that sees sustainable technology, sustainable consumption, and market-based solutions (indeed "sustainable capitalism") as providing the answers.[8] Here, social scientists parallel the stance of mainstream environmental technocrats—such as Thomas Friedman, Fred Krupp of the Environmental Defense Fund, Ted Nordhaus and Michael Shellenberger from the Breakthrough Institute, as well as Newt Gingrich—who propose that a green industrial revolution, rooted in technological innovation and efficiency, will produce a green society. For this group, new "green markets" will enhance economic growth, which remains the real objective.[9]

Thus as natural scientists have become more concerned about the detrimental effects of the economic system on the environment, and correspondingly radicalized, asking more and more root questions, social scientists have increasingly turned to the existing economic system as the answer. Indeed, it is no longer surprising to see a major European social scientist, such as Ulrich Beck (the originator of the "risk society" concept), writing of "the global consensus on climate protection that is now within reach [and which] is also creating new markets.... Under a regime of 'green capitalism' composed of transnationally structured ecological enforced markets, ecology no longer represents a hindrance to the economy. Rather, the opposite holds: ecology and climate protection could soon represent a direct route to profits." For Beck, capitalism—removed from its reality as a system of capital accumulation without end—can be seen as fully compatible with sustaining the earth, in sharp opposition to decades of green analysis that argued precisely the opposite. Likewise environmental sociologist Arthur Mol points to what he calls "promising developments and prospects in the taming of transnational capitalism" offered by "the European Union and to a lesser extent NAFTA," which as supra-national power structures are said to be in a position "to counteract the environmental side effects of global capitalism."[10]

How do we explain this growing quiescence of social science with respect to environmental problems (explicitly including environmental social science) even as the problem itself, as natural scientists insist, is rapidly accelerating? Answering this question requires that we look at some of the persistent weaknesses that permeate social science, and how this relates to the ecological crisis specifically. Social science has been in many ways hamstrung in our society precisely because its object is the *social*, and hence both its analysis and what is deemed acceptable/unacceptable tends to be filtered through the dominant institutions and structures of the prevailing hierarchical social order. The stagnation that has so often characterized contemporary social science is thus in many ways a built-in function of the system's commitment to stasis in its fundamental social-property relations. Social scientists have often displayed amazing, ingenious techniques for getting around this problem and raising critical ideas despite the limitations imposed by the hegemonic culture. Important observations and discoveries are made. But more often than not, such challenges are directed at what from the standpoint of the social system as a whole are marginal issues, thereby more readily tolerated. Where social science goes beyond this and addresses the problem of power head on, most of these contributions, no matter how singular, end up being treated as isolated discoveries, which, lacking any meaningful relation to the dominant social practice, are quickly forgotten.

In the mid-twentieth century, leading British scientist and Marxist social critic, J. D. Bernal, provided a useful starting point for a discussion of the weaknesses of social science in a class society, in his monumental book *Science in History* (1954). Considering some of the reasons commonly given for what he called the "backwardness" of the social sciences in relation to the natural sciences in the twentieth century, Bernal dismissed two of these as illusory: (1) the supposed impossibility of experimentation in the social sciences; and (2) the notion that the advance of social science is seriously inhibited by the fact that it involves value judgments. He also dismissed three other commonly offered reasons as of very limited explanatory power, in terms of explaining the failures of social science: (1) the reflexive nature of the social sciences, whereby human beings are both the subject and object of study; (2) the sheer complexity of human society, viewed as much more than a mere aggregation of the complex psychological attributes of its members; and (3) the changing nature of society, which does not allow for fixed laws. For Bernal, these factors contributed to the distinctiveness of the social sciences but did not significantly block their advance or explain their underdevelopment.

Rather, the real obstacles facing the social sciences, he suggested, could be attributed almost entirely to the fact that they were seriously circumscribed by and often directly subservient to the established order of power, and specifically to the dominant social/property relations (in ways only indirectly applicable to the natural sciences). Despite important advances and revolutionary developments, social science in "normal times" has been more about maintaining/managing a given social order than encouraging the historical changes necessary to human society, where social capacities and challenges keep evolving. The imperative of those in power to maintain hegemonic control was so compelling, Bernal explained, that Plato "deliberately constructed myths instead of rational explanations for the common people in the *Republic*," as a means of defending his ideal aristocratic order. Today similar ideological means, although not always so deliberate, are employed. "In short," Bernal wrote, "the backwardness and emptiness of the social sciences are due to the overriding reason that in all class societies they are inevitably corrupt."[11]

Social science thus often enters a relatively dormant state once a new system of power is established. A new class-social order, once it surpasses its initial revolutionary stage and consolidates itself, demands nothing so much as "the bad conscience and evil intent of apologetics"—since the main goal from then on is to maintain its position of power/hegemony.[12] Circumscribed in this way by the power structure, social science in normal—non-revolutionary—periods is unable to develop in a rational direction that would allow knowledge to interact in meaningful ways with social practice, particularly of a democratic kind.

The "corruption" referred to here is clearly not about such petty academic crimes as plagiarism, falsification of data, and being coopted directly by private interests—all of which are common enough—but the much more pervasive problem of the widespread capitulation to the status quo, and evasion of all alternative perspectives, even at the cost of the abandonment of rational analysis and meaningful social practice. Getting ahead in the academy (as well as in the media, the government, and other places in which social scientists are to be found) all too often involves self-censorship, a narrow focus on the relatively inconsequential, and leaving the big stuff—in terms of social change—off the table. Hence, "social science becomes an accumulation of harmless platitudes with disconnected empirical additions."[13] The more powerful a set of insights offered by a given social theory, say, Marxian theory—or the more penetrating insights of particular thinkers, say, Rousseau, Hegel, Weber, Veblen, Schumpeter, and Keynes—the more likely they are to be discarded in essence, to be winnowed down or bastardized and replaced by frameworks more conducive to the mere perpetuation of the status quo. A good example of this is the rise of what Joan Robinson famously called "bastard

Keynesianism," better known as the "neoclassical synthesis," in economics after the Second World War, in which all the main elements of the Keynesian revolution were discarded, since they were too threatening to the established order.[14]

It is no accident, then, that the greatest achievements in social science have occurred during periods of social disruption. Indeed, the social sciences as we know them today are largely a product of the bourgeois revolutions in Europe from the seventeenth to the nineteenth centuries, and are associated with the birth of capitalism. This period gave rise to both modern liberal and modern socialist conceptions. By the twentieth century, however, social science was increasingly consolidated and entrenched as part of a relatively stable period of global capitalist hegemony. Although there were revolutionary challenges to the system, emanating primarily from the periphery rather than the core of the world system, the hegemonic social science within the system's center remained largely untouched—despite its growing disconnection from world trends. Mainstream social science was therefore marked by a number of contradictory tendencies: (1) its increasingly static, ahistorical character; (2) its reductionism and abstract empiricism (mimicking in crude versions of positivism the worst tendencies of natural science); (3) its increasing relativism, culturalism, and irrationalism (in the form of a complementary anti-historical tradition emanating from the humanities); and (4) its anti-naturalism (in the sense of the divorce from the wider context of existence, the natural prerequisites of life).[15]

More and more the human sciences have taken on the dominant attitudes and incapacities of "liberal practicality," involving scattered attention to "innumerable factors" of individual milieus. The result is an inability to conceive of adequate causes, which are invariably structural, operating "behind the backs" of individuals. In its most extreme forms, such a distorted approach to social science leads to an abandonment of any pretense to realism altogether, and hence an emphasis not so much on innumerable factors as on innumerable discourses and cultural constructions.[16] The result is the kind of confused "pure empiricism," in which, as Hegel wrote, "everything has equal rights with everything else; one characteristic is as real as another, and none has precedence."[17]

Although these various proclivities of contemporary social science are seemingly diverse and often appear diametrically opposed to one another, they share an incapacity to connect a critical historical perspective on human society with the forms of social practice necessary to carry out meaningful social change. The result is an effectual capitulation to the status quo. The space of real action envisioned within mainstream social science (insofar as such action remains an object) is relegated to a kind of "pragmatic" managerialism appropriate to a bureaucratic ethos—a philosophy of simply "making do" with all the main parameters of society predetermined.

The Dehistoricization of Society

"Nothing in biology," the geneticist Theodosius Dobzhansky famously declared, "makes sense except in the light of evolution."[18] Similarly, little or nothing in human society makes sense except in the light of history. It is therefore telling that the most important feature distinguishing establishment social science/ideology from its more radical counterparts is its attempt to depict *the present as non-history*, particularly where society's productive forms and relations are concerned.[19] For the former, historical development may have occurred in the past but only in the sense that it led to the present as the *end of history*, that is, the culmination of historical development. The ideas that characterize the present era are removed from their relation to material existence and generalized as the universal ideas of all human community—or the ones to which it was always headed up to this point—so as to allow for no real, fundamental improvement.[20] Francis Fukuyama succinctly summed up this general tendency by describing contemporary liberalism as the era of *The End of History and the Last Man*, in which there are no more revolutions to be expected in social relations, only marginal improvements.[21] Social science thus loses, with respect to the analysis of the present, what John Stuart Mill characterized as its primary purpose. "The fundamental problem . . . of the social science," he wrote, "is to find the [historical] laws according to which any state of society produces the state which succeeds it and takes its place."[22] Since the future will be essentially like the present, this "fundamental problem" no longer exists. The systematic disregard of historical influences, required by the hegemonic culture, cripples the social sciences in nearly every respect.

Social scientists of a more positivist bent, as in economics, search for the bases of social existence, but do so in a context in which history is largely negated, in conformity with the view that the present is non-history. As a result, there is no real attention to the *historical specificity* of present conditions or the contradictory and transitory relations in which we live.[23] A narrow spectrum of time in which social conditions have seemed to be relatively stable is frequently translated into a set of permanent conditions, and consequently these conditions/parameters disappear from the analysis since they are, in effect, naturalized. Factors in a historical era that are constant for long periods of time can be rationally labeled "background conditions." Because of their ubiquity and constancy they become, like gravity, invisible to most empirical analyses. We can say that such "background conditions" act like "social gravity" or law-like emergent social properties particular to a specific historical era. They influence all aspects of society while often being unnoticed by social observers. In "conventional economic analysis," as Fred Block observes, fundamental "social relations are relegated to the category of background factors, which are assumed to remain constant over time," and thus remain unanalyzed, even though they constitute the foundations of social life.[24] This is the proverbial problem of the fish that cannot see the water that surrounds and permeates its existence. It is not recognized that these fundamental conditions can change: the water can be polluted, the stream can be drained.[25]

Indeed, the historical specificity that defines the present means that conditions, which are commonly seen as simply given, as constituting a kind of permanent equilibrium, can in fact be abruptly transformed. Often the failure to see this is due to too short a historical reference point or lack of a historical consciousness—both of which plague contemporary social science. For example, former Federal Reserve Chairman Alan Greenspan, testifying to Congress on the Great Financial Crisis of 2007–2009, declared that the defect in the Federal Reserve Board's computer modeling of the financial economy under his leadership was that the data had been drawn from too narrow a spectrum of time, thereby leaving out the real variability of change. As Greenspan himself put it in October 2008, "The whole intellectual edifice . . . collapsed in the summer of last year because the data integrated into the risk management models generally covered only the past two decades, a period of euphoria. Had instead the models been fitted more appropriately to historic periods of stress, capital requirements would have been much higher and the financial world would be in far better shape today."[26] In other words, having excluded not only the Great Depression but also the major economic crises of the mid-1970s and the early 1980s from the data going into the computer models, the results generated by these models showed, not surprisingly, that a serious economic and financial crisis was out of the picture, and that the financial system essentially went up but never down. Important historical tendencies endemic to a capitalist economy (major periodic crises of accumulation) had been exorcised from the analysis in this way, producing faulty results.

In avoiding such dehistoricized, reified understandings, historical materialism is an indispensable guide. The Marxian tradition is highly skeptical of the claim that most observed social phenomena can be explained by spatiotemporally invariant laws, as in the case of distinguished sociological theorist Jonathan Turner, who is strongly committed to what he calls "the goal of positivism [which] is to formulate and then test laws that apply to *all* societies in *all* places at *all* times." For Turner, "Marxists and others make a fundamental mistake in assuming that the laws of social organization are time bound, such that the laws governing the operation of feudalism are somehow different than those directing capitalism."[27]

Marxian theorists, in contrast to such exponents of invariant spatiotemporal laws, highlight the dramatic changes in social structures and patterns that have occurred throughout human history and argue that what appear to be invariant laws to observers in any particular period, may in fact be transient tendencies unique to that historical era, emerging from the dialectical interaction of an ensemble of social and natural processes. Historicity—the notion that social "laws" (if not natural laws as well) vary across different historical periods—is one of the most basic concepts of the historical materialist tradition. This approach does not deny the existence of social laws, but rather historicizes them, seeing them as tendencies (tendential laws) arising out of historically specific conditions.

Despite the fact that many social laws seem immutable within a given era, history demonstrates that these can be swept away, often in a strikingly short span of time. Such major transformations, changes in the social gravity, are especially true during a period of structural crisis of an entire social system. As the conservative nineteenth-century cultural historian Jacob Burckhardt observed, a "historical crisis" occurs when "a crisis in the whole state of things is produced, involving whole epochs and all or many peoples of the same civilization.... The historical process is suddenly accelerated in terrifying fashion. Developments which otherwise take centuries seem to flit by like phantoms in months or weeks, and are fulfilled."[28]

Given the irrationalities of a social system—in which those on top have a vested interest in blocking fundamental change—a "crackpot realism," which suggests that anything but the present unacceptable reality is "utopian," may be designed to preclude precisely those revolutionary or transformative actions that are most urgently needed by humanity as a whole.[29]

Indeed, history teaches us that there are times when society is ready to "burst asunder" and the failure to carry out drastic change can mean the demise of civilization or worse—potentially worse today because we are threatened with the demise of the planet as we know it.[30] People make their own history, but not exactly as they please. Rather, they must struggle under conditions established by the complex coevolution of nature and human production, and the relations of power and ideology, inherited from the past.[31] At certain historical moments the equilibrium that is thought to constitute a relatively stable society is punctuated by rapid developments. It is precisely this kind of historic moment that now confronts humanity, and represents the greatest challenge to social change since the Industrial Revolution.

In a deservedly famous passage in *The Communist Manifesto*, Karl Marx and Frederick Engels wrote:

> The bourgeoisie, during its rule of scarce one hundred years, has created more massive and more colossal productive forces than have all preceding generations together. Subjection of nature's forces to man's machinery, application of chemistry to industry and agriculture, steam-navigation, railways, electric telegraphs, clearing of whole continents for cultivation, canalization of rivers, whole populations conjured out of the ground. What earlier century had even a presentiment that such forces slumbered in the lap of social labor?[32]

When this was written in 1847, capitalism still held sway only in a small corner of the world. Since then it has more and more been transformed into a world system, or globalized. The development of science and technology, industry and agriculture has expanded worlds beyond the level pictured by Marx and Engels. Indeed, the transformations in the human relation to the environment since the Industrial Revolution and particularly since 1945—nuclear power, modern organic chemistry, the conquest of the air and space exploration, digital technology, and biotechnology—have so transformed our relation to the planet as to raise the question of the Anthropocene. In all of this, capitalism has remained essentially (if not more so) what it was from the beginning: an enormous engine for the ceaseless accumulation of capital, propelled by the competitive drive of individuals and groups seeking their own self-interest in the form of private gain. Such a system recognizes no absolute limits to its own advance. The race to accumulate, the real meaning of economic growth under the system, is endless.[33]

Joseph Schumpeter, the great conservative economist, famously characterized capitalism as a system of "creative destruction."[34] In our time, however, its destructive features are coming more and more to the fore. Economically, the system, although driven by and dependent on growth, has demonstrated itself to be prone to deep stagnation tendencies—a slowdown in economic growth, arising from increasing inequality associated with a waning of many of the historical factors (most notably, the buildup of industry from scratch) that propelled its earlier "golden ages." This has resulted, as a compensatory mechanism, in the financialization of the accumulation process, producing more and more irrationalities.[35] Ecologically, the system draws ever more destructively on the limited resources and absorptive capacity of nature, as the economy continually grows in scale in relation to the planetary system. The result is emerging and expanding ecological rifts that are turning into planetary chasms.

The essential nature of the problem resides in the fact that there is no way out of this dilemma within the laws of motion of a capitalist system, in which capital accumulation is the primary goal of society. As Schumpeter stated, and as recognized by all of the leading economists—Adam Smith, David Ricardo, Karl Marx, Thorstein Veblen, Alfred Marshall, John Maynard Keynes—down to the present, "Capitalism is a process [of accumulation and growth], stationary capitalism would be a *contradictio in adjecto*."[36] There is no conceivable capitalist economics compatible with a "steady-state economy," a system that abandons endless growth of the economy as its central feature. Practically without exception, mainstream economists praise economic growth, only conceding today that changes may need to be made to ensure that this growth is sustainable—but, as always, still conceived in terms of economic growth.

By abandoning the critique of capitalism, and increasingly dropping it from their analytical vision altogether, mainstream social scientists (even environmental sociologists in some cases) have removed from their analytic vision the essential problem now facing planetary society. Once the notion of "green capitalism" is accepted—as if capitalism was not a system of self-expanding value, or that endless accumulation was somehow compatible with environmental sustainability—then the environmental problem becomes merely a question of management and markets. There is then no need to address the relentless drive that constitutes the global system of monopoly-finance capital and its processes of rapacious exploitation of the earth, propelled forward by a speculative system of asset-based accumulation.[37] Rather than confronting capitalism as a real, historical entity, arising over centuries, which must be explained organically in terms of its own presuppositions and development, we are more and more encouraged to see the economic order in which we live in purely technocratic terms as a "market system." Such reified versions of reality involve nothing so much as "a forgetting."[38]

Central to the constitution of environmental sociology as a field is a dispute that has arisen between those who believe in "green capitalism," or the *ecological modernization school*, and those who see the dominant social and economic system as driven by a commitment to the expansion of production at the expense of all other, particularly environmental, needs, or the *treadmill of production school*. Ecological modernization proponents strip from the concept of capital all determinant content, reducing it to little more than a "market system." Hence Arthur Mol, a leading representative of this view within sociology, refers to "capitalism" on the first page of his *Globalization and Environmental Reform* "as simply 'a catchword'" of "the late 1960s and 1970s"—one "much like" that of "globalization" today. Thereby, the focus, as he observes elsewhere, is "on redirecting and transforming 'free market Capitalism' in such a way that it less and less obstructs, and increasingly contributes to, the preservation of society's sustenance base in a fundamental/structural way."

Likewise, the main opponents of the *ecological modernization school* (the dialectical opposites of the former), the *treadmill of production school*, all too often steer away from a thoroughgoing critique of capitalism as a historical mode of production, preferring to pose the problem simply in mechanistic terms—as if the metaphor of an economic treadmill was an adequate substitute for the historical reality of capitalism.[39]

This abandonment of the historical critique of capital, at a time when it is more than ever needed, is the greatest single fault of contemporary social science, including ecological social science. As John

Kenneth Galbraith pointedly observed in his chapter on "The Renaming of the System" in his last book, *The Economics of Innocent Fraud*:

> When capitalism, the historical reference, ceased to be acceptable, the system was renamed. The new term was benign but without meaning. . . . In reasonably learned expression there came "the market system." There was no adverse history here, in fact no history at all. It would have been hard, indeed, to find a more meaningless designation—this a reason for the choice. . . . So it is of the market system we teach the young. It is this, as I've said, that sophisticated political leaders, compatible journalists and many scholars speak. No individual or firm is dominant. No economic power is evoked. There is nothing here from Marx or Engels. There is only the impersonal market, a not wholly innocent fraud.[40]

This "renaming of the system" systematically eludes the rift in society associated with the historical concept of capitalism. For Galbraith this is a form of "innocent fraud," for Bernal it was a "corruption" of social science built into class society. As Marxian social scientists have argued (particularly since Lukács), social reality is "reified," that is, social-productive relations are turned into relations between things. There is a process of dehistoricization, erasing the social relations and the social gravity that shape material reality.[41] The historical system of capitalism, with its social class characteristics rooted in production, is turned into a mere "market," a seemingly concrete entity but without any real definition—an example of what Alfred North Whitehead called "the fallacy of misplaced concreteness."[42]

The Dehistoricization of Nature

Social science today is crippled not only by its growing failure to confront the historical specificity (and thus the hegemonic structures) of present-day society, but also by its repeated refusal to engage critically with the reality of the natural world. Thus the social sciences and the humanities (in particular fields such as economics, political science, sociology, cultural anthropology, philosophy, and cultural studies) are all characterized to varying degrees by their radical separation from nature—from the concerns that preoccupy natural science, and more particularly from notions of natural history or evolution. The long revolt against positivism and the domination of the social sciences by natural-scientific conceptions has of course been crucial to the development of today's social sciences/humanities. However, the anthropocentrism and culturalism, and the extreme neglect of natural/material conditions, that this has given rise to, has proven increasingly debilitating, especially in the present age of planetary ecological crisis.

The word *nature*, we should interject here, is one of the more complex words in modern language, standing as it does in different contexts for the material world and even the universe; the most fundamental domain of existence; the elemental drives of life; the object of natural science; certain timeless, immutable laws; evolution; the non-human and non-social; the non-intellectual and non-spiritual, and so on. So various and yet indispensable are its usages that Max Weber referred to the fundamental "ambiguity of the concept of 'nature,'" arguing that the most we can say is that in each of its many usages it refers to "a complex of certain kinds of *object*s, a complex that is distinguished from another complex of *objects*, which have different properties."[43] Indeed, the concept of nature can be seen as perhaps the prime example of what Fredric Jameson has referred to as a fundamental "ontological rift" in existence, posing dialectical oppositions that can be fathomed but never fully bridged.[44]

The dialectical recognition of this ontological rift—and attempts to work through it in various ways from a consistent materialist, naturalist, realist perspective—has led in recent years to advances in critical and dialectical realism, which recognize the possibility of a far-reaching naturalism, that is, the unification at some level of higher principle of the natural and social sciences.[45] Nevertheless, for the most part the human sciences today seem to be committed to an ever stricter divide between what C. P. Snow called the "two cultures"—represented by the humanities/social sciences and the natural sciences. Thus we see the social sciences erecting ever higher walls, constructing deeper moats, all directed at separating themselves off from the objects (if not the methods) of natural science.[46]

In declaring its independence from nature (viewed as the object of natural science) social science has all too often reacted to an earlier Newtonian mechanism that saw nature primarily in terms of timeless, immutable laws. Here the resistance is often to nature as "essentialism" in its various forms, whereby human beings/society are reduced to mere biological entities/byproducts (often in grossly distorted ways as in classical racist and sexist ideologies). From this perspective, nature stands for what is fixed and unchanging, or changing too slowly to be of direct relevance to human society. Theodor W. Adorno observed that "losing its genesis," as a natural-historical phenomenon, nature is transformed into "something which in principle . . . is unalterable."[47]

It thus became customary in the social sciences to view the realm of humanity/society/culture/the mind as a realm constructed apart from nature. Such anthropocentric views were reinforced by the so-called conquest of nature associated with modern science and technology, feeding a "Human-Exemptionalist Paradigm," or the notion that human beings were not only exempt from nature's general laws but could transcend them in almost infinite ways, given ingenuity. Nature was taken for granted, as the social world existed outside the bounds and limits of natural influences. It was assumed that whatever *social* problems arose in relation to nature, scientific-technological fixes could be employed to maintain the existing social order. As a result, the environmental sustainability of human societies was not a problem.[48]

Although denying any need to address the natural conditions of human society, social science, in its more abstract-empiricist form, has often tried to replicate the methodological successes of natural science, by searching by means of a crude positivism for immutable laws of social science paralleling those discovered for nature itself. This has almost invariably meant, however, the dehistoricizing of both nature and society—modeling all of human society (and nature itself) on the basis of either an unchanging status quo or a structural-functionalist and teleological notion of "modernism." What has *become* is treated as absolute, as its own final cause. At most it is seen as the unfolding of a predetermined result—a view far removed from the contingency of Darwinian evolution or Marxian historical materialism. All of the past that does not fit within such an artificial, "Whiggish" view of history is simply discarded as wrong or failed. This approach has given rise to what Adorno called "the paradox . . . whereby the prevalent empiricism is amputating, precisely, experience" through its denial of real history.[49]

Moreover, recent dissenters from this dominant, modernist tradition have not so much held on to history as amputated it by other means, through irrealism, relativism, and a postmodernism without content. For example, exhibiting the ontological fallacy (the denial of any independent existence to nature apart from thought) so characteristic of the skeptical "strong programme" in the sociology of science, Keith Tester declared: "A fish is only a fish if it is socially classified as one. . . . Animals are indeed a blank paper which can be inscribed with any message, and symbolic meaning, that the social wishes."[50]

Sociology has had a particularly tortured legacy in this respect. Its early devastating encounter with social Darwinism, racism, and crude positivism, which summed up much nineteenth-century sociology, resulted in a subsequent rejection in the twentieth century of all views that sought to connect nature and the biological sciences with social development. Hence, environmental sociology, insofar as it has taken ecological questions seriously, has found itself in continual, sharp opposition to the dominant sociological paradigm of human exmptionalism—which, in denying the human dependence on (or even interdependence with) nature, has presented a radical constructivist view in which the physical world and evolution scarcely exist.[51]

The irony in all of these developments is that since at least the mid-nineteenth century, with the rise of Darwin's theory of evolution, natural science (as opposed to social science) has tended to become more, not less, historical and dialectical in its outlook. Since the early twentieth century, the biosphere (a notion developed by V. I. Vernadsky in conjunction with the discovery that life had played a role in the construction of the atmosphere) and the universe itself have been seen increasingly in temporal terms.[52] As Hegel once wrote, "The true is the whole." But this only tells us that it cannot be understood outside of its "becoming."[52] The development of ecology and today's earth system (including climate) science reflects the movement toward complex, historical, materialist, holistic forms of analysis, taking account of contingency—very far removed from the supposed mechanistic laws of Newtonian science. Indeed, the growing planetary ecological crisis has given rise to an understanding of how fast nature can change under certain conditions and of the coevolution of humanity and nature.[54] Peter M. Vitousek, an ecologist, notes that humans are forcing qualitative and historical changes on the world that will "alter the structure and function of Earth as a system." Hence, it is of utmost importance for the social sciences to take up these issues.[55]

Although natural processes have often been viewed as operating according to principles of geological time that are too slow and gradual to affect human history, and nature therefore has been frequently treated as the realm of the permanent, this is now rapidly changing. Nature as history (that is, natural history) is increasingly being impressed on our consciousness, since it is now more and more subject to the forces of human history.[56] The development of the human economy in the Anthropocene has acted as a catalyst for the unprecedented acceleration of changes in the atmosphere, the climate, the ocean, and the earth's ecosystems. The end of the Holocene due to anthropogenically induced global warming means that suddenly geological-scale change has entered human history itself.

Indeed, perhaps the greatest danger of climate change to life is the accelerating tempo of the change in the earth system, overwhelming natural-evolutionary processes and even social adaptation, and thus threatening the mass extinction of species and even human civilization itself. Human-induced global warming is occurring at rates far faster, for example, than the warming that normally occurs at the end of ice ages.[57]

One way in which species are attempting to adapt to climate change is by moving toward the polar regions with their cooler temperatures (and also into higher alpine regions). But the warmer temperature zones are in effect moving toward the poles faster than species. As James Hansen explains in *Storms of My Grandchildren*,

> Studies of more than one thousand species of plants, animals, and insects (including butterfly ranges charted by members of the public) found an average migration rate toward the north and south poles of about four miles per decade in the second half of the twentieth century. That is not fast enough. During the past thirty years the lines marking the regions in which a given average temperature prevails ("isotherms") have been moving poleward at a rate of change of about thirty-five miles per decade. . . .
>
> As long as the total movement of isotherms toward the poles is much smaller than the size of the habitat, or the ranges in which animals live, the effect on species is limited. But now the [isotherm] movement is inexorably toward the poles and totals more than one hundred miles over the past several decades. If greenhouse gases continue to increase at business-as-usual rates, then the rate of isotherm movement will double in this century to at least seventy miles per decade.

In effect, species trying to outrun global warming are losing the race. The tempo of temperature change is simply too great. The worst fate awaits polar and alpine species, which are in effect being "pushed off the planet."[58]

In eighteenth-century France, a leading geological scientist Horace-Bénédict de Saussure (1740–99) used the term "revolution," as was common in the time, to describe massive changes in the past history of the earth, or "nature's revolutions." Indeed, as the great historian of geohistory, Martin J. S. Rudwick, has written in *Bursting the Limits of Time*, eighteenth-century geologists used the word "revolution" in the same general sense as in political history, applying it "to the world of nature, and particularly to the past history of the earth. . . . What they all agreed, in the face of evidence as disparate as mountains, fossils, and volcanoes, was that the earth had somehow undergone massive changes; and that those changes had happened over a timescale that was certainly vast in relation to human lives, perhaps unimaginably so in relation to the whole of human history." They represented structural forces that had built up over time and then burst asunder.[59]

This notion of nature's revolutions was part of the discovery of geological time that was occurring in the eighteenth and nineteenth centuries, when it was determined that the world was not just a few centuries old but millions of years old. It thus impressed itself on the consciousness of society. Yet the doctrine of gradualism, associated with Charles Lyell's uniformitarian geology and Charles Darwin's evolutionary biology (the product of the quieter climate of nineteenth-century England), tended to efface somewhat the importance of "nature's revolutions" and to generate a view of nature as relatively "passive," no longer a significant actor *from the standpoint of human history and social science.*[60]

The failure of social scientists to recognize both the radical historicity of human society and the radical historicity of nature thus leads to a failure to address the ecological crisis of our time with the realism, dialectical understanding, urgency, and commitment to revolutionary transformations in human society that it requires. Natural scientists tell us that U.S. society needs to cut its carbon emissions practically to zero while social scientists tell us—rather absurdly in this context and completely removed from the most pressing ecological relations—that we need to eco-modernize our shopping habits.[61] The structural significance and scale of the ecological crisis is not reflected in solutions of a corresponding significance and scale. This failure of both imagination and social practices is in many ways a product of a double alienation: from nature and within human society itself. Not only has this generated inertia with respect to social change—indeed a tendency to fiddle while Rome burns—but it has also led to the belief that the crisis can be managed by essentially the same social institutions that brought it into being in the first place.

The danger lies not so much in the vast majority of social scientists (economists, sociologists, political scientists, anthropologists, geographers), who have essentially ignored, ostrich-like, the ecological crisis of our time, and still need to awaken to the reality, as in those who propose to manage the crisis (environmental economists, environmental sociologists, environmental political scientists), who profess they can "green" capitalism and green "modernity," all the time refusing to recognize that capitalism is not an immutable condition of human existence and that nature, far from being stabilized, is *for all practical purposes* being destroyed by this very system. Common notions of ecological modernization, for example, are increasingly being rendered irrelevant or useless in the face of the ongoing development of the ecological rift.

The proposition that unlimited economic growth under capitalism can and should be managed so as to generate a system of sustainable capitalist development (a view we call "Capitalism in Wonderland") rejects at one and the same time an understanding of capitalism as a historical system and the notion that nature itself is historically complex and contingent in ways that we are only beginning to understand.[62] The great geological eras in the history of the planet are separated by massive die-downs in species. *Homo sapiens* under the present economic and social system are destroying natural habitat, which is driving the sixth mass extinction.[63] Given the alienation and reification that are today so pervasive, ecological destruction has simply become a way of life in an era dominated by the interests of capital.[64] The only thing that can save us is a revolution in the constitution of human society itself. Without such revolutionary changes in the human metabolic relation to the earth (in the material relations of production), the future of the world, like bourgeois human nature, will be "nasty, brutish, *and short*."[65]

Planetary Boundaries and the Drive to Amass Capital

The critical problem of our time is the breaching of planetary boundaries, which have been discovered precisely because the system of production now threatens to overshoot them, and in three instances—climate change, the nitrogen cycle, and biological diversity—already has. This problem has become a life-and-death issue for the earth system as we know it—so much so that the threat to these planetary *bound-*

aries from the human economy has been singled out by scientists as the defining feature of what is now being referred to as the Anthropocene.

What makes matters so serious is the inability of our social system to respond effectively to this planetary crisis. It is an inner characteristic of the capitalist economy that it is essentially limitless in its expansion. It is a grow-or-die system. The "drive to amass capital" recognizes no physical boundaries. All obstacles are treated as mere barriers to be surmounted in an infinite sequence.[66] Capital is thus, from a wider social and ecological standpoint, a *juggernaut*, an unstoppable, crushing force.[67] It is a system not concerned directly with the expansion of use value, but rather with the expansion of exchange value—a merely quantitative element, which derives its meaning only from its exponential increase. As István Mészáros, the great Marxist philosopher, explains, "Quantity rules absolute in the capital system."[68] Qualitative social relations, including those with the natural conditions of life, are not part of its system of accountancy.

Marx famously explained this in terms of his M-C-M′ formula—whereby capital is understood as the "continuous transformation of capital-as-money into capital-as-commodities, followed by a retransformation of capital-as-commodities into capital-as-more-money."[69] It is exchange value, which knows only quantitative increase—not use value, which relates to the qualitative aspects of production—which drives the system. Thus capital constantly metamorphosizes into more capital, which includes surplus value, or profits, the generation of which is "the absolute law of this mode of production."[70] "The strength of the idea of private enterprise," E. F. Schumacher wrote, "lies in its terrifying simplicity. It suggests that the totality of life can be reduced to one aspect—profits." And profit by its nature is quantitative increase.[71]

We can understand the implications of this more fully if we turn to Hegel, who distinguished dialectically between *barriers* and *boundaries* in the life of any organic entity. "A boundary" that is posited as something that it is "essential to overcome," he wrote, "is not merely [a] boundary as such but [a] barrier." Boundaries thus refer to real limits; and barriers are mere obstacles to be surmounted. "A sentient creature" faced with an obstacle, which it is essential for it to overcome, "hunger, thirst, etc.," will see this, Hegel argues, not as a firm boundary but will seek "to go beyond its limiting barrier."[72]

Employing this Hegelian distinction, Marx described capital in the *Grundrisse* as

> the endless and limitless drive to go beyond its limiting barriers. Every boundary is and has to be a [mere] barrier for it. Else it would cease to be capital—money as self-reproductive. If ever it perceived a certain boundary not as a barrier, but became comfortable with it as a boundary, it would itself have declined from exchange value to use value, from the general [abstract] form of wealth to a specific, substantial mode of the same. Capital as such creates a specific surplus value because it cannot create an infinite one all at once; but it is the constant movement to create more of the same. The quantitative boundary of the surplus value appears to it as a mere natural barrier, as a necessity which it constantly tries to violate and beyond which it constantly seeks to go.

He adds in a footnote: "A barrier to capital's advance appears [to it] as an accident which has to be conquered. . . . If capital increases from 100 to 1,000, then 1,000 is now the point of departure, from which the increase has to begin; the tenfold multiplication, by 1,000%, counts for nothing."[73] And so the process continues, simply establishing a new starting point for its expansion on an ever-larger scale.

It should come as no surprise, then, that orthodox or neoclassical economists, who are mainly concerned with articulating the needs of the capitalist system, have adamantly resisted (or played down) the notion that there are insurmountable physical boundaries to economic growth beyond which the ecological viability of the planet is compromised. Like the corporate world, economists' principal interest is in the expansion of the economic system—production, profits, accumulation, concentrated wealth—and only secondarily (if then) in the quality of existence. Hence, the dominant economic approach to climate change, even in the case of those concerned about the problem, such as Nicholas Stern, chief author of *The Stern Review*, is to advocate constraints on greenhouse gas emissions in order to save the planet—*except where such restraints would lead to a significant decline in economic growth (capital accumulation)*.[74]

The notion of "sustainable development," though an essential concept in the context of growing ecological crisis insofar as it emphasizes the need for ecological sustainability, has often been used as a category to reinforce the need for *sustaining economic growth*. An example is leading British environmental economist David Pearce, author of the UK government's Pearce Report, *Blueprint for a Green Economy*, who has stated that "sustainable economic development . . . [is] fairly simply defined. It is continuously rising, or at least non-declining, consumption per capita, or GNP, or whatever the agreed indicator of development is. And this is how sustainable development has come to be interpreted by most economists addressing the issue."[75]

Ecological modernization theorists simply added to this the corporate greenwashing claim that the eco-modernizing tendencies intrinsic to capitalism or "the market system" allowed the "expansion of the limits" of growth. Ecological modernization, according to its leading advocate, Arthur Mol, is the belief that "an environmentally sound society" can be created without reference to "a variety of other social criteria and goals, such as the scale of production, the capitalist mode of production, workers' influence, equal allocation of economic goods, gender criterion, and so on. Including the latter set of criteria might result in a more radical programme (in the sense of moving away from the present social order), but not necessary a more ecologically radical programme." As another leading ecological modernization theorist, Maarten Hajer, has acknowledged, ecological modernization "does not call for any structural change but is, in this respect, basically a modernist and technocratic approach to the environment that suggests that there is a techno-institutional fix for the present problems." For this reason, ecological modernization sees no reason to address the reality of capitalism. In Hajer's words, "It is . . . obvious that ecological modernization . . . does not address the systemic features of capitalism that make the system inherently wasteful and unmanageable."[76]

This rigid notion of ecological modernization as a mere *correction* in the original modernizing tendency of society leaves little room for considerations of social inequality. Additionally, ecological modernization thinkers do not normally address the larger problems of the global ecological crisis, such as global warming, or the forms taken by human-nature interactions.[77] Rather than engaging in an overarching critique of the historical relation between society and nature, the increasingly dominant ecological modernization perspective takes all of this for granted. It begins and ends with the notion of *technique* (technics), which is both cause and effect, problem and solution: at most a question of technological innovation coupled with the appropriate forms of ecological management.

Ecological modernization is thus all about the development and management of green technologies (techniques), displacing the old, environmentally harmful operations. Entrepreneurs are deemed to be an important driver of this transition, as they respond to increasing environmental consciousness among the public and pursue important innovations as far as products and technologies. "The basic notion of the ecological modernization processes," as the principal sociological advocates of this perspective state, is "aimed at 'regaining one of the crucial design faults of modernity'" through technological innovation.[78] The standard way in which to square the expanding circle (or spiral) of capitalist production is to bring in the black box of technology as constituting the solution to all problems.

Yet, technology cuts both ways. "The assumption of some critics that technological change is exclusively a part of the solution and no part of the problem," Herman Daly writes, "is ridiculous on the face of it and totally demolished by the work of Barry Commoner [in *The Closing Circle*] (1971). We need not accept Commoner's extreme emphasis on the importance of the problem-causing nature of post–World War II technology (with the consequent downplaying of the roles of population and affluence) in order to recognize that recent technological change has been more a part of the problem than of the solution."[79]

To be sure, technological change is a necessary part of any ecological solution. But ecological modernizers in sociology and sustainable-developers in mainstream economics go beyond this by arguing that technology can work magic: "dematerializing" economic production so that the capitalist economy can then walk on air (or create a "weightless society"), thereby continuing its relentless expansion—but with a rapidly diminishing effect on the environment. Needless to say, such technological fantasies have no basis in reality.[80]

Still, technological optimism is pervasive in the ecological literature (and especially among ecological modernization theorists). All sorts of "positive-sum" and "win-win" technical fixes are proposed. Hajer speaks confidently of the "*technicisation of ecology*" as the answer to the ecological crisis. In this view, "microelectronic technologies are presented as the solution for the 'juggernaut effect'" of capitalism.[81] Technological change is promoted in an attempt to argue that social relations (of power and property) can remain the same—whereas it is merely values, consciousness, and knowledge that change, and that direct technological innovation. Such views are worse than those of necromancers of old, since they wish away all pretenses to a scientific understanding in the name of science. Not only are the basic physics of thermodynamics set aside, but the way in which technology is embedded within the social system is also ignored.[82]

The notion that economic production in general under the present system can continually expand without ecological waste and degradation (the dematerialization hypothesis) goes against the basic laws of physics. As the brilliant ecological economist Nicholas Georgescu-Roegen wrote: "Had economics recognized the entropic nature of the economic process, it might have been able to warn its co-workers for the betterment of mankind—the technological sciences—that 'bigger and better' washing machines, automobiles, and superjets must lead to 'bigger and better' pollution."[83] Although new technologies (and indeed much older technologies) can accomplish great things in terms of reducing the environmental impact per unit of production, the scale effects of economic expansion generally override any energy/environmental savings (a phenomenon known as the Jevons Paradox).[84] Since

1975 the amount of energy expended per dollar of GDP in the United States has decreased by half, marking an increase in energy efficiency by that amount. But at the same time the overall consumption of energy by U.S. society has risen by some 40 percent.[85]

New environmental technologies are adopted not on the basis of their value in creating a sustainable relation to the environment but on the basis of the profit considerations of corporations, which rarely converge with ecological requirements. As economist Juliet Schor notes, "Firms are reluctant to install technologies whose gains they cannot capture. A decentralized system of solar and wind, for example, may have technical superiorities such as avoiding the power loss that accompanies long-distance power generation in centralized facilities. But if the technologies are small-scale and easy to replicate, large firms have difficulty capturing the profits that make investments desirable."[86]

Indeed, the single-minded goal of technological innovation under capitalism is expansion of production, profits, accumulation, and wealth for those at the top, not protection of the environment. According to Donella Meadows and her co-authors in *The Limits to Growth: The 30-Year Update*: "If a society's implicit goals are to exploit nature, enrich the elites, and ignore the long term, then that society will develop technologies and markets that destroy the environment, widen the gap between the rich and the poor, and optimize for short-term gains. In short, that society develops technologies and markets that hasten a collapse instead of preventing it."[87]

Although the proverbial efficiency of the market is often brought in as an "invisible hand" that will solve such environmental problems, ecological degradation ranks along with inequality and poverty as powerful evidence of "market failure." The capitalist market system is geared at all times to the concentration of economic surplus and wealth together with the displacement of the majority of costs onto society and the environment. It provides a distorted accounting of human and environmental welfare in its gross national income statistics. It is responsible for the most profligate forms of waste.[88] As Charles Lindbloom has indicated, what is most remarkable about "the market system" of developed capitalism is the full extent of its "allocative inefficiencies."[89] One example of such aggregate allocative inefficiency is the short-term time horizon built into such a system that works almost exclusively on individual greed. The heavy discounting of the future in the modern profit system means, as mainstream economist Lester Thurow grimly observed, that "good capitalists" will "decide to do nothing" no matter how bad the environmental problems in the distant future will be due to their immediate actions. Eventually a generation will come along that no longer has the opportunity to act, because the conditions of the irrevocable deterioration of "earth's altered environment" will already have been set in place. "Each generation makes good capitalistic decisions, yet the effect is collective social suicide."[90]

One way to look at this is to see capitalism as a bubble economy, which uses up environmental resources and the absorptive capacity of the environment while displacing the costs back on Earth itself, thus incurring an enormous ecological debt. As long as the system is relatively small and can keep on expanding outwardly, this ecological debt is displaced, often without any recognition of the costs that have been incurred. Once the economic system begins to approach not just its regional boundaries but planetary boundaries, the mounting ecological debt will become ever more precarious, threatening an ecological crash. To stop this requires nothing less than an ecological revolution aimed at bringing the social relations of production in line with the conditions of ecological sustainability.[91]

The New Whole Human

In the nineteenth century, Karl Marx introduced the notion of the "metabolic rift," or a rift in the metabolic exchange between humanity and nature. The context was the robbing of the soil of the countryside of nutrients and the sending of these nutrients to the cities in the form of food and fiber, where they ended up contributing to pollution. This rupture in the soil nutrient cycle undermined the regenerative capacities of the ecosystem. Marx argued that it was necessary to "restore" the soil metabolism to ensure environmental sustainability for the generations to come. Such transformation in the metabolic relation required a society directed by associated producers, who regulated the qualitative and quantitative interchange between society and the conditions of life.[92] Nature can be seen as a web or a fabric made up of innumerable processes, relations, and interactions, the tearing of which ultimately results in a crash of the ecological system. Metabolic analysis serves as a means to study these complex relationships of ecological degradation and sustainability. Hence, Marx's concept of socio-ecological metabolism and the emergence under capitalism of a metabolic rift will be central to this book.

Marx's analysis, although primarily related to the nitrogen and phosphorus (also potassium) cycles, can be seen as a key to the whole problem of planetary boundaries. Ecosystems and natural processes are complex and exist in multiple layers, ultimately serving as the determinants for life. The cutting down of an ancient/old growth forest (as in the Pacific Northwest of North America or the Brazilian Amazon) that numerous species depend on often produces what looks like a crazy quilt of isolated patches when viewed from the air. Green exists where trees remain standing; brown marks where the flora has been removed. The forest appears to be literally cut to pieces. At some point, logging creates rifts in the forest, as the integrity of the entire ecosystem—usually measured in terms of the decline of "keystone" species—is destroyed. Attempts to substitute a mono-cropped industrial tree plantation for original forestland is simply the replacement of a diverse ecological system with a relatively sterile (and often toxic due to contamination with pesticides) human-made environment. Such an industrial forest is all the more extreme because it is turned into a pure commodity: so many board feet of standing timber.[93]

This same process of being torn to pieces and reorganized in line with the human economy is happening in various ways to the earth as a whole—in ways we have barely begun to understand. Nevertheless, it is clear that the general nature of the "division of nature" under capitalism is such that it simplifies what was formerly complex.[94] This produces severe ecological breaks, which then get displaced or shifted around until the point is reached where this is no longer possible, as in the crossing of planetary boundaries.[95]

This ecological rift is, at bottom, the product of a social rift: the domination of human being by human being. The driving force is a society based on class, inequality, and acquisition without end. At the global level it is represented by what L. S. Stavrianos in *Global Rift*—a history of the third world—described as the imperialist division between center and periphery, North and South, rich and poor countries.[96] This larger world of unequal exchange is as much a part of capitalism as the search for profits and accumulation. The notorious imperialist Cecil Rhodes once said "I would annex the planets if I could."[97] As it was, he was only able to seize some of the richer parts of Africa, enslaving people and mining the earth for gold and diamonds. No solution to the world's ecological problem can be arrived at that does not take the surmounting of capitalism, as an imperialist world system, as its object.

It is time to take back the planet for sustainable human development. Already such an ecological revolution, emanating first and foremost from the global South, is emerging in our age, providing new bases for hope. An environmental proletariat, representing a new struggle for ecological hegemony, is arising to challenge both the ecological degradation and social exploitation associated with capitalist expansion. In 1998, the Bolivian government, in conjunction with the World Bank, privatized water in Cochabamba. Bechtel, a U.S. engineering firm, received the contract. Water prices tripled and service was eliminated to those who could not afford water. People took to the streets in a series of water wars, challenging the privatization of water and the domination of corporate interest. In April 2000, Bechtel was forced out of the country. Throughout Bolivia similar struggles took place, as people fought against the Suez water company that was given control of water in La Paz. This groundswell helped usher in Bolivia's socialist and indigenous president Evo Morales, who proclaimed that water must be provided for free and cannot be run by private business.[98]

On April 22, 2010, the tenth anniversary of the victory of the water wars in Cochabamba, *The Cochabamba Protocol*, or *The People's Agreement on Climate Change and the Rights of Mother Earth* was released in Bolivia under the leadership of Morales. In his opening speech he began with the slogan "Planet or Death, We Shall Overcome!" *The Cochabamba Protocol* declared: "Humanity confronts a great dilemma: to continue on the path of capitalism, depredation, and death, or to choose the path of harmony with nature and respect for life." Among other things it insisted that 6 percent of the GDP of the rich countries (roughly equivalent to the real share of military spending in U.S. GDP) be devoted to helping the poorer countries adapt to climate change. And it demanded a 50 percent reduction in the greenhouse gas emissions of developed countries from the 1990s over the next decade.[99]

The goal of human development, as Morales has articulated it, should not be living better, but living well.[100] This means that the guiding principle is "enough," requiring economic development for the poorest countries along with the stabilization of world economic output overall. This can only be achieved with much greater focus on substantive equality along with qualitative human development. "A free life," Epicurus wrote, "cannot acquire great wealth, because the task is not easy without slavery to the mob or those in power. . . . And if it does somehow achieve great wealth, one could easily share this out in order to obtain the good will of one's neighbors."[101]

Given the "closing circle" of planetary boundaries, the world must develop an economic system that is not based on endless accumulation, but rather on providing for the needs of people while protecting ecosystems. However, at present there is a contradiction between ecology and the dominant economic system. As Elmar Altvater explains in *The Future of the Market*: "The 'steady-state principle' is . . . rational within the ecological system. And yet, what is rational in the ecological system is irrational in terms of [capitalist] market economics: an economy without profit."[102] There is only one possible solution to this contradiction: an ecological and social revolution that will rid us of the narrow profit system and replace it with a sustainable and just society.

"Socialism is of interest," Schumacher observed in his *Small Is Beautiful*, precisely because of "the possibility it creates for the overcoming of the religion of economics," that is, "the modern trend towards total quantification at the expense of the appreciation of qualitative differences."[103]

If the Holocene of the last ten to twelve millennia stood for the "New Whole" epoch in geological evolution, and the Anthropocene of the last two centuries stands for the "New Human" epoch (as marked ironically by the crisis in the human domination of the planet), what we need to strive for is a Holoanthropocene—an epoch of the "New Whole Human" based on transcending the alienation of humanity and nature. The rift in ecology and society can only be healed through a new revolutionary transformation in human social and ecological relations. "Nature's revolution" and social revolution need to be made one. Humanity must at long last reach a new stage in its real historical development, in which the earth is a boundary and life is respected.

PART ONE

Capitalism and Unsustainable Development

1. The Paradox of Wealth

Today orthodox economics is reputedly being harnessed to an entirely new end: saving the planet from the ecological destruction wrought by capitalist expansion. It promises to accomplish this through the further expansion of capitalism itself, cleared of its excesses and excrescences. A growing army of self-styled "sustainable developers" argues that there is no contradiction between the unlimited accumulation of capital—the credo of economic liberalism from Adam Smith to the present—and the preservation of the earth. The system can continue to expand by creating a new "sustainable capitalism," bringing the efficiency of the market to bear on nature and its reproduction. In reality, this vision amounts to little more than a renewed strategy for profiting on planetary destruction.

Behind this tragedy-cum-farce is a distorted accounting deeply rooted in the workings of the system that sees wealth entirely in terms of value generated through exchange. In such a system, only commodities for sale on the market really count. External nature—water, air, living species—outside this system of exchange is viewed as a "free gift." Once such blinders have been put on, it is possible to speak, as the leading U.S. climate economist William Nordhaus has, of the relatively unhindered growth of the economy a century or so from now, under conditions of business as usual—despite the reality that leading climate scientists see following the identical path over the same time span as absolutely catastrophic both for human civilization and life on the planet as a whole.[1]

Such widely disparate predictions from mainstream economists and natural scientists are due to the fact that, in the normal reckoning of the capitalist system, both nature's contribution to wealth and the destruction of natural conditions are largely invisible. Insulated in their cocoon, orthodox economists either implicitly deny the existence of nature altogether or assume that it can be completely subordinated to narrow, acquisitive ends.

This fatal flaw of received economics can be traced back to its conceptual foundations. The rise of neoclassical economics in the late nineteenth and early twentieth centuries is commonly associated with the rejection of the labor theory of value of classical political economy and its replacement by notions of marginal utility/productivity. What is seldom recognized, however, is that another critical perspective was abandoned at the same time: the distinction between wealth and value (use value and exchange value). With this was lost the possibility of a broader ecological and social conception of wealth. These blinders of orthodox economics, shutting out the larger natural and human world, were challenged by figures inhabiting what John Maynard Keynes called the "underworlds" of economics. This included critics such as James Maitland (Earl of Lauderdale), Karl Marx, Henry George, Thorstein Veblen, and Frederick Soddy. Today, in a time of unlimited environmental destruction, such heterodox views are having a comeback.[2]

The Lauderdale Paradox

The ecological contradictions of the prevailing economic ideology are best explained in terms of what is known in the history of economics as the "Lauderdale Paradox." James Maitland, the eighth Earl of Lauderdale (1759–1839), was the author of *An Inquiry into the Nature and Origin of Public Wealth and into the Means and Causes of Its Increase* (1804). In the paradox with which his name came to be associated, Lauderdale argued that there was an inverse correlation between public wealth and private riches such that an increase in the latter often served to diminish the former. "Public wealth," he wrote, "may be accurately defined,—*to consist of all that man desires, as useful or delightful to him*." Such goods have use value and thus constitute wealth. But private riches, as opposed to wealth, required something additional (had an added limitation), consisting "*of all that man desires as useful or delightful to him; which exists in a degree of scarcity*."

Scarcity, in other words, is a necessary requirement for something to have value in exchange, and to augment private riches. But this is not the case for public wealth, which encompasses all value in use, and thus includes not only what is scarce but also what is abundant. This paradox led Lauderdale to argue that increases in scarcity in such formerly abundant but necessary elements of life as air, water, and food would, if exchange values were then attached to them, enhance individual private riches, and indeed the riches of the country—conceived of as "the sum-total of individual riches"—but only at the expense of the common wealth. For example, if one could monopolize water that had previously been freely available by placing a fee on wells, the measured riches of the nation would be increased at the expense of the growing thirst of the population.

"The common sense of mankind," Lauderdale contended, "would revolt" at any proposal to augment private riches "by creating a scarcity of any commodity generally useful and necessary to man." Nevertheless, he was aware that the bourgeois society in which he lived was already, in many ways, doing something of the very sort. He explained that Dutch colonialists in particularly fertile periods burned "spiceries" or paid natives to "collect the young blossoms or green leaves of the nutmeg trees" to kill them off; and that in plentiful years "the tobacco-planters in Virginia," by legal enactment, burned "a certain proportion of tobacco" for every slave working their fields. Such practices were designed to increase scarcity, augmenting private riches (and the wealth of a few) by destroying what constituted public wealth—in this case, the produce of the earth. "So truly is this principle understood by those whose interest leads them to take advantage of it," Lauderdale wrote, "that nothing but the impossibility of general combination protects the public wealth against the rapacity of private avarice."[3]

From the beginning, wealth, as opposed to mere riches, was associated in classical political economy with what John Locke called "intrinsic value," and what later political economists were to call "use value."[4] Material use values had, of course, always existed, and were the basis of human existence. But commodities produced for sale on the market under capitalism also embodied something else: exchange value (value). Every commodity was thus viewed as having "a twofold aspect," consisting of use value and exchange value.[5] The Lauderdale Paradox was nothing but an expression of this twofold aspect of wealth/value, which generated the contradiction between total public wealth (the sum of use values) and the aggregation of private riches (the sum of exchange values).

David Ricardo, the greatest of the classical-liberal political economists, responded to Lauderdale's paradox by underscoring the importance of keeping wealth and value (use value and exchange value) conceptually distinct. In line with Lauderdale, Ricardo stressed that if water, or some other natural resource formerly freely available, acquired an exchange value due to the growth of absolute scarcity, there would be "an actual loss of wealth" reflecting the loss of natural use values—even with an increase of private riches.[6]

In contrast, Adam Smith's leading French follower, Jean Baptiste Say, who was to be one of the precursors of neoclassical economics, responded to the Lauderdale Paradox by simply defining it away. He argued that wealth (use value) should be subsumed under value (exchange value), effectively obliterating the former. In his *Letters to Malthus on Political Economy and Stagnation of Commerce* (1821), Say thus objected to "the definition of which Lord Lauderdale gives of *wealth*." It was absolutely essential, in Say's view, to abandon altogether the identification of wealth with use value. As he wrote:

> *Adam Smith*, immediately having observed that there are two sorts of values, one *value in use*, the other *value in exchange*, completely abandons the first, and entirely occupies himself all the way through his book with *exchangeable value* only. This is what you yourself have done, Sir [addressing Malthus]; what Mr. Ricardo has done; what I have done; what we have all done: for this reason that there is no other value in political economy. . . . [Consequently,] wealth consists in the value of the things we possess; confining this word *value* to the only admitted and exchangeable value.

Say did not deny that there were "things indeed which are natural wealth, very precious to man, but which are not of that kind about which political economy can be employed." But political economy was to encompass in its concept of value—which was to displace altogether the concept of wealth—nothing but exchangeable value. Natural or public wealth, as opposed to value in exchange, was to be left out of account.[7]

Nowhere in liberal political economy did the Lauderdale Paradox create more convolutions than in what Marx called the "shallow syncretism" of John Stuart Mill.[8] Mill's *Principles of Political Economy* (1848) almost seemed to collapse at the outset on this basis alone. In the "Preliminary Remarks" to his book, Mill declared (after Say) that "wealth, then, may be defined, [as] all useful or agreeable things which possess exchangeable value"—thereby essentially reducing wealth to exchange value. But Mill's characteristic eclecticism and his classical roots led him also to expose the larger irrationality of this, undermining his own argument. Thus we find in the same section a penetrating treatment of the Lauderdale Paradox, pointing to the conflict between capital accumulation and the wealth of the commons. According to Mill:

> Things for which nothing could be obtained in exchange, however useful or necessary they may be, are not wealth in the sense in which the term is used in Political Economy. Air, for example, though the most absolute of necessaries, bears no price in the market, because it can be

> obtained gratuitously: to accumulate a stock of it would yield no profit or advantage to any one; and the laws of its production and distribution are the subject of a very different study from Political Economy. But though air is not wealth, mankind are much richer by obtaining it gratis, since the time and labour which would otherwise be required for supplying the most pressing of all wants, can be devoted to other purposes. It is possible to imagine circumstances in which air would be a part of wealth. If it became customary to sojourn long in places where the air does not naturally penetrate, as in diving-bells sunk in the sea, a supply of air artificially furnished would, like water conveyed into houses, bear a price: and if from any revolution in nature the atmosphere became too scanty for the consumption, or could be monopolized, air might acquire a very high marketable value. In such a case, the possession of it, beyond his own wants, would be, to its owner, wealth; and the general wealth of mankind might at first sight appear to be increased, by what would be so great a calamity to them. The error would lie in not considering that, however rich the possessor of air might become at the expense of the rest of the community, all persons else would be poorer by all that they were compelled to pay for what they had before obtained without payment.[9]

Mill signaled here, in line with Lauderdale, the possibility of a vast rift in capitalist economies between the narrow pursuit of private riches on an increasingly monopolistic basis and the public wealth of society and the commons. Yet despite these deep insights, Mill closed off the discussion with these "Preliminary Remarks," rejecting the Lauderdale Paradox in the end, by defining wealth simply as exchangeable value. What Say said with respect to Smith in the *Wealth of Nations*—that he entirely occupied "himself all the way through his book [after his initial definitions] with exchangeable value only"—therefore applied also to Mill in his *Principles of Political Economy*.[10] Nature was not to be treated as wealth but as something offered "gratis," as a free gift from the standpoint of capitalist value calculation.

Marx and the Lauderdale Paradox

In opposition to Say and Mill, Marx, like Ricardo, not only held fast to the Lauderdale Paradox but also made it his own, insisting that the contradictions between use value and exchange value, wealth and value, were intrinsic to capitalist production. In *The Poverty of Philosophy*, he responded to Proudhon's confused treatment (in *The Philosophy of Poverty*) of the opposition between use value and exchange value by pointing out that this contradiction had been explained most dramatically by Lauderdale, who had "founded his system on the inverse ratio of the two kinds of value." Indeed, Marx built his entire critique of political economy in large part around the contradiction between use value and exchange value, indicating that this was one of the key components of his argument in *Capital*. Under capitalism, he insisted, nature was rapaciously mined for the sake of exchange value: "The earth is the reservoir, from whose bowels the use-values are to be torn."[11]

This stance was closely related to Marx's attempt to look at the capitalist economy simultaneously in terms of its economic-value relations and its material transformations of nature. Thus Marx was the first major economist to incorporate the new notions of energy and entropy, emanating from the first and second laws of thermodynamics, into his analysis of production.[12] This can be seen in his treatment of the metabolic rift—the destruction of the metabolism between human beings and the soil, brought on by the shipment of food and fiber to the city, where nutrients withdrawn from the soil, instead of returning to the earth, ended up polluting the air and the water. In this conception, both nature and labor were robbed, since both were deprived of conditions vital for their reproduction: not "fresh air" and water but "polluted" air and water, Marx argued, had become the mode of existence of the worker.[13]

Marx's analysis of the destruction of the wealth of nature for the sake of accumulation is most evident in his treatment of capitalist ground rent and its relation to industrial agriculture. Ricardo had rooted his agricultural rent theory in "the original and indestructible powers of the soil"; Marx replied that "the soil has no 'indestructible powers'"—in the sense that it could be degraded, that is, subject to conditions of ecological destruction. It is here in Marx's treatment of capitalist agriculture that the analysis of the metabolic rift and the Lauderdale Paradox are brought together within his overall critique. It is here, too, that he frequently refers to sustainability as a material requirement for any future society—the need to protect the earth for "successive generations." A condition of sustainability, he insisted, is the recognition that no one (not even an entire society or all societies put together) owns the earth—which must be preserved for future generations in accordance with the principles of good household management. For a sustainable relation between humanity and the earth to be possible under modern conditions, the metabolic relation between human beings and nature needs to be rationally regulated by the associated producers in line with their needs *and* those of future generations. This means that the vital conditions of life and the energy involved in such processes need to be conserved.[14]

Few things were more important, in Marx's view, than the abolition of the big private monopolies in land that divorced the majority of humanity from: (1) a direct relation to nature, (2) the land as a means of production, and (3) a communal relation to the earth. He delighted in quoting at length from Herbert Spencer's chapter in his *Social Statics* (1851), "The Right to the Use of the Earth." There, Spencer openly declared: "Equity . . . does not permit property in land, or the rest would live on the earth by sufferance only. . . . It is impossible to discover any mode in which land can become private property. . . . A claim to the exclusive possession of the soil involves land-owning despotism." Land, Spencer insisted, properly belongs to "the great corporate body—society." Human beings were "co-heirs" to the earth.[15]

Although Marx usually looked at nature from an exclusively human perspective, in terms of sustaining use values, he also referred at times to nature's right not to be reduced to a mere commodity. Thus, he quoted Thomas Müntzer's famous objection that, in the developing bourgeois society, "all creatures have been made into property, the fish in the water, the birds in the air, the plants on the earth—all living things must also become free."[16]

Ecology and the Labor Theory of Value

Ironically, green thinkers (both non-socialist and socialist) frequently charge that the labor theory of value, to which Marx adhered in his critique of capitalism, put him in direct opposition to the kind of ecologically informed value analysis that is needed today. In *Small Is Beautiful*, E. F. Schumacher observed that in modern society there is an inclination "to treat as valueless everything that we have not made ourselves. Even the great Dr. Marx fell into this devastating error when he formulated the so-called 'labour theory of value.'" Luiz Barbosa, a contributor to *Twenty Lessons in Environmental Sociology* (2009), has written that Marx "believed raw materials are given to us gratis (for free) by nature and that it is human labor that gives it value. Thus, Marx failed to notice the intrinsic value of nature." Eco-socialist Jean-Paul Deléage has complained that, in making labor the only source of value, Marx "attributes no intrinsic value to natural resources." Social ecologist Mathew Humphrey gives credence to the view that "Marx's attachment to the labour theory of value in which non-human nature is perceived as valueless" can be taken as an indication of "his anthropocentric outlook."[17]

Here it is important to understand that certain conceptual categories that Marx uses in his critique of political economy, such as nature as a "free gift" and the labor theory of value itself, were inventions of classical-liberal political economy that were integrated into Marx's *critique* of classical political economy—insofar as they exhibited the real tendencies and contradictions of the system. Marx employed these concepts in an argument aimed at transcending bourgeois society and its limited social categories. The idea that nature was a "free gift" for exploitation was explicitly advanced by the physiocrats, and by Adam Smith, Thomas Malthus, David Ricardo, and John Stuart Mill—well before Marx.[18] Moreover, the idea was perpetuated in mainstream economics long after Marx. Although accepting it as a reality of bourgeois political economy, Marx was nevertheless well aware of the social and ecological contradictions embedded in such a view. In his *Economic Manuscripts of 1861–63*, he repeatedly attacked Malthus for falling back on this "physiocratic notion" of the environment as "a gift of nature to man," while failing to recognize that the concrete appropriation of nature for production—and the entire value framework built upon this in capitalist society—was in fact associated with historically specific social relations.[19] For Marx, with his emphasis on the need to protect the earth for future generations, the capitalist expropriation of the environment as a free object simply pointed to the contradiction between natural wealth and a system of accumulation of capital that systematically "robbed" it.

Nevertheless, since the treatment of nature as a "free gift" was intrinsic to the workings of the capitalist economy, it continued to be included as a *basic proposition underlying neoclassical economics*. It was repeated as an axiom in the work of the great late-nineteenth-century neoclassical economist Alfred Marshall, and has continued to be advanced in orthodox economic textbooks. Hence, the tenth edition of *Economics* (1987), a widely used introductory textbook by Campbell McConnell, states the following: "Land refers to all natural resources—all 'free gifts of nature'—which are useable in the production process." And further along in the same book we find: "Land has no production cost; it is a 'free and nonreproducible gift of nature.'"[20] Indeed, so crucial is this notion to neoclassical economics that it continues to live on in mainstream environmental economics. For example, Nick Hanley, Jason F. Shogren, and Ben White state in their influential *Introduction to Environmental Economics* (2001) that "natural capital comprises all [free] gifts of nature."[21]

Green critics, with only the dimmest knowledge of classical political economy (or of neoclassical economics), often focus negatively on Marx's adherence to the labor theory of value—the notion that only labor generated value. Yet it is important to remember that the labor theory of value was not confined to Marx's critique of political economy but constituted the *entire basis* of classical-liberal political economy. Misconceptions pointing to the anti-ecological nature of the labor theory of value arise due to conflation of the categories of *value* and *wealth*—since, in today's received economics, these are treated synonymously. It was none other than the Lauderdale Paradox, as we have seen, that led Say, Mill, and others to abandon the autonomous category of wealth (use value)—helping to set the stage for the neoclassical economic tradition that was to follow. In the capitalist logic, there was no question that nature was valueless (a free gift). The problem, rather, was how to jettison the concept of wealth, as distinct from value, from the core framework of economics, since it provided the basis of a critical—and what we would now call "ecological"—outlook.

Marx, as noted, strongly resisted the jettisoning of the wealth-value distinction, going so far as to criticize other socialists if they embraced the "value equals wealth" misconception. If human labor were one source of wealth, he argued—one that became the basis of value under capitalism—nature was another indispensable source of wealth. Those who—falling prey to the commodity fetishism of capitalist value analysis—saw labor as the sole source of wealth were thus attributing "supernatural creative power" to it. "Labour," Marx pronounced at the beginning of the *Critique of the Gotha Programme*, "is *not the source* of all wealth. *Nature* is just as much the source of use values (and it is surely of such that material wealth consists!) as is labour, which itself is only the manifestation of a natural force, human labour power." In the beginning of *Capital*, he cited William Petty, the founder of classical political economy, who had said, "labour is the father of material wealth, the earth is its mother."[22] "Man and nature," Marx insisted, were "the two original agencies" in the creation of wealth, which "continue to cooperate." Capitalism's failure to incorporate nature into its value accounting, and its tendency to confuse value with wealth, were *fundamental contradictions of the regime of capital itself*. Those "who fault Marx for not ascribing value to nature," Paul Burkett has written, "should redirect their criticisms to capitalism itself."[23]

As with Lauderdale, only with greater force and consistency, Marx contended that capitalism was a system predicated on the accumulation of value, even at the expense of real wealth (including the social character of human labor itself). The capitalist, Marx noted, adopted as his relation to the world: "*Après moi le déluge*!"[24] Or, as he would frequently observe, capital had a vampire-like relation to nature—representing a kind of living death maintained by sucking the blood from the world.[25]

Unworldly Economists and Their Critics

Nevertheless, the whole classical conception of wealth, which had its highest development in the work of Ricardo and Marx, was to be turned upside down with the rise of neoclassical economics. This can be seen in the work of Carl Menger—one of the founders of the Austrian School of Economics and of neoclassical economics, more generally. In his *Principles of Economics* (1871, published only four years after Marx's *Capital*), Menger attacked the Lauderdale Paradox directly (indeed, the reference to it as a "paradox" may have originated with him), arguing that it was "exceedingly impressive at first glance," but was based on false distinctions. For Menger, it was important to reject both the use value/exchange value and wealth/value distinctions. Wealth was based on exchange, which was now seen as rooted in subjective utilities. Replying to both Lauderdale and Proudhon, he insisted that the deliberate production of scarcity in nature was beneficial (to capital). Indeed, standing Lauderdale on his head, he contended that it would make sense to encourage "a long continued diminution of abundantly available (non-economic) goods [such as, air, water, natural landscapes, since this] must finally make them scarce in some degree—and thus components of wealth, which is thereby increased." In the same vein, Menger claimed that mineral water could conceivably be turned eventually into an economic good due to its scarcity. What Lauderdale presented as a paradox or even a curse—the promotion of private riches through the destruction of public wealth—Menger, one of the precursors of neoliberalism in economics, saw as an end in itself.[26]

This attempt to remove the paradox of wealth from economics led to scathing indictments by Henry George, Thorstein Veblen, and Frederick Soddy, along with others within the underworld of economics. In his best-selling work, *Progress and Poverty* (1879), George strongly stressed the importance of retaining a social concept of wealth:

> Many things are commonly spoken of as wealth which in taking account of collective or general wealth cannot be considered as wealth at all. Such things have an exchange value . . . insomuch as they represent as between individuals, or between sets of individuals, the power of obtaining wealth; but they are not truly wealth [from a social standpoint], inasmuch as their increase or decrease does not affect the sum of wealth. Such are bonds, mortgages, promissory notes, bank bills, or other stipulations for the transfer of wealth. Such are slaves, whose value represents merely the power of one class to appropriate the earnings of another class. Such are lands, or other natural opportunities, the value of which is but the result of the acknowledgement in favor of certain persons of an exclusive right to their use, and which represents merely the power thus given to the owners to demand a share of the wealth produced by those who use them. . . . By enactment of the sovereign political power debts might be canceled, slaves emancipated, and land resumed as the common property of the whole people, without the aggregate wealth being diminished by the value of a pinch of snuff, for what some would lose others would gain.[27]

Carefully examining the changing definitions of wealth in economics, George roundly condemned Say, Mill, and the Austrian School for obliterating the notion of use value and defining wealth entirely in terms of exchange value. Produced wealth, he argued, was essentially the result of "exertion impressed on matter" and was to be associated with producible use values. Value came from labor. Like Marx, he drew upon the basic tenets of Greek materialism (most famously extolled by Epicurus and Lucretius), arguing that nothing can be created merely by labor; "nothing can come out of nothing."[28]

Other economic dissidents also challenged the narrow orthodox economic approach to wealth. Veblen contended that the main thrust of capitalist economics under the regime of absentee ownership was the seizure of public wealth for private benefit. Calling this the "American plan" because it had "been worked out more consistently and more extensively" in the United States "than elsewhere," he referred, in Lauderdale-like terms, to it as "a settled practice of converting all public wealth to private gain on a plan of legalised seizure"—marked especially by "the seizure of the fertile soil and its conversion to private gain." The same rapacious system had its formative stages in the United States in slavery and in "the debauchery and manslaughter entailed on the Indian population of the country."[29]

Soddy, the 1921 Nobel Prize winner in chemistry, was an important forerunner of ecological economics. He was an admirer of Marx—arguing that it was a common error to think that Marx saw the source of all wealth as human labor. Marx, Soddy noted, had followed Petty and the classical tradition in seeing labor as the father of wealth, the earth as the mother.[30] The bounty of nature was part of "the general wealth" of the world. Reviving the Lauderdale Paradox, in his critique of mainstream economics, Soddy pointed out that

> the confusion enters even into the attempt of the earlier [classical] economists to define . . . "wealth," though the modern [neoclassical] economist seems to be far too wary a bird to define even that. Thus we find that wealth consists, let us say, of the enabling requisites of life, or something equally unequivocal and acceptable, but, if it is to be had in unlimited abundance, like sunshine or oxygen or water, then it is not any longer wealth in the economic sense, though without either of these requisites life would be impossible.

In this, Soddy wrote, "the economist, ignorant of the scientific laws of life, has not arrived at any conception of wealth," nor given any thought to the costs to nature and society, given the degradation of the environment.[31] Turning to Mill's contorted treatment of the Lauderdale Paradox, Soddy referred to the "curious inversions" of those who, based on making market exchange the sole criterion of value/wealth, thought that the creation of scarcity with respect to food, fuel, air, and so on made humanity richer. The result was that "the economist has effectually impaled himself upon the horns of a very awkward dilemma."[32]

Despite the devastating criticisms arising from the underworld of economics, however, the dominant neoclassical tradition moved steadily away from any concept of social/public wealth, excluding the whole question of social (and natural) costs—within its main body of analysis. Thus, as ecological economist K. William Kapp explained in his landmark *Social Costs of Private Enterprise* in 1950, despite the introduction of an important analogue to the orthodox tradition with the publication of Pigou's *Economics of Welfare*, it remained true that the "analysis of social costs is carried on not within the main body of value and price theory but as a separate system of so-called welfare economics." Kapp traced the raising of the whole problem of social wealth/social costs to none other than Lauderdale, while viewing Marx as one of the most devastating critics of capitalism's robbing of the earth.[33]

The Return of the Lauderdale Paradox

Today Lauderdale's paradox is even more significant than it was when originally formulated in the early nineteenth century. Water scarcities, air pollution, world hunger, growing fuel shortages, and the warming of the earth are now dominant global realities. Moreover, attempts within the system to expand private riches by exploiting these scarcities, such as the worldwide drive to privatize water, are ever-present. Hence, leading ecological economist Herman Daly has spoken of "The Return of the Lauderdale Paradox"—this time with a vengeance.[34]

The ecological contradictions of received economics are most evident in its inability to respond to the planetary environmental crisis. This is manifested both in repeated failures to apprehend the extent of the danger facing us and in the narrow accumulation strategies offered to solve it. The first of these can be seen in the astonishing naïveté of leading orthodox economists—even those specializing in environmental issues—arising from a distorted accounting that measures exchange values but largely excludes use values, that is, issues of nature and public wealth. Thus William Nordhaus was quoted in *Science* magazine in 1991 as saying: "Agriculture, the part of the economy that is sensitive to climate change, accounts for just 3% of national output. That means there is no way to get a very large effect on the U.S. economy" just through the failure of agriculture. In this view, the failure of agriculture in the United States would have little impact on the economy as a whole! Obviously, this is not a contradiction of nature, but of the capitalist economy—associated with its inability to take into account material realities. Oxford economist Wilfred Beckerman presented the same myopic view in his book *Small Is Stupid* (1995), claiming that "even if the net output of [U.S.] agriculture fell by 50 per cent by the end of the next century this is only a 1.5 per cent cut in GNP." This view led him to conclude elsewhere that global warming under business as usual would have a "negligible" effect on world output. Likewise, Thomas Schelling, winner of the Bank of Sweden's Nobel Memorial Prize in Economic Sciences, wrote in *Foreign Affairs* in 1997: "Agriculture [in the developed world] is practically the only sector of the economy affected by climate, and it contributes only a small percentage—three percent in the United States—of national income. If agricultural productivity were drastically reduced by climate change, the cost of living would rise by one or two percent, and at a time when per capita income will likely have doubled."[35]

The underlying assumption here—that agriculture is the only part of the economy that is sensitive to climate change—is obviously false. What is truly extraordinary in such views, however, is that the blinders of these leading neoclassical economists effectively prevent even a ray of common sense from getting through. GDP measurements become everything, despite that such measurements are concerned only with economic value added and not with the entire realm of material existence. There is no understanding here of production as a system, involving nature (and humanity), outside of national income accounting. Even then, the views stated are astonishingly naïve—failing to realize that a decrease by half of agricultural production would necessarily have an extraordinary impact on the price of food. Today, with a "tsunami of hunger sweeping the world," and at least one billion people worldwide lacking secure access to food, these statements of only a decade ago by leading mainstream environmental economists seem criminal in their ignorance.[36]

The same distorted accounting, pointing to "modest projected impacts" on the economy from global warming, led Nordhaus in 1993 to classify climate change as a "second-tier issue," and to suggest that "the conclusion that arises from most economic studies is to impose modest restraints, pack up our tools, and concentrate on more pressing problems." Although he acknowledged that scientists were worried about the pending environmental catastrophe associated with current trends, the views of most economists were more "sanguine."[37]

None of this should surprise us. Capitalism's general orientation with respect to public welfare, as is well known, is a kind of trickle-down economics, in which resources and human labor are exploited intensively to generate immeasurable affluence at the top of society. This is justified by the false promise that some of this affluence will eventually trickle down to those below. In a similar way, the ecological promises of the system could be called "trickle-down ecology." We are told that by allowing unrestrained accumulation the environment will be improved through ever-greater efficiency—a kind of secondary effect. The fact that the system's celebrated efficiency is of a very restricted, destructive kind is hardly mentioned.

A peculiarity of capitalism, brought out by the Lauderdale Paradox, is that it feeds on scarcity. Hence, nothing is more dangerous to capitalism as a system than abundance. Waste and destruction are therefore rational for the system. Although it is often supposed that increasing environmental costs will restrict economic growth, such costs continue to be externalized under capitalism on nature (and society) as a whole. This perversely provides new prospects for private profits through the selective commodification of parts of nature (public wealth).

All of this points to the fact that there is no real feedback mechanism, as commonly supposed, from rising ecological costs to economic crisis, that can be counted on to check capitalism's destruction of the biospheric conditions of civilization and life itself. By the perverse logic of the system, whole new industries and markets aimed at profiting on planetary destruction, such as the waste management industry and carbon trading, are being opened up. These new markets are justified as offering partial, ad hoc "solutions" to the problems generated nonstop by capital's laws of motion.[38]

The growth of natural scarcity is seen as a golden opportunity in which to further privatize the world's commons. This tragedy of the privatization of the commons only accelerates the destruction of the natural environment, while enlarging the system that weighs upon it. This is best illustrated by the rapid privatization of fresh water, which is now seen as a new mega-market for global accumulation. The drying up and contamination of fresh water diminishes public wealth, creating investment opportunities for capital, while profits made from selling increasingly scarce water are recorded as contributions to income and riches. It is not surprising, therefore, that the UN Commission on Sustainable Development proposed, at a 1998 conference in Paris, that governments should turn to "large multinational corporations" in addressing issues of water scarcity, establishing "open markets" in water rights. Gérard Mestrallet, CEO of the global water giant Suez, has openly pronounced: "Water is an efficient product. It is a product which normally would be free, and our job is to sell it. But it is a product which is absolutely necessary for life." He further remarked: "Where else [other than in the monopolization of increasingly scarce water resources for private gain] can you find a business that's totally international, where the prices and volumes, unlike steel, rarely go down?"[39]

Not only water offers new opportunities for profiting on scarcity. This is also the case with respect to fuel and food. Growing fuel shortages, as world oil demand has outrun supply—with peak oil approaching—has led to increases in the prices of fossil fuels and energy in general, and to a global shift in agriculture from food crops to fuel crops.

This has generated a boom in the agrofuel market (expedited by governments on the grounds of "national security" concerns). The result has been greater food scarcities, inducing an upward spiral in food prices and the spiking of world hunger. Speculators have seen this as an opportunity for getting richer quicker through the monopolization of land and primary commodity resources.[40]

Similar issues arise with respect to carbon-trading schemes, ostensibly aimed at promoting profits while reducing carbon emissions. Such schemes continue to be advanced even though experiments in this respect have thus far failed to reduce emissions. Here, the expansion of capital trumps actual public interest in protecting the vital conditions of life. At all times, ruling-class circles actively work to prevent radical structural change in this as in other areas, since any substantial transformation in social-environmental relations would mean challenging the treadmill of production and launching an ecological-cultural revolution.

Indeed, from the standpoint of capital accumulation, global warming and desertification are blessings in disguise, increasing the prospects of expanding private riches. We are thus driven back to Lauderdale's question: "What opinion," he asked, "would be entertained of the understanding of a man, who, as the means of increasing the wealth of... a country should propose to create a scarcity of water, the abundance of which was deservedly considered one of the greatest blessings incident to the community? It is certain, however, that such a projector would, by this means, succeed in increasing the mass of individual riches."[41]

Numerous ecological critics have, of course, tried to address the contradictions associated with the devaluation of nature by designing new green accounting systems that would include losses of "natural capital."[42] Although such attempts are important in bringing out the irrationality of the system, they run into the harsh reality that the current system of national accounts *does* accurately reflect capitalist realities of the non-valuation/undervaluation of natural agents (including human labor power). To alter this, it is necessary to transcend the system. The dominant form of valuation in our age of global ecological crisis is a true reflection of capitalism's mode of social and environmental degradation—causing it to profit on the destruction of the planet.

In Marx's critique, value was conceived of as an alienated form of wealth.[43] Real wealth came from nature and labor power and was associated with the fulfillment of genuine human needs. Indeed, "it would be wrong," Marx wrote, "to say that labour which produces use-values is the *only* source of the wealth produced by it, that is of material wealth. . . . Use-value always comprises a natural element. . . . Labour is a natural condition of human existence, a condition of material interchange [metabolism] between man and nature." From this standpoint, Lauderdale's paradox was not a mere enigma of economic analysis but rather the supreme contradiction of a system that, as Marx stressed, developed only by "simultaneously undermining the original sources of all wealth—the soil and the worker."[44]

2. Rifts and Shifts

Humans depend on functioning ecosystems to sustain themselves, and their actions affect those same ecosystems. As a result, there is a necessary "metabolic interaction" between humans and the earth that influences both natural and social history. Increasingly, the state of nature is being defined by the operations of the capitalist system, as anthropogenic forces are altering the global environment on a scale that is unprecedented. The global climate is rapidly changing due to the burning of fossil fuels and deforestation. No area of the world's ocean is unaffected by human influence, as the accumulation of carbon, fertilizer runoff, and overfishing undermine biodiversity and the natural services it provides. The Millennium Ecosystem Assessment explains that the majority of the world's ecosystems are overexploited and/or polluted. Environmental problems are increasingly interrelated. James Hansen, the leading climatologist in the United States, warns that we are dangerously close to pushing the planet past its tipping point, setting off cascading environmental problems that will radically alter the conditions of nature.[1]

Although the ecological crisis has captured public attention, the dominant economic forces are attempting to seize the moment by assuring us that capital, technology, and the market can be employed so as to ward off any threats without a major transformation of society. For example, numerous technological solutions are proposed to remedy global climate change, including agrofuels, nuclear energy, and new coal plants that will capture and sequester carbon underground. The ecological crisis is thus presented as a technical problem that can be fixed within the current system, through better ingenuity, technological innovation, and the magic of the market. In this view, the economy will be increasingly dematerialized, reducing demands placed on nature.[2] The market will ensure that new avenues of capital accumulation are created in the very process of dealing with environmental challenges.

This line of thought ignores the drivers of the ecological crisis. The social metabolic order of capitalism is inherently anti-ecological, since it systematically subordinates nature in its pursuit of endless accumulation and production on ever-larger scales. Technical fixes to socio-ecological problems typically have unintended consequences and fail to address the root of the problems: the political-economic order. Rather than acknowledging metabolic rifts, natural limits, and/or ecological contradictions, capital seeks to play a shell game with the environmental problems it generates, moving them around rather than addressing the root causes.

One obvious way capital shifts around ecological problems is through simple geographic displacement—once resources are depleted in one region, capitalists search far and wide to seize control of resources in other parts of the world, whether by military force or markets. One of the drivers of colonialism was clearly the demand for more natural resources in rapidly industrializing European nations.

However, expanding the area under the control of global capitalism is only one of the ways in which capitalists shift ecological problems around. There is a qualitative dimension as well, whereby one environmental crisis is "solved" (typically only in the short term) by changing the type of production process and generating a different crisis, such as how the shift from the use of wood to plastic in the manufacturing of many consumer goods replaced the problems associated with wood extraction with those associated with plastics production and disposal.

Thus one problem is transformed into another—a shift in the type of rift. We illustrate these issues here by focusing on the soil crisis Marx identified in his time, which continues to the present, and our contemporary energy and climate crisis.

The Expanding Social Metabolic Order of Capital and Ecological Crisis

A metabolic relationship involves regulatory processes that govern the interchange of materials. Karl Marx noted that natural systems, such as the nutrient cycle, had their own metabolism, which operated independently of and in relation to human society, allowing for their regeneration and/or continuance. He employed the concept of social metabolism to refer to "the complex, dynamic interchange between human beings and nature" of matter and energy, which recognized how both "nature-imposed conditions" and human actions transform this process. Each mode of production creates a particular social metabolic order that determines the interchange between society and nature. Such interactions influence the ongoing reproduction of society and ecosystems.[3]

István Mészáros explains that a fundamental change in the social metabolism took place with the onset of capitalism, as a new social metabolic order came to dominate the material interchange between society and nature. Capitalism imposes a particular form of "productive interchange of human beings with nature," given that its very logic of operation is a "'*totalizing*' framework of control into which everything else, including human beings, must be fitted, and prove thereby their 'productive viability,' or perish if they fail to do so." Capitalists pursue their own interests to maximize profit, above and beyond any other interests, subsuming all natural and social relationships to the drive to accumulate capital. Natural cycles and processes are subjected to the whims of the economic cycle, given that "the only modality of time which is directly meaningful to capital is *necessary labor time* and its operational corollaries, as required for securing and safeguarding the conditions of *profit-oriented time-accountancy* and thereby the realization of capital on an extended scale." The competition of capital produces an "*ultimately uncontrollable mode of social metabolic control*" running roughshod over the regulatory processes that govern the complex relationships of interchange within natural systems and cycles.[4]

Paul Sweezy explained that the capitalist economic system "is one that never stands still, one that is forever changing, adopting new and discarding old methods of production and distribution, opening up new territories, subjecting to its purposes societies too weak to protect themselves." Thus the tendency of capital is to violate the natural conditions that ensure nature's vitality, undermining the base on which ecological and human sustainability depends. In part, this is because capital freely appropriates nature and its bounty—it is "purely a matter of utility." The exploitation of nature and labor serve "as a means to the paramount ends of profit-making and still more capital accumulation." Hence, the expansion and intensification of the social metabolic order of capital generates rifts in natural cycles and process, forcing a series of shifts on the part of capital, as it expands environmental degradation.[5]

Marx and the Metabolic Rift in Soil Nutrients

Capitalism's destructive metabolic relation to nature came into focus in the nineteenth century. The German chemist Justus von Liebig, in the 1850s and '60s, employed the concept of metabolism in his studies of soil nutrients. He explained that British agriculture, with its intensive methods of cultivation to increase yields for the market, operated as a system of robbery, destroying the vitality of the soil. Liebig detailed how the soil required specific nutrients—nitrogen, phosphorus, and potassium—to maintain its ability to produce crops. As crops grew they took up these nutrients. In earlier societies, the produce of nature was often recycled back to the land, fertilizing it. But the concentration of land ownership, which involved the depopulation of rural areas, and the increasing division between town and country, changed this process. Food and fiber were shipped from the countryside to distant markets. In this, the nutrients of the soil were transferred from the country to the city where they accumulated as waste and contributed to the pollution of the cities, rather than being returned to the soil. This caused a rupture in the nutrient cycle.

Marx, who was influenced by Liebig's work, recognized that soil fertility and the conditions of nature were bound to the historical development of social relations. Through his studies of soil science, Marx gained insights in regard to the nutrient cycle and how soil exhaustion was caused. On this basis he provided a materialist critique of modern agriculture, describing how capitalist operations inevitably produced a metabolic rift, as the basic processes of natural reproduction were undermined, preventing the return to the soil of the necessary nutrients.[6]

The transfer and loss of nutrients was tied to the accumulation process. Marx described how capital creates a rupture in the "metabolic interaction" between humans and the earth, one that is only intensified by large-scale agriculture, long-distance trade, and massive urban growth. With these developments, the nutrient cycle was interrupted and the soil continually impoverished. He explained that the drive for the accumulation of capital "reduces the agricultural population to an ever decreasing minimum and confronts it with an ever growing industrial population crammed together in large towns; in this way it produces conditions that provoke an irreparable rift in the interdependent process of social metabolism, a metabolism prescribed by the natural laws of life itself. The result of this is a squandering of the vitality of the soil, which is carried by trade far beyond the bounds of a single country."[7]

The development of capitalism, whether through colonialism, imperialism, or market forces, expanded the metabolic rift to the global level, as distant regions across the oceans were brought into production to serve the interests of capitalists in core nations. While incorporating distant lands into the global economy—a form of geographical displacement—helped relieve some of the demands placed on agricultural production in core nations, it did not serve as a remedy to the metabolic rift. The systematic expansion of production on a larger scale subjected more of the natural world to the dictates of capital. The consequence of this, as Marx noted, is that "it disturbs the metabolic interaction between man and the earth, i.e. it prevents the return to the soil of its constituent elements consumed by man in the form of food and clothing; hence it hinders the operation of the eternal natural condition for the lasting fertility of the soil."[8]

Rifts, Shifts, and the Soil Crisis

The metabolic rift in the nutrient cycle and degradation of the soil hastened the concentration of agricultural production among a small number of proprietors who adopted ever more intensive methods of production to further expand and enhance production. The logic of capital and competition drives "bourgeois production out of its old course and . . . compels capital to intensify the productive forces of labour." It gives "capital no rest," Marx noted, "and continually whispers in its ear: 'Go on! Go on!'"[9] This sets off a series of rifts and shifts, whereby metabolic rifts are continually created and addressed—typically only after reaching crisis proportions—by shifting the type of rift generated. To the myopic observer, capitalism may appear at any one moment to be addressing some environmental problems, since it does on occasion mitigate a crisis. However, a more far-sighted observer will recognize that new crises spring up where old ones are supposedly cut down. This is unavoidable given that capital is propelled constantly to expand.

One of the consequences of the metabolic rift and declining soil fertility in core nations in the 1800s was the development of an international guano/nitrate trade. Guano (bird droppings) from islands off the coast of Peru with large seabird colonies had high concentrations of phosphate and nitrogen. At the time, guano was recognized as one of the best fertilizers, both enriching the soil and increasing the yield. This new fertilizer sparked an international scramble to claim islands that had guano deposits. Millions of tons of guano were dug up by imported Chinese "coolies" in Peru under conditions worse than slave labor and exported to the United States and European nations. The necessity to import fertilizer reflected a crisis in capitalist agriculture, but it did not mend the metabolic rift. Rather, it redirected a natural resource, which had been used for centuries to enrich the soils of Peru, to the global market, rapidly diminishing the reserves on the islands. The nitrate trade pitted Peru and Bolivia against Chile in the War of the Pacific, a war encouraged and supported by nitrate investors in Britain. It was a war between poor countries struggling to control guano and nitrate fields that were used to meet the fertilizer demands of core nations. Nonetheless, the metabolic rift in regard to soil degradation continued to plague core capitalist nations. It was only when Fritz Haber, a German chemist and nationalist, devised a process—just before the First World War—for fixing nitrogen from the air that a radical shift took place in agriculture, as artificial nitrogen fertilizer was produced in large quantities and applied to soils to sustain yields.[10]

The social metabolic order of capital undermined the natural cycles and processes that allow for the regeneration and use of soil nutrients. Various geological and technological shifts were introduced to maintain production, but they created new rifts while not alleviating old ones. Increasingly, industrial processes were incorporated into agricultural practices, intensifying the social metabolism of society. As a result, agriculture became increasingly dependent upon industrial operations and materials—such as the industrial fixation of nitrogen— in order to continue. Even in his day, Marx recognized the transformations taking place in agriculture and noted: "Agriculture no longer finds the natural conditions of its own production within itself, naturally, arisen, spontaneous, and ready to hand, but these exist as an independent industry separate from it—and, with this separateness the whole complex set of interconnections in which this industry exists is drawn into the sphere of the conditions of agricultural production."[11]

Marx explained that the expansion of capitalist industrialized operations increased the scale of exploitation and environmental degradation, subjecting nature to the rapacious logic of capital:

> Large-scale industry and industrially pursued large-scale agriculture have the same effect. If they are originally distinguished by the fact that the former lays waste and ruins labour-power and thus the natural power of man, whereas the latter does the same to the natural power of the soil, they link up in the later course of development, since the industrial system applied to agriculture also enervates the workers there, while industry and trade for their part provide agriculture with the means of exhausting the soil.[12]

Technology is not neutral, given that it embodies capitalist relations, whether it is to facilitate the division of labor or to increase production through the exploitation of labor and nature. Technological innovations serve as an additional means to enlarge and expand the social metabolic order of capital. In regard to capitalist agriculture, Marx explicated: "All progress in capitalist agriculture is a progress in the art, not only of robbing the worker, but of robbing the soil; all progress in increasing the fertility of the soil for a given time is a progress towards ruining the more long-lasting sources of that fertility. . . . Capitalist production, therefore, only develops the techniques and the degree of combination of the social process of production by simultaneously undermining the original sources of all wealth—the soil and the worker."[13]

Chemical processes and inputs were initiated in agriculture to duplicate, replace, and/or reproduce natural operations and what they produce. Synthetic fertilizer was widely introduced to sustain and increase agricultural production, but it did not resolve the metabolic rift in the nutrient cycle. Karl Kautsky, drawing upon the work of Marx and Liebig, explained that artificial fertilizers:

> allow the reduction in soil fertility to be avoided, but the necessity of using them in larger and larger amounts simply adds a further burden

> to agriculture—not one unavoidably imposed on nature, but a direct result of current social organization. By overcoming the antithesis between town and country . . . the materials removed from the soil would be able to flow back in full. Supplementary fertilisers would then, at most, have the task of enriching the soil, not staving off its impoverishment. Advances in cultivation would signify an increase in the amount of soluble nutrients in the soil without the need to add artificial fertilisers.[14]

Kautsky identified the creation of a fertilizer treadmill, whereby a continuous supply of artificial fertilizer was needed to produce high yields on land that was exhausted. Diminished natural conditions, such as depleted soils, forced capital to shift its operations in order to continue production. Rather than solving the problem, this shift and the ones that followed created additional environmental problems, escalating the magnitude of the ecological crisis.

Food production has increased through expanding agricultural production to less fertile land—depleting the nutrients in these areas—and through the incorporation in the agricultural process of large quantities of oil, used in the synthesis of chemical fertilizers and pesticides, contributing to global climate change as well as a myriad of other environmental problems. Modern agriculture has become the art of turning oil into food.[15] Constant inputs are needed simply to sustain this operation, given the depletion of the soil. Genetically modified crops are developed to grow in arid, depleted soils with the help of artificial fertilizer. Each step is an attempt to overcome barriers for the sake of accumulation, regardless of the ecological implications.

The incorporation of the "technological fix" of artificial nitrogen fertilizer has created additional ecological rifts and other environmental problems. The production of synthetic fertilizer produces airborne nitrogen compounds that increase global warming. Nitrogen runoff overloads marine ecosystems with excess nutrients, which compromise natural processes that generally remove nutrients from the waterways. The increased concentration of nutrients within the water causes eutrophication. This leads to oxygen-poor water and the formation of hypoxic zones—otherwise known as "dead zones" because crabs and fishes suffocate within these areas.

Hence the shifting logic of ecological destruction spreads rifts throughout the system. The drive to increase agricultural production, the separation of town and country, and the loss of soil nutrients produce a metabolic rift in the soil nutrient cycle. In an attempt to overcome natural limits, capital engages in a series of shifts to sustain production, importing natural fertilizers and producing artificial fertilizer. As a result, the social metabolism is intensified, as more of nature is subjected to the demands of capital, and additional ecological problems are created.

Energy and Climate Crisis

The development of energy production technologies provides one of the best examples of rifts and shifts, as technological fixes for energy problems create new ecological crises in the attempt to alleviate old ones. Biomass, particularly wood, has, of course, been one of the primary energy sources humans have depended on throughout their history. The development of more energy intensive processes, such as the smelting of metals, was therefore connected with greater pressure on forests, as trees were fed to the fires. By the time the Industrial Revolution began to emerge in Europe, vast regions of the continent had already been deforested, particularly in areas close to major sites of production, and much of this deforestation was driven by the demand for fuel. As industrialization advanced, new sources of power were desired to fuel the machines that allowed for production to take place on a growing scale. Whole forests could be devoured at an unprecedented rate, making wood ever more scarce. The tension between the desire of the capitalist owners of the new industrial technologies for expanding the accumulation of capital and the biophysical limits of Earth were apparent from the start of the Industrial Revolution. However, capitalists did not concern themselves with the internal/external contradictions of capitalism, except insofar as they were barriers to be transcended. Thus efforts to achieve what we would today call sustainability were not even considered by the elite. Rather, coal, and subsequently other fossil fuels, quickly became the standard fuel of industry, temporarily sidestepping the fuelwood crisis—although forests continued to fall due to the many demands placed on them—but laying the foundations for our current global climate change crisis by dramatically increasing the emission of carbon dioxide.[16]

The pattern has remained similar to how it was in the early years of the Industrial Revolution. Oil was quickly added to coal as a fuel source and a variety of other energy sources were increasingly exploited. Among these was hydropower, the generation of which requires damming rivers, and thus destroying aquatic ecosystems. For example, the expansion of hydropower over the twentieth century in the Pacific Northwest was the primary force leading to the widespread depletion and extinction of salmon runs. Nuclear power was the most controversial addition to the power mix. Despite initial claims that it would provide clean, unlimited power that would be too cheap to meter, it turned out to be an expensive, risky power source that produced long-lived highly radioactive waste for which the development of truly safe, sufficiently long-term, storage sites has proven elusive.

Now, in the twenty-first century, with global climate change finally being recognized by the elite as a serious problem, the proposed solutions are to shift the problem from one form of energy to a new form of energy. Nuclear power, despite its drop in popularity toward the end of the last century, due to high costs and widespread public opposition, is now very much back on the agenda, with new promises of how the new nuclear plants are safer—never mind the issue of radioactive waste. We are also regaled with promises of agrofuels, ironically bringing us back to the pre-coal energy crisis. Recent scientific reports note that growing crops for agrofuel to feed cars may actually increase the carbon emitted into the atmosphere.[17] But even this ignores the fact that the production of agrofuel would be based on unsustainable agricultural practices that demand massive inputs of fertilizers and would only further the depletion of soil nutrients, bringing us back to the metabolic rift that Marx originally addressed.

Two recent examples of technical approaches to mitigating climate change are particularly illustrative of how technological optimism distracts us from the political-economic sources of our environmental problems. Nobel Laureate Paul Crutzen, who admirably played a central role in identifying and analyzing human-generated ozone depletion in the stratosphere, has argued that climate change can be avoided by injecting sulfur particles into the stratosphere to increase the albedo of Earth and reflect more of the sun's energy back into space, which would counter the warming stemming from rising concentrations of greenhouse gases. Although doubtless offered sincerely and out of desperation stemming from the failure of those in power to adequately address the mounting climate crisis, the technical framing of the climate change issue makes it easy for political and business leaders to avoid addressing greenhouse gas emissions, since they can claim that technical fixes make it unnecessary to take action to preserve forests and curtail the burning of fossil fuels. Engineering the atmosphere on this scale is likely to have many far-reaching consequences (acid rain being only the most obvious), many of which have not been anticipated.

In a similar vein, well-known physicist Freeman Dyson recently suggested that we can avoid global climate change by replacing one-quarter of the world's forests with genetically engineered carbon-eating trees. The ecological consequences of such an action would likely be extraordinary.

Both of these so-called solutions avoid addressing an economic system that is largely structured around burning fossil fuels and must constantly renew itself on a larger scale as it runs roughshod over nature. Often techno-solutions are proposed as if completely removed from the world as it operates, without any sense of the social and economic relations of power. The irony is that such narrowly conceived "solutions" would only serve as a means to prop up the very forces driving ecological degradation, allowing those forces to continue to operate, as they create additional ecological rifts.[18]

Toward a New Social Metabolic Order

The pursuit of profit is the immediate pulse of capitalism as it reproduces itself on an ever-larger scale. A capitalist economic system cannot function under conditions that require accounting for the reproduction of nature, which may include timescales of a hundred years or more, not to mention maintaining the particular, integrated natural cycles that help sustain living conditions. The social metabolic order of capital is characterized by rifts and shifts, as it freely appropriates nature and attempts to overcome, even if only temporarily, whatever natural and social barriers its confronts. In this, Marx noted, capital turns to problems with "the land only after its influence has exhausted it and after it has devastated its natural qualities." And at this point, it only makes shifts or proposes technological fixes to address the pressing concern, without addressing the fundamental crisis, the force driving the ecological crisis—capitalism itself. Mészáros warns: "In the absence of miraculous solutions, capital's arbitrarily self-asserting attitude to the objective determinations of causality and time in the end inevitably brings a bitter harvest, at the expense of humanity [and nature itself]."[19]

The global reach of capital is creating a planetary ecological crisis. A fundamental structural crisis cannot be remedied within the operations of the system. Marx explained that the future could be ruined and shortened as a result of a social metabolism that exhausted the conditions of life. Capital shows no signs of slowing down, given its rapacious character. The current ecological crisis has been in the making for a long time, and the most serious effects of continuing with business as usual will not fall on present but rather future generations. But, as James Hansen warns, the time that we have to respond and change the forces driving the ecological crisis is getting shorter and shorter. Each delay in taking decisive action compounds the problem and makes the necessary intervention that much larger.[20]

Capitalism is incapable of regulating its social metabolism with nature in an environmentally sustainable manner. Its very operations violate the laws of restitution and metabolic restoration. The constant drive to renew the capital accumulation process intensifies its destructive social metabolism, imposing the needs of capital on nature, regardless of the consequences to natural systems. Capitalism continues to play out the same failed strategy again and again. The solution to each environmental problem generates new environmental problems (and often does not curtail the old ones). One crisis follows another, in an endless succession of failure, stemming from the internal contradictions of the system. If we are to solve our environmental crises, we need to go to the root of the problem: the social relation of capital itself, given that this social metabolic order undermines "the vital conditions of existence."[21]

Resolving the ecological crisis requires in the end a complete break with the logic of capital and the social metabolic order it creates, which does not mean we cannot take beneficial actions *within* the present system—although these will necessarily go against the *internal logic* of the system. Marx proposed that a society of associated producers served as the basis for potentially bringing the social metabolism in line with the natural metabolism in order to sustain the inalienable condition for the existence and reproduction of the chain of human generations. Given that human society must always interact with nature, concerns regarding the social metabolism are a constant, regardless of the society. But a mode of production in which associated producers can regulate their exchange with nature in accordance with natural limits and laws, while retaining the regenerative properties of natural processes and cycles, is fundamental to an environmentally sustainable social order.

As Frederick Engels stressed: "Freedom does not consist in the dream of independence from natural laws, but in the knowledge of these laws." In fact, "real human freedom" requires living "an existence in harmony with the laws of nature that have become known."[22]

A multitude of environmental problems, each with its own dynamics (although increasingly interrelated), are causing the ecosystems on which lives depend to collapse. We confront an ecological crisis at the global level generated by particular social forces. Rather than perpetuating a social metabolic order that generates metabolic rifts and ecological crises, merely attempting to shift the problems around, we need to transcend this system, to create a social metabolism that allows nature to replenish and restore itself within timescales relevant to its continued reproduction.

3. Capitalism in Wonderland

In a 2008 essay in *Nature*, "Economics Needs a Scientific Revolution," physicist Jean-Philippe Bouchaud, a researcher for an investment management company, asks rhetorically, "What is the flagship achievement of economics?" Bouchaud's answer: "Only its recurrent inability to predict and avert crises."[1] Although his discussion is focused on the current worldwide financial crisis, his comment applies equally well to mainstream economic approaches to the environment—where, for example, ancient forests are seen as non-performing assets to be liquidated, and clean air and water are luxury goods for the affluent to purchase at their discretion. The field of economics in the United States has long been dominated by thinkers who unquestioningly accept the capitalist status quo and, accordingly, value the natural world only in terms of how much short-term profit can be generated by its exploitation. As a result, the inability of received economics to cope with or even perceive the global ecological crisis is alarming in its scope and implications.

Bouchaud penetratingly observes, "The supposed omniscience and perfect efficacy of a free market stems from economic work done in the 1950s and '60s, which with hindsight looks more like propaganda against communism than plausible science." The capitalist ideology that undergirds economics in the United States has led the profession to be detached from reality, rendering it incapable of understanding many of the crises the world faces. Mainstream economics' obsession with the endless growth of GDP—a measure of "value added," not of human well-being or the intrinsic worth of ecosystems and other species—and its failure to recognize the fundamental ecological underpinnings of the economy has led to more than simply an inability to perceive the deterioration of the global environment. The problem goes much deeper. Orthodox economics, like the capitalist system that it serves, leads to an "*Après moi le déluge!*" philosophy that is anything but sustainable in orientation. As Naomi Klein has said, there is something perversely "natural" about "disaster capitalism."[2]

Economists in Wonderland

The inherent incapacity of orthodox or neoclassical economics to take ecological and social costs into account was perhaps best exemplified in the United States by the work of Julian Simon. In articles and exchanges in *Science* and *Social Science Quarterly* and in his book *The Ultimate Resource* published at the beginning of the 1980s, he insisted that there were no serious environmental problems, that there were no environmental constraints on economic or population growth, and that there would never be long-term resource shortages. For example, he infamously claimed that copper (an element) could be made from other metals and that only the mass of the universe, not that of the earth, put a theoretical limit on how much copper could be produced. The free market if left unfettered, he contended, would ensure continuous progress into the distant future. These and other dubious assertions led ecologist Paul Ehrlich to refer to Simon as "an economist in Wonderland."[3]

Apologists for capitalism continue to occupy Wonderland, because it is only in Wonderland that environmental problems either do not really exist or can be solved by capitalism, which can also improve the quality of life for the mass of humanity. Bjørn Lomborg, a Danish statistician and political scientist (now an adjunct professor at the Copenhagen Business School), picked up Simon's torch, publishing his salvo aimed at environmentalism, *The Skeptical Environmentalist*, in 2001. Lomborg argued, for example, that attempting to prevent climate change would cost more and cause more harm than letting it happen. Lomborg's book was immediately praised to the skies by the mass media, which was looking for a new anti-environmental crusader. Soon after the publication of *The Skeptical Environmentalist*, environmental scientists documented the countless flaws (not all of them inadvertent) in Lomborg's reasoning and evidence. *Scientific American* devoted part of an issue to four articles by leading scientists sharply criticizing Lomborg. As a result of its serious flaws, the book was rejected by the scientific community. Yet despite the adamant rejection of *The Skeptical Environmentalist* by natural scientists, all of this seemed only to add to Lomborg's celebrity within the corporate media system. *The Economist* touted the book and its conclusions, proclaiming it to be "one of the most valuable books on public policy," having dispelled the notion of "looming environmental disaster" and "the conviction that capitalism is self-destructive."[4] *Time* magazine in 2004 designated Lomborg as one of the 100 most influential people in the world; and in 2008 Britain's *Guardian* newspaper labeled him as one of the "50 people who could save the planet."

In 2003 Lomborg organized what he called the "Copenhagen Consensus" to rank the world's leading problems. This was carried out through the writing of a number of reports on various global priorities by a group of hand-picked, mainly economic authorities, and then the subsequent ranking of these problems by eight "experts"—all economists, since economists were declared to be the only experts on "economic prioritization," that is, decisions on where to put society's resources. The eight Copenhagen Consensus economists not surprisingly all ranked climate change at or near the bottom of the world's agenda, backing up Lomborg's position.[5]

Lomborg's 2007 book *Cool It: The Skeptical Environmentalist's Guide to Global Warming* was an extended attack on the Kyoto Protocol and all attempts to carry out substantial cuts in greenhouse gas emissions. For Lomborg the essential point was that, "all major peer-reviewed economic models agree that little emissions reduction is justified." He relied particularly on the work of Yale economist William Nordhaus, a leading economic contributor to the discussion of global warming, who has opposed any drastic reductions in greenhouse gases, arguing instead for a slow process of emissions reduction, on the grounds that it would be more economically justifiable.[6]

Economists versus Natural Scientists

Needless to say, establishment economists, virtually by definition, tend to be environmental skeptics. Yet they have an outsized influence on climate policy as representatives of the dominant end of capitalist society, before which all other ends are subordinated. (Social scientists other than economists either side with the latter in accepting accumulation as the primary goal of society or are largely excluded from the debate.) In sharp contrast, natural and physical scientists are increasingly concerned about the degradation of the planetary environment, but have less direct influence on social policy responses.

Mainstream economists are trained in the promotion of private profits as the singular "bottom line" of society, even at the expense of larger issues of human welfare and the environment. The market rules over all, even nature. For Milton Friedman the environment was not a problem since the answer was simple and straightforward. As he put it: "Ecological values can find their natural space in the market, like any other consumer demand."[7]

Natural scientists, as distinct from economists, however, typically root their investigations in a materialist conception of nature and are engaged in the study at some level of the natural world, the conditions of which they are much more disposed to take seriously. They are thus much less inclined to underrate environmental problems.

The conflict between economists and natural scientists on global warming came out in the open as a result of an article by Nordhaus that appeared in the leading natural science journal, *Science*, in 1993.

Nordhaus projected that the loss to gross world output in 2100 due to continuation of global warming trends would be insignificant (about 1 percent of GDP in 2100). His conclusion clearly conflicted with the results of natural science since these same business-as-usual trends could lead, according to the UN Intergovernmental Panel on Climate Change (IPCC) scenarios at the time, to as much as a 5.8°C (10.4°F) increase in average global temperature, which for scientists was nothing less than catastrophic for civilization and life itself. Nordhaus had concluded in his article that attempts at emissions stabilization would be worse than inaction. This led to a number of strong replies by noted natural scientists (in letters to *Science*), who viewed Nordhaus's analysis as patently absurd.

Nordhaus subsequently defended his views by surveying a number of influential economists and scientists, asking them for their best guesstimates, and publishing his results in the *American Scientist* in 1994. The economists he chose to survey agreed with him that climate change would have little effect on the economy. Yet the natural scientists saw the consequences as potentially catastrophic. One physical scientist responded by claiming that there was a 10 percent chance under present trends of the complete destruction of civilization—similar views would likely be even more common today. Nordhaus observed that those who knew most about the economy were optimistic. Stephen Schneider, a Stanford biologist and climate scientist (and a leading critic of both Lomborg and Nordhaus), retorted that those who knew most about the environment were worried. As Schneider summed up the debate in 1997 in his *Laboratory Earth*: "Most conventional economists . . . thought even this gargantuan climate change [a rise in average global temperature of 6°C]—equivalent to the scale of change from an ice age to an interglacial epoch in a hundred years, rather than thousands of years—would have only a few percent impact on the world economy. In essence, they accept the paradigm that society is almost independent of nature."[8]

Orthodox economists, it is true, often project economic costs of global warming in 2100 to be only a few percentage points and therefore hardly significant, even at levels of climate change that would endanger most of the "higher" species on the planet and human civilization itself, costing hundreds of millions, if not billions, of human lives.

The failure of economic models to count the human and ecological costs of climate change should not surprise us. Bourgeois economics has a carefully cultivated insensitivity to human tragedy (not to mention natural catastrophe) that has become almost the definition of "man's inhumanity to man." Thomas Schelling, a recipient of the Bank of Sweden's Nobel Memorial Prize in Economic Sciences, and one of Lomborg's eight experts in the Copenhagen Consensus, is known for arguing that since the effects of climate change will fall disproportionately on the poorer nations of the global South, it is questionable how much in the way of resources the rich nations of the global North should devote to the mitigation of climate trends. (Schelling in his Copenhagen Consensus evaluation ranked climate change at the very bottom of world priorities.)[9] Here one can't help but be reminded of Hudson Institute planners, who in the process of proposing a major dam on the Amazon in the early 1970s in effect contended—as one critic put it at the time—that "if the flooding drowns a few tribes who were not evacuated because they were supposed to be on higher ground, or wipes out a few forest species, who cares?"[10] Similarly, Lawrence Summers, now Obama's top economic advisor, wrote an internal memo while chief economist of the World Bank, in which he stated: "The economic logic behind dumping a load of toxic waste in the lowest-wage country is impeccable and we should face up to that." He justified this by arguing: "The measurement of the costs of health-impairing pollution depends on the foregone earnings from increased morbidity and mortality. From this point of view a given amount of health-impairing pollution should be done in the country with the lowest cost, which will be the country of the lowest wages."[11]

Discounting the Future

Nordhaus—who ranks as one of the most influential mainstream economists on global warming today and is a cut above figures like Simon and Lomborg—has proposed, in his 2008 book *A Question of Balance: Weighing the Options on Global Warming Policies*, a go-it-slow strategy on combating greenhouse emissions.[12] Nordhaus demonstrates here that despite impressive credentials he remains hobbled by the same ideology that has crippled other mainstream economists. In essence this comes down to the belief that capitalism offers the most efficient response to questions of resource use, and indeed a sufficient answer to the world's problems.

A Question of Balance presents a fairly standard economic argument about how to address global climate change, although it is backed by Nordhaus's own distinctive analyses using sophisticated modeling techniques. He acknowledges that global climate change is a real problem, and is human generated, arguing that it is necessary to slowly move away from carbon-emitting energy sources. Nevertheless, the central failures of his approach are that it assigns value to the natural environment and human well-being using standard economic measures that are fundamentally inadequate for this purpose, and that it fails properly to incorporate the possibility that an ecological collapse could utterly undermine the economy, and indeed the world as we know it. These failures, which are those of mainstream economics, are clearly apparent in his approach to discounting for purposes of estimating how much effort should be put into reducing carbon emissions. Nordhaus argues we should only invest a modest amount of effort in reducing carbon emissions in the short term and slowly increase this over time—and justifies this by introducing a high discount rate.

The issue of discounting may seem esoteric to most people but it is not to economists, and deserves some examination. Discounting is fundamentally about how we value the future relative to the present—insofar as it makes any sense at all to attach numbers to such valuations. The "discount rate" can be thought of as operating in inverse relation to compound interest. While "compounding measures how much present-day investments will be worth in the future, discounting measures how much future benefits are worth today."[13] Estimation of the discount rate is based on two moral issues. First, there is the issue of how we value the welfare of future generations relative to present ones (the time discount rate). As Nordhaus states, "A zero discount rate means that all generations into the indefinite future are treated the same; a positive discount rate means that the welfare of future generations is reduced or 'discounted' compared with nearer generations." A catastrophe affecting humanity fifty years from now, given a discount rate of 10 percent, would have a "present value" less than 1 percent of its future cost. Second, there is the issue of how wealthy future generations will be relative to present ones and whether it is appropriate to shift costs from the present to the future. If we assume a high rate of economic growth into the indefinite future, we are more likely to avoid investing in addressing problems now, because we assume that future generations will be wealthier than we are and can better afford to address these problems, even if the problems become substantially worse.[14]

The difficulty of the discount rate, as environmental economist Frank Ackerman has written, is that "it is indeed a choice; the appropriate discount rate for public policy decisions spanning many generations cannot be deduced from private market decisions today, or from economic theory. A lower discount rate places a greater importance on future lives and conditions of life. To many, it seems ethically necessary to have a discount rate at or close to zero, in order to respect our descendants and create a sustainable future."[15] Indeed, the very notion of sustainability is about maintaining the environment *for future generations*.

Economic growth theorist Roy Harrod argued in the 1940s that discounting the future based on a "pure time preference" (the myopic preference for consumption today apart from all other considerations) was a "polite expression for rapacity." A high discount rate tends to encourage spending on policies/projects with short-term benefits and long-term costs as opposed to ones with high up-front costs and long paybacks. It therefore encourages "wait-and-see" and "go-it-slow" approaches to impending catastrophes, such as climate change, rather than engaging in strong preventive action.[16]

Nordhaus, like most mainstream economists, through his support of a high discount rate, places a low value on the welfare of future generations relative to present ones, and assumes, despite considerable uncertainty in this regard, that future generations will be much wealthier than present ones. This leads him to argue against large immediate investments in curtailing climate change. He advocates putting a tax on carbon of $30 to $50 per ton and increasing this to about $85 by mid-century. Taxing carbon at $30 a ton would increase the price of gasoline by a mere seven cents a gallon, which gives one a sense of the low level of importance Nordhaus places on curtailing climate change as well as the future of humanity and the environment. Nordhaus has tripled his estimate of the loss to global economic output due to climate change in 2100, moving from his earlier estimate of almost 1 percent to nearly 3 percent in his latest study.[17] Still, such losses are deemed insignificant, given a high discount rate, in comparison to the costs that would be incurred in any attempt to curtail drastically climate change today, leading Nordhaus to advocate a weak-kneed response.

Nordhaus is particularly interested in countering the arguments presented in *The Economics of Climate Change* (commonly known as *The Stern Review*), the report written by Nicholas Stern (former chief economist of the World Bank) for the British government, which advocates immediate and substantial investments aimed at reducing carbon emissions. Stern, deviating from the practice of most orthodox economists, uses a low discount rate, arguing that it is morally inexcusable to place low value on the welfare of future generations and to impose the costs of the problems we generate on our descendants. Nordhaus discounts the future at roughly 6 percent a year; Stern by 1.4 percent. This means that for Stern having a trillion dollars a century from now is worth $247 billion today, while for Nordhaus it is only worth $2.5 billion.[18] Due to this, Stern advocates imposing a tax on carbon of greater than $300 per ton and increasing it to nearly $1,000 before the end of the century.[19] Lomborg in the *Wall Street Journal* characterized the *Stern Review* as "fear-mongering," and referred to it in *Cool It* as a "radical report," comparing it unfavorably to Nordhaus's work.[20]

The Unworldly Economists

It is important to recognize that the difference displayed here between Nordhaus and Stern is fundamentally a moral, not technical, one. Where they primarily differ is not on their views of the science behind climate change but on their value assumptions about the propriety of shifting burdens to future generations. This lays bare the ideology embedded in orthodox neoclassical economics, a field which regularly presents itself as using objective, even naturalistic, methods for modeling the economy. However, past all of the equations and technical jargon, the dominant economic paradigm is built on a value system that prizes capital accumulation in the shortterm, while devaluing everything else in the present *and everything altogether in the future.*

Some of the same blinders are common in varying degrees to both Nordhaus and Stern. Nordhaus proposes what he calls an "optimal path" in economic terms aimed at slowing down the growth of carbon emissions. In his "climate policy ramp" emissions reductions would start slow and get bigger later but would nonetheless lead eventually (in the next century) to an atmospheric carbon concentration of nearly 700 parts per million (ppm). This would present the possibility of global average temperature increases approaching 6°C (10.8°F) above preindustrial levels—a level that Mark Lynas in his *Six Degrees* compares to the sixth circle of hell in Dante's *Inferno*.[21]

Indeed, with a level of carbon concentration much less than this, 500 ppm (associated with global warming on the order of 3.5°C or 6.3°F), the effects both on the world's biological diversity and on human beings themselves would be disastrous. "A conservative estimate for the number of species that would be exterminated (committed to extinction)" at this level, according to James Hansen, director of NASA's Goddard Institute for Space Studies, "is one million."

Moreover, rising sea levels, the melting of glaciers, and other effects could drastically affect hundreds of millions, conceivably even billions, of people. Hansen, the world's most famous climatologist, argues that in order to avoid catastrophic change it is necessary to *reduce* atmospheric carbon to a level of 350 ppm.[22]

Yet, the *Stern Review*, despite being designated as a "radical" and "fear-mongering" report by Lomborg, targets an atmospheric carbon concentration stabilization level of around 480 ppm (550 ppm in carbon dioxide equivalent), which—if never reaching Nordhaus's near 700 ppm peak (over 900 ppm carbon dioxide equivalent)—is sure to be disastrous if the analysis of Hansen and most other leading climatologists is to be believed.[23] Why such a high atmospheric carbon target?

The answer is provided explicitly by the *Stern Review* itself, which argues that past experience shows that anything more than a 1 percent average annual cut in carbon emissions in industrial countries would have a significant negative effect on economic growth. Or as the *Stern Review* puts it, "It is difficult to secure emission cuts faster than about 1 percent a year except in instances of recession."[24] So the atmospheric carbon target in the perspective of the *Stern Review* is to be determined not according to what is necessary to sustain the global environment, protect species, and ensure the sustainability of human civilization, but by what is required to keep the capitalist economy itself alive.

The starting point that led to Summers's conclusion in his 1992 World Bank memo is the same that underlies the analyses of both Nordhaus and Stern. Namely, human life in effect is worth only what each person contributes to the economy as measured in monetary terms. So if global warming increases mortality in Bangladesh, which it appears likely it will, this is only reflected in economic models to the extent that the deaths of Bengalis hurt the economy. Since Bangladesh is very poor, economic models of the type Nordhaus and Stern use would not estimate it to be worthwhile to prevent deaths there since these losses would show up as minuscule in the measurements. Nordhaus, according to his discount analysis, would go a step beyond Stern and place an even slighter value on the lives of people if they are lost several decades in the future. This economic ideology extends beyond just human life, of course, such that all of the millions of species on earth are valued only to the extent they contribute to GDP. Thus ethical concerns about the intrinsic value of human life and of the lives of other creatures are completely invisible in standard economic models. Increasing human mortality and accelerating the rate of extinction are to most economists only problems if they undermine the "bottom line." In other respects they are invisible: as is the natural world as a whole.

From any kind of rational perspective, that is, one not dominated exclusively by the narrow economic goal of capital accumulation, such views would seem to be entirely irrational, if not pathological. To highlight the peculiar mindset at work it is useful to quote a passage from Lewis Carroll's *Through the Looking Glass*:

> "The prettiest are always further!" [Alice] said at last, with a sigh at the obstinacy of the rushes in growing so far off, as, with flushed cheeks and dripping hair and hands, she scrambled back into her place, and began to arrange her new-found treasures.
>
> What mattered it to her just then that the rushes had begun to fade, and to lose all their scent and beauty, from the very moment that she picked them? Even real scented rushes, you know, last only a very little while—and these, being dream-rushes, melted away almost like snow, as they lay in heaps at her feet—but Alice hardly noticed this, there were so many other curious things to think about.[25]

A society that values above all else the acquisition of abstract value-added, and in the prospect lays waste to nature in an endless quest for further accumulation, is ultimately an irrational society. What matters it to capital what it leaves wasted at its feet, as it turns elsewhere in its endless pursuit of more?

Mainstream economics, ironically, has never been a materialist science. There is no materialist conception of nature in what Joseph Schumpeter called its "preanalytic vision."[26] It exists in almost complete ignorance of physics (constantly contravening the second law of thermodynamics) and the degradation of the biosphere. It sees the world simply in terms of an endless, enlarging "circular flow" of economic relations.

The ecological blinders of neoclassical economics, which serves to exclude the planet from its preanalytic vision, are well illustrated by a debate that took place within the World Bank, related by ecological economist Herman Daly. As Daly tells the story, in 1992 (when Summers was chief economist of the World Bank and Daly worked for the Bank) the annual *World Development Report* was to focus on the theme *Development and the Environment*:

> An early draft contained a diagram entitled "The Relationship Between the Economy and the Environment." It consisted of a square labeled "economy," with an arrow coming in labeled "inputs" and an arrow going out labeled "outputs"—nothing more. I suggested that the picture failed to show the environment, and that it would be good to have a large box containing the one depicted, to represent the environment. Then the relation between the environment and the economy would be clear—specifically, that the economy is a subsystem of the environment both as a source of raw material inputs and as a "sink" for waste outputs.
>
> The next draft included the same diagram and text, but with an unlabeled box drawn around the economy like a picture frame. I commented that the larger box had to be labeled "environment" or else it was merely decorative, and that the text had to explain that the economy is related to the environment as a subsystem within the larger ecosystem and is dependent on it in the ways previously stated. The next draft omitted the diagram altogether.[27]

To be sure, not all economics is as resolutely unworldly as this. Nicholas Georgescu-Roegen, an economist critical of the anti-ecological orientation of economics—and the founder of the heterodox tradition known as ecological economics, which builds into its preanalytic vision the notion that the economy is materially limited by physics and ecology—explained that the drive for continuous social wealth and economic profit increased the ecological demands placed on nature, expanding the scale of environmental degradation. He highlighted the error of pretending that the economy could be separated from ecology. Others, like Herman Daly, and Paul Burkett in the Marxist tradition, have pushed forward this notion of ecological economics.[28] Yet, these ecological economists remain on the margins, excluded from major policy decisions and academic influence.

The Juggernaut of Capital

Mainstream economists see themselves as engaged in the science of economic growth. Nevertheless, the assumption of endless economic growth, as if this were the purpose of society and the way of meeting human needs, seems naïve at best. As Daly says, "an ever growing economy is biophysically impossible."[29] The Wonderland nature of such an assumption is particularly obvious in light of the fact that the very underpinning of the economy, the natural environment itself, is being compromised.

Marx did not miss the importance of this social-ecological relationship. He pointed out that humans are dependent upon nature, given that it provides the energy and materials that make life possible. As capitalists focused on exchange value and short-term gains, Marx explained that the earth is the ultimate source of all material wealth, and that it needed to be sustained for "successive generations." The "conquest of nature" through the endless pursuit of capital, which necessitated the constant exploitation of nature, disrupted natural cycles and processes, undermining ecosystems and causing a metabolic rift. Frederick Engels warned that such human actions left a particular "stamp . . . upon the earth" and could cause unforeseen changes in the natural conditions that exact the "revenge" of nature.[30]

Today carbon dioxide is being added to the atmosphere at an accelerating rate, much faster than natural systems can absorb it. Between 2000 and 2006, according to Josep G. Canadell and his colleagues, in an article in the *Proceedings of the National Academy of Sciences*, the emissions growth rate increased as the global economy grew and became even more carbon intensive, meaning that societies emitted more carbon per unit of economic activity at the beginning of the new millennium than they did in the past. At the same time, the capacity of natural sinks to absorb carbon dioxide has declined, given environmental degradation such as deforestation. This contributed to a more dramatic upswing in carbon accumulation in the atmosphere than was anticipated.[31] The juggernaut of capital overexploits both the resource taps and waste sinks of the environment, undermining their ability to operate and provide natural services that enhance human life.

There are many good reasons to think that the patterns and processes that held for the past one hundred years—for example, economic growth—may not hold for the next one hundred, a point on which the present economic crisis should perhaps focus our attention. Justifying shifting costs from the present to the future based on the assumption that future generations will be richer than present ones is highly dubious. In relation to the economy as well as the ecology the future is highly uncertain, though current trends clearly point to disaster. If global climate change, not to mention the many other interconnected environmental problems we face, has some of the more catastrophic effects that scientists predict, economic growth may not only be hampered, but the entire economy may be undermined, not to mention the conditions of nature on which we depend. Therefore, future generations may be much poorer than present ones and even less able to afford to fix (if still possible) the problems we are currently creating.

The growth mania of neoclassical economists focuses on the kinds of things, mainly private goods reflecting individual interests, that comprise GDP, while collective goods and the global commons are devalued in comparison. It therefore encourages an economic bubble approach to the world's resources that from a deeper and longer perspective cannot be maintained.

For all of these reasons, the current economic order tends to mismeasure the earth and human welfare. Capitalism, in many respects, has become a failed system in terms of the ecology, economy, and world stability. It can hardly be said to deliver the goods in any substantive sense, and yet in its process of unrestrained acquisition it is undermining the long-term prospects of humanity and the earth.[32]

If we cannot rely on orthodox economists to avert crises in financial markets, an area that is supposedly at the core of their expertise, why should we rely on them to avert ecological crises, the understanding of which requires knowledge of the natural environment that is not typically covered in their training? Nor is such an awareness compatible with the capitalist outlook that is embedded in received economics. As Ehrlich noted, "Most economists are utterly ignorant of the constraints placed upon the economic system by physical and biological factors," and they fail to "recognize that the economic system is completely and irretrievably embedded in the environment," rather than the other way around. Due to these problems, he stated pointedly, "It seems fair to say that most ecologists see the growth-oriented economic system and the economists who promote that system as the gravest threat faced by humanity today." Furthermore, "The dissociation of economics from environmental realities can be seen in the [false] notion that the market mechanism completely eliminates the need for concern about diminishing resources in the long run."[33]

Plan B: The Technological Wonderland

The demonstrated failure of received economics to offer a solution to the environmental problem compatible with a capitalist economy has recently resulted in a Plan B in which technological silver bullets would carry out a "green revolution" without altering the social and economic relations of the system. Often this plan is presented in terms of an "investment strategy" geared to new Schumpeterian epoch-making innovations of an environmental nature that will somehow save the day for both the economy and ecology, while restoring U.S. empire. Orthodox economists assume that the resource problems of today will force prices up tomorrow and that these higher prices will force the creation of new technology. The new army of environmental technocrats claims that the innovations that will solve all problems are simply there waiting to be developed—if only a market is created, usually with the help of the state.

Such views have been promoted recently by figures like Thomas Friedman, Newt Gingrich, Fred Krupp of the Environmental Defense Fund, and Ted Nordhaus and Michael Shellenberger of the Breakthrough Institute. Krupp and Miriam Horn present this as a question of a competitive race between nations to be first in the green technologies and markets that will save the world. "The question," they write, "is no longer just how to avert the catastrophic impacts of climate change, but which nations will produce—and export—the green technologies of the twenty-first century."[34] These analyses tend to be big on the wonders of technology and the market, while setting aside issues of physics, ecology, the contradictions of accumulation, and social relations. They assume that it mostly comes down to energy efficiency (and other technical fixes) without understanding that in a capitalist system, growth of efficiency normally leads to an increase in scale of the economy (and further rifts in ecological systems) more than negating any ecological gains made.[35]

Like the establishment economists with whom they are allied the technocrats promise to solve all problems while keeping the social relations intact. The most ambitious schemes involve massive geoengineering proposals to combat climate change, usually aimed at enhancing the earth's albedo (reflectivity). These entail schemes like using high-flying aircraft, naval guns, or giant balloons to launch reflective materials (sulfate aerosols or aluminum oxide dust) into the upper stratosphere to reflect back the rays of the sun. There are even proposals to create "designer particles" that will be "self-levitating" and "self-orienting" and will migrate to the atmosphere above the poles to provide "sunshades" for the Polar Regions.[36] Such technocrats live in a Wonderland where technology solves all problems, and where the *Sorcerer's Apprentice* has never been heard of. All of this is designed to extend the conquest of the earth rather than to make peace with the planet.

Ecological Revolution

If there is a definite beginning to the modern ecological revolution, it can be traced back to Rachel's Carson's *Silent Spring*. In attempting to counter what she called the "sterile preoccupation with things that are artificial, the alienation from the sources of our strength," that has come to characterize the capitalist Wonderland, Carson insisted that it was necessary to cultivate a renewed "sense of wonder" toward the world and living beings.[37] Yet it was not enough, as she was to demonstrate through her actions, merely to *contemplate* life. It was necessary also to *sustain* it, which meant actively opposing the "gods of profit and production"—and their faithful messengers, the dominant economists of our time.

4. The Midas Effect

Climatologist James Hansen warns that global climate change today constitutes a "planetary emergency." Existing trends threaten to set in motion irreversible climate transformations, proceeding "mostly under their own momentum," thereby fundamentally transforming the conditions of life on earth.[1] It is becoming increasingly evident that capitalism, given its insatiable drive for accumulation, is the main engine behind impending catastrophic climate change. Unfortunately, mainstream economics, although now acknowledging the importance of environmental issues, remains hamstrung by its adherence to the existing system of economic relations. It therefore relies increasingly on what can be called transmutation myths—referred to here as "the Midas Effect"—as a way out of the global environmental crisis. In contrast, our argument in this chapter suggests that nothing less than an ecological revolution—a fundamental reordering of relations of production and reproduction to generate a more sustainable society—is required to prevent a planetary disaster.

The 350 Imperative

Human activities, primarily fossil fuel combustion and deforestation, are unequivocally responsible for the observed warming of the earth's atmosphere.[2] In the 1990s, global carbon emissions increased 0.9 percent per year, but in 2000–2008 they increased by 3.5 percent per year, presenting a scenario outside of the range of possibilities considered in the 2007 IPCC report.[3] This recent escalation has been due to economic growth, rising carbon intensity, and the continuing degradation of ecosystems that serve as natural carbon sinks.[4] At the IPCC meeting held in Copenhagen in March 2009, several researchers noted how global climate conditions had gone from bad to worse: "Emissions are soaring, projections of sea level rise are higher than expected, and climate impacts around the world are appearing with increasingly frequency."[5]

The carbon dioxide concentration in the atmosphere has increased from the preindustrial level of 280 parts per million (ppm) to 390 ppm in 2010 (higher than ever before during recorded human history), with an average rate of growth of 2 ppm per year. Climatologists had previously indicated that an increase above 450 ppm would be extremely dangerous, given that various positive feedbacks would be set in motion, furthering climate change. But 450 ppm is now seen as too high, given that—because of inadequate knowledge—most climate models failed to consider "slow" climate feedback processes such as the disintegration of ice sheets and the release of greenhouse gases from soils and the tundra.[6]

Hansen and his colleagues warn that "if humanity wishes to preserve a planet similar to that on which civilization developed and to which life on Earth is adapted, paleoclimate evidence and ongoing climate change suggest" that carbon dioxide must be reduced to "at most 350 ppm."[7] Thus it is imperative to act now, since we have already surpassed the limit, and the longer we exceed this point and the further we push up these numbers, the greater the threat of creating irreversible environment changes with dire consequences. Global temperature is already at the warmest it has been during the Holocene (the last 12,000 years, which includes the rise of human civilization). Climate change has shifted the habitat zones for animals and plants and influenced the hydrologic cycle. Specific positive feedbacks have been set in motion, so that even if carbon dioxide emissions do not increase further, significant additional warming would still occur.

Society, through its expanding production and the resulting carbon emissions, is already in the process of racing off the cliff. For instance, the thawing of the tundra will release massive quantities of the potent greenhouse gas methane. The melting of ice and snow throughout the planet will reduce the earth's reflectivity, accelerating the warming process. Drought conditions will cause "the loss of the Amazon rainforest," greatly diminishing natural sequestration.[8] Other related trends include a rapidly increasing extinction rate, growing severity of weather events, rising sea levels, and expanding numbers of ecological refugees throughout the world.

Under these circumstances of what can be called, without hyperbole, threatened apocalypse, it is critically important to assess what forces are driving the ecological crisis, especially the accumulation of carbon in the atmosphere.[9] What is abundantly clear at this point is that the logic of capital accumulation runs in direct opposition to environmental sustainability. The motor of capitalism is competition, which ensures that each firm must grow and reinvest its "earnings" (surplus) in order to survive. By its nature, capital is self-expanding value, and accumulation is its sole aim. Hence, capitalism as a system does not adhere to nor recognize the notion of *enough*. Joseph Schumpeter observed that "stationary capitalism would be a *contradictio in adjecto*."[10] The capitalist economy must increase in scale and intensity. The earth and human labor are systematically exploited/robbed to fuel this juggernaut. Today we are threatened by the transformation of the entire atmosphere of the earth as a result of economic processes.

The Orthodox Economics of Climate Change

Although mitigation of and/or adaptation to climate change is now definitely on the global agenda, there remains a real danger that it will be hijacked by mainstream economics, which plays a critical role in constraining possible social responses. The threatening implications of this are clearly revealed in the work of Nicholas Stern and William Nordhaus, who represent the limits of variance that exist within the neoclassical economics mainstream on the issue of climate change.

In the most progressive neoclassical treatment of global warming, Stern argues that carbon dioxide *equivalent* concentration in the atmosphere (which includes other greenhouse gases as well) should be stabilized at 550 ppm.[11] This corresponds to an atmospheric carbon dioxide concentration of 480 ppm and a rise in global temperature of 3–4°C (6.1–7.2°F) above preindustrial levels.[12] Even though this exceeds atmospheric carbon targets proposed by climatologists, Stern insists that efforts to limit greenhouse gases to levels below this should not be attempted, given that they "are unlikely to be economically viable" and would threaten the economic system.[13] In other words, the level of atmospheric carbon is not to be determined by ecological considerations in this conception, but by what the present economic system will permit.

Nordhaus, the most prominent U.S. economic analyst of climate change, suggests that only modest reductions in greenhouse gas emissions should be implemented in the short term, and in the long term more ambitious reductions could be put into place.[14] In support of this "climate-policy ramp," he argues against drastic attempts to stabilize emissions this century. Instead he insists on an "optimal path" that would slow the growth of carbon emissions, peaking at about 700 ppm by 2175, with a global average temperature approaching 6°C (10.8°F) above preindustrial levels. This way the economy will be permitted to grow, allowing for various investments in welfare-enhancing areas of the economy to address whatever risks may arise from climate changes. Taking strong measures to reduce carbon levels, even to the extent proposed by Stern, is seen by Nordhaus as being too economically costly.

Both of these options, offered by orthodox economists who are seen as taking pro-environment positions, would lead to atmospheric carbon dioxide stabilization goals that many scientists see as catastrophic. Thus the mainstream economics of climate change directs us toward an ecologically unsustainable target—one that climatologists believe would imperil human civilization itself, and could result in deaths in the millions, even billions, plus the loss of countless numbers of species.[15]

The Midas Effect

The critical issue that clearly arises here is the unworldliness (in the sense of ecological blinders) of received economics and of the capitalist system it serves. To find an appropriate comparison one has to enter the misty realm of mythology. Indeed, the characteristic relation of orthodox economics to the environment, we suggest, can be best described as "the Midas Effect." We use this term to refer to a set of transmutation myths or ecological alchemy, whereby economics, in addressing environmental problems, constantly seeks to transmute ecological values into economic ones. In the Greek and Roman myth of Midas, as told by Ovid in his *Metamorphoses*, the god Bacchus (Dionysus) offers King Midas of Phrygia his choice of whatever he wishes for, in return for aid he had given to the satyr Silenus, Bacchus's tutor and foster father. Midas chooses the gift of having everything that he touches turn to gold. Bacchus grants him his wish, and Midas rejoices in his new power. Everywhere he tests:

> the efficacy of his gift by touching
> one thing and another: even he
> could scarcely credit it, but when he snapped
> a green twig from the low branch of an oak,
> the twig immediately turned to gold;

> he picked a stone up, and it did the same;
> he touched a clod, and at his potent touch,
> the piece of earth became a lump of ore;
> ripe wheat-heads plucked produced a golden harvest. . . .
> *All turns to gold!* He scarcely could imagine![16]

Nature itself—branch, stone, earth, grain, stream—thus became gold at his mere touch. The folly of Midas's choice, however, materializes when he discovers that his food and drink also turns to gold at his mere touch, leaving him hungry and thirsty. In one version of the myth he turns his daughter into gold. Midas therefore pleads with Bacchus to free him from his curse, and the god shows him mercy. Thereafter Midas scorns wealth and becomes a worshiper of Pan, the god of nature.

Cursed by their own gods of profit and production, today's mainstream economists see the challenge of the environmental limits to growth as surmountable due to three transmutations: (1) the universal *substitutability* of everything in nature so that nothing natural is irreplaceable or irreversible; (2) *dematerialization*, or the decoupling of the economy from actual resource use; and (3) the conversion of nature into *natural capital*, whereby everything in nature is assigned an economic value. By such fantastic means today's dominant economists dream of turning the earth into money in order to overcome the external limits to economic expansion. "The commodity" economy of capitalism, Elmar Altvater wrote in *The Future of the Market*, "is narcissistic: it sees only itself reflected in gold."[17]

In such a commodified world, ecological alchemy prevails. Anti-environmentalist economist Julian Simon proclaimed a number of times that if the world ran out of copper, then copper, an element, could be produced artificially.[18] Resorting to philosophical idealism to defend this position, he later declared: "You see, in the end copper and oil come out of our minds. That's really where they are."[19]

Mainstream environmental economists, though rarely as crude in their rejection of environmental issues as Simon, typically adopt what is called the "weak sustainability hypothesis," that is, the notion that everything in nature if exhausted (or exterminated) can be substituted for with the help of technology. This means that the natural environment presents no actual limits or critical thresholds to the infinite transmutation of the world into gold—the cash nexus.[20] As Robert Solow, a winner of the Bank of Sweden's Nobel Memorial Prize in the Economic Sciences, once wrote in criticism of the "limits to growth" perspective: "If it is very easy to substitute other factors for natural resources, then there is in principle no 'problem.' The world can, in effect, get along without natural resources, so exhaustion is just an event, not a catastrophe."[21]

Similarly, one of Britain's leading mainstream environmental economists, David Pearce, author of the UK government's *Blueprint for a Green Economy*, has stated: "Sustainable economic development. . . . is continuously rising, or at least non-declining, consumption per capita, or GNP." He stresses that "most economists" addressing sustainable development approach it this way.[22] Sustainability is thus defined entirely in terms of economic growth, monetary wealth, and consumption, without any direct reference to the environment. Given the assumption of substitutability, nature simply disappears. Only money matters.

In another transmutation, mainstream environmental sociologists and ecological modernization proponents have repeatedly turned to the notion of dematerialization. This is the view that the growth of economic value and even the production of goods can be decoupled from the consumption of nature's resources, through ever-greater efficiency. Production can be so transformed to create a "weightless economy."[23] So far, however, all such dreams have proven illusory. Even where greater efficiency in the use of energy and materials is attained, the efficiency gains, under a capitalist system, are used to expand the scale of the system, outweighing any "dematerializing" tendencies—a phenomenon known as the Jevons Paradox.[24]

Others, like Paul Hawken, Amory Lovins, and L. Hunter Lovins, claim that the solution is to reconceptualize nature as natural capital, and thus to extend capitalism to all of nature.[25] It is assumed that the proverbial efficiency of the market will then take over, safeguarding environmental values. However, conceiving forests as so many millions of board feet of standing timber (thereby as natural capital) has historically done very little to preserve forest ecosystems. Putting price tags on species and ecosystems will only serve in the end to subsume nature to the endless growth of production and profits.

The Midas Effect thus stands for the inability of received economics and the capitalist system itself to recognize that there are intrinsic values and critical thresholds in nature that we ignore at our own cost. Midas turned his daughter into gold in his mad search for wealth. Today's economics threatens to destroy the lives of future generations as well as those of innumerable other species in like fashion.

Perhaps the most potent example of the ecological blinders of mainstream economics is to be found in figures like Stern and Nordhaus who argue for levels of atmospheric carbon dioxide that threaten the planet as we know it, along with human civilization. This is justified, as we have seen, on the basis that a serious program for the control of greenhouse gases to save the planet would imperil capitalist economic growth—as if capitalism exists at a more basic level than the planet. Biologist and climate scientist Stephen Schneider, highlighting the absurdity of this position, asserts that mainstream economists simply assume that society (and the economy) is not bound by natural conditions.[26]

The Revolution for Enough

The ancient Greek philosopher Epicurus observed: "Nothing is enough to someone for whom enough is little."[27] This statement contains the germ of a materialist ecological critique of the current system, and indicates what must be transcended in order to pursue environmental sustainability. The goal of society needs to shift radically, as Marx emphasized in the nineteenth century, from the endless pursuit of private profit and accumulation to sustainable human development for the sake of "successive generations."[28]

Recognizing the incompatibility between a capitalist system geared to exponential growth and the goal of sustaining the earth for future generations, influential environmentalist James Gustave Speth has recently written: "Capitalism as we know it today is incapable of sustaining the environment."[29] Others within the Marxist tradition have gone even further in their ecological criticisms of capitalism. Writing two decades ago on "Capitalism and the Environment," U.S. Marxist economist Paul Sweezy concluded that seriously addressing the ecological crisis required "a reversal, not merely a slowing down, of the underlying trends of the last few centuries."[30] As Evo Morales, socialist president of Bolivia, stated: "Competition and the thirst for profit without limits of the capitalist system are destroying the planet. Under Capitalism we are not human beings but consumers. Under capitalism mother earth does not exist, instead there are raw materials. . . . The earth is much more important than stock exchanges of Wall Street."[31]

A full-fledged ecological revolution means that the human relations with nature would need to be completely restructured. It is our contention that an *elementary triangle of ecology* (related to Hugo Chávez's "elementary triangle of socialism"), prescribed by the natural laws of life itself, constitutes the necessary foundation of the new society: (1) social use, not ownership, of nature; (2) rational regulation by the associated producers of the metabolism between human beings and nature; and (3) the satisfaction of communal needs—not only of present but also future generations.[32] What is needed, in other words, is a green cultural revolution, in which humanity as a whole radically redefines its needs in relation to community, equality, and sustainability.

Transition Strategies

Some argue today that the speed and intensity of the ecological threat leaves us with no choice but to stick with the existing system and embrace its limited and myopic solutions to environmental problems: such strategies as "cap-and-trade" carbon markets and market-driven technological silver bullets. The fantastic nature of these strategies reflects their adherence to the Midas Effect of mainstream economics: environmental change must kowtow to the "bottom line" of capital accumulation.

Where adopted, carbon markets have accomplished little to reduce carbon emissions. This has to do with numerous factors, not least of all provisions for nations to buy out of the actual reductions in various ways. The idea that technology can solve the global environmental problem, as a kind of *deus ex machina* without changes in social relations, belongs to the area of fantasy and science fiction. Thomas Friedman provides a vision of green industrial revolution in his *Hot, Flat, and Crowded* in which he repeatedly tells his readers that if given "abundant, clean, reliable, and cheap electrons," we could move the world and end all ecological problems.[33] Gregg Easterbrook, in what he calls environmental "realism," argues that even if we destroy this biosphere we can "terraform" Mars—so humanity's existence is not necessarily impaired by environmental destruction.[34]

The very desperation of such establishment arguments, which seek to address the present-day environmental problem without confronting the reality of capitalism, highlights the need for more radical measures in relation to climate change and the ecological crisis as a whole. Especially noteworthy in this respect is Hansen's carbon tax proposal, and global contraction-convergence strategies. In place of carbon markets, which invariably include various ways to buy out of emissions reductions (registering reductions while actually increasing emissions), Hansen proposes a carbon tax for the United States to be imposed at well-head and point of entry, which is aimed at bringing carbon dioxide emissions down to near zero, with 100 percent of the revenue from the tax being deposited as monthly dividends directly into the bank accounts of the public on a per person basis (with children receiving half-shares).[35] Not all carbon taxes of course are radical measures. But Hansen's emergency strategy, with its monthly dividends, is designed to keep carbon in the ground and at the same time appeal to the general public. It explicitly circumvents both the market and state power in order to block those who desire to subvert the process. In this, the hope is to establish a mass popular constituency for combating climate change by promoting social redistribution of wealth toward those with smaller carbon footprints (the greater part of the population).

Hansen insists that any serious attempt to protect the climate means going against Big Coal. An important step would be to declare a moratorium on new coal-fired power stations, which he describes as "death factories" since the carbon emissions they produce contribute to escalating extinction rates (as well as polluting regional environments and directly impairing human health).[36] He argues that we need to leave as much coal as possible in the ground and to close existing coal-fired power stations if we are to prevent catastrophic environmental change.

From a global standpoint, ecological degradation is influenced by the structure and dynamics of a world system hierarchically divided into numerous nation-states, competing with each other both directly and via their corporations. In an attempt to counter carbon imperialism, Anil Agarwal and Sunita Narain propose that carbon emissions of nations should be determined on an equal per capita basis, rooted in what is allowable within the shared atmosphere.[37] The global North, with its relatively smaller population than the South, has used a disproportionate amount of the atmospheric commons, given its immense carbon emissions. Tom Athanasiou and Paul Baer and other climate justice activists thus propose a process of contraction and convergence.[38] The rich nations of the North would be required to reduce (contract) their emissions of greenhouse gases to appropriate levels as determined by the atmospheric carbon target. Given global inequalities, the nations of the South would be allowed to increase their emissions gradually to a limited extent—but only if a nation had a per capita carbon emission rate below the acceptable level established by the target. This would create a world converging toward "equal and low, per capita allotments."[39] Today contraction and convergence would necessarily aim at stabilizing atmospheric carbon dioxide at 350 ppm, in conformity with scientific indications.

Such a proposal would mean that the rich nations would have to reduce their carbon emissions very rapidly by levels approaching 100 percent, and a massive global effort would be needed to help countries in the global South move toward emissions stabilization as well, while not jeopardizing sustainable human development. Such a process of contraction and convergence would require that the global North pay the ecological debt that it has accrued through using up the bulk of the atmospheric commons by carrying the main cost of mitigation globally and aiding nations of the South in adapting to negative climate effects.

Ecological Revolution

In reality, the radical proposals discussed above, although ostensibly transition strategies, present the issue of revolutionary change. Their implementation would require a popular revolt against the system itself. A movement (or movements) powerful enough to implement such changes on the necessary scale might well be powerful enough to implement a full-scale social-ecological revolution. Humanity cannot expect to reach 350 ppm and avoid planetary climatic disaster except through a major global social transformation, in line with the greatest social revolutions in human history. This would require not simply a change in productive forces but also in productive relations, necessitating a green cultural revolution. The answer to today's social and environmental crisis, as Lewis Mumford argued in *The Condition of Man*, lies in the creation of a new "organic person," and a system of sustainable human development.[40] This means the creation of cultural forms that present the opportunity for balance in the human personality. Rather than promoting the asocial traits of humanity, the emphasis would be on nurturing the social and collective characteristics. Each human being would be "in dynamic interaction with every part of his environment."

For revolutionary environmental thinker-activists, the first condition of sustainability is the restoration of genuine human community (and communities of communities). The concept of community, as Herman Daly and John Cobb insisted in *For the Common Good*, points to a social order with definite "communal" characteristics.[41] It involves extensive collective participation in decision making, and thus necessitates, at its highest level of development, what the early communist François Babeuf called "a society of equals," that is, a system of substantive equality.[42] A society that is actively *communal* in this sense can arise only out of a strong collective bond, dissolving mere individual economic exchange. Moreover, a sustainable community requires both the cultivation of a sense of place and the extension of the community ethic to what Aldo Leopold referred to as a "land ethic," incorporating the surrounding ecology.[43]

It is only at this point in human history, if it were to be reached, that we could speak of the implementation in full of the elementary triangle of ecology. The sustainable development of each would be the key to the sustainable development of all—with both *the each* and *the all* now extended to the earth itself. Such a vital, humanistic-naturalistic community would require for its emergence, however, an ecological revolution against capitalism—in other words, the fall of Midas.

5. Carbon Metabolism and Global Capital Accumulation

The capacity of humans to transform nature in detrimental ways has long been known. However, it is only recently that the long-term survival of human civilization has been called into question, given the scale of environmental problems and the crossing of planetary ecological boundaries.[1] One of the most pressing environmental challenges is global climate change. The Intergovernmental Panel on Climate Change (IPCC) indicates that the observed increases in average global temperatures are unequivocally due to human activities, such as fossil fuel combustion and deforestation, leading to the accumulation of carbon dioxide in the atmosphere.[2] Climate scientists stress that "business as usual" cannot continue if we wish to preserve the conditions of life.

Capitalism—since the late fifteenth century—has been the global hegemonic economic system, influencing human interactions with nature, shaping the particular organization of material exchange. Thus it is important to grapple directly with how global climate change is related to the historical era of capitalism, which serves as the background condition influencing social development. Through understanding the logic of capital, it is possible to assess how such a socioeconomic system confronts natural systems and affects their ability to sustain human life. In what follows, we present how the accumulation of carbon dioxide in the atmosphere is tied to the accumulation of capital, how ongoing environmental destruction contributes to climate change, and how the structural conditions under the current economic system limit the ecological benefits of technological development.

Karl Marx employed a metabolic analysis to study the environmental problems of his day, which involved assessing the metabolism of natural systems, the relations between organisms and their surroundings, and the material exchange in these relationships.[3] Metabolism, which is one of the foundational concepts in ecology, provides an avenue for grappling with both qualitative and quantitative dimensions of relationships. Marx's metabolic analysis serves as a means for studying the empirical reality of the nature–society relationship. Furthermore, it helps establish the theoretical framework to deal with both sides of the dialectic between society and nature, considering the processes that take place in each realm, as well as examining how these positions interact and transform each other.[4]

We draw upon the strength of Marx's metabolic analysis for studying the nature–society dialectic. We extend its application to examine global climate change, including human influence on the carbon cycle and its consequences. Broadly, our discussion consists of linking three major ideas: (1) the utility of metabolic analysis for comprehending recent anthropogenic changes in the global carbon cycle—or more specifically, the creation of a metabolic rift in the carbon cycle—given capitalist development; (2) the Jevons Paradox, where improvements in efficiency actually increase the use of natural resources under capitalist relations, therefore, diminishing the potential for developing ecological sustainability based on technological fixes; and (3) the dialectic between the flooding and destruction of carbon sinks and the endless pursuit of capital accumulation.

Metabolism and the Metabolic Rift

Metabolic analysis draws upon the historical development of the concept within the natural sciences, as well as how Marx used it to study environmental problems. German physiologists in the 1830s and '40s adopted the term "metabolism" (which was introduced around 1815) to describe the "material exchanges within the body, related to respiration."[5] Justus von Liebig, the great German chemist, applied the term on a wider basis, using it to refer to metabolic processes in relation to "tissue degradation" and as a key concept for understanding the processes at both "the cellular level and in the analysis of entire organisms."[6]

In the mid-1800s, agricultural chemists and agronomists in Britain, France, Germany, and the United States alerted people to the loss of soil nutrients—such as nitrogen, phosphorus, and potassium—through the transfer of food and fiber from the country to the cities. In contrast to traditional agricultural production where essential nutrients were returned to the soil, capitalist agriculture transported nutrients essential for replenishing the soil, in the form of food and fiber, hundreds, even thousands, of miles to urban areas, where they ended up as waste. In 1859, Liebig determined that the intensive methods of British agriculture were a system of robbery, as opposed to rational agriculture.[7] The soil was depleted continually of its necessary nutrients, decreasing the productive potential of the land.

Marx, drawing upon Liebig's research, employed the concept of metabolism to refer to "the complex, dynamic interchange between human beings and nature."[8] For Marx, there is a necessary "metabolic interaction" between humans and the earth.[9] Marx contended that "man *lives* on nature" and that in this dependent relationship "nature is his *body*, with which he must remain in continuous interchange if he is not to die."[10] Thus, a sustainable social metabolism is "prescribed by the natural laws of life itself."[11] Labor is the process in which humans interact with nature through the exchange of organic matter.[12] In this metabolic relationship, humans both confront the nature-imposed conditions of the processes found in the material world and affect these processes through labor (and the associated structure of production). Marx, in studying the work of soil chemists, recognized that Liebig's critique of modern agriculture complemented and paralleled his own critique of political economy.[13]

The natural conditions found in the world, such as soil fertility and species of plants in a country, are, in part, "bound up with the social relations of the time."[14] The degradation of the soil contributed to a greater concentration of agricultural land among a smaller number of proprietors who adopted even more intensive methods of production, including the application of artificial fertilizers, which placed demands on other natural resources. An "irreparable rift" (rupture) emerged in the metabolic interaction between humans and the earth, one that was only intensified by large-scale agriculture, long-distance trade, massive urban growth, and large and growing synthetic inputs (chemical fertilizers) into the soil. The pursuit of profit sacrificed reinvestment in the land, causing the degradation of nature through depleting the soil of necessary nutrients and despoiling cities with the accumulation of waste as pollution.[15] The metabolic rift was deepened and extended with time, as capitalism systematically violated the basic conditions of sustainability on an increasingly large scale (both internally and externally), through soil intensification and global transportation of nutrients, food, and fiber.[16]

Marx noted that the metabolic interaction of humans with nature serves as the "regulative law of social production."[17] Yet capitalist agriculture is unable to maintain the conditions necessary for the recycling of nutrients. As a result, it creates a rift in our social metabolism with nature. To make matters worse, the ongoing development of capitalism continues to intensify the rift in agriculture and creates rifts in other realms of the society–nature relationship, such as the introduction of artificial fertilizers. Incidentally, food production has increased through expanding agricultural production to less fertile land—depleting the nutrients in these areas—and through the incorporation of large quantities of oil used in the synthesis of chemical fertilizers and pesticides, contributing to the carbon rift. Constant inputs are needed simply to sustain this operation, given the depletion of the soil.[18] Here attempts to "solve" the metabolic rift (related to the soil) created additional rifts and failed to solve the primary problem, given the continuation of production based on the accumulation of capital. Marx argued that the "systematic restoration" of this metabolic relation, through a system of associated producers, was required to govern and regulate the material interchange between humans and nature.[19]

Metabolic analysis has become a powerful approach for analyzing human interactions with nature and ecological degradation, especially regarding agricultural production.[20] Marx's conception of the metabolic rift under capitalism illuminates social-natural relations and the degradation of nature in a number of ways: (1) "The decline in the natural fertility of the soil due to the disruption of the soil nutrient cycle" while transferring nutrients over long distances to new locations; (2) new scientific and technological developments, under capitalist relations, increase the exploitation of nature, intensifying the degradation of the soil, expanding the rift; and (3) the nutrients transferred to the city accumulate as waste and become a pollution problem.[21]

In this chapter, we extend this metabolic analysis to the carbon cycle and global climate change. In the extension, we utilize the general properties provided by Marx's model of a metabolic rift as well as the processes of the carbon cycle, where "climate is intimately embedded within human ecosystems."[22] A "metabolic rift" refers to an ecological rupture in the metabolism of a system. The natural processes and cycles (such as the soil nutrient cycle) are interrupted. The division between town and country is a particular geographical manifestation of the metabolic rift, in regard to the soil nutrient cycle. But the essence of a metabolic rift is the rupture or interruption of a natural system. Our analysis of the carbon cycle and climate change follows this logic, extending the metabolic rift to a new realm of analysis. The metabolic rift also entails the division of nature, which is tied to the division of labor, as the world is subdivided to enhance the accumulation of capital.[23] Materials and energy are transformed into new forms, generally associated with the system of commodity production. In this process, environmental degradation takes place, leading to the accumulation of pollution.[24] Lastly, attempts to remedy metabolic rifts, without systematic change to the current political-economic system, compound the problems associated with rifts between the social metabolism and natural metabolism.

Metabolic rifts, like any social issue, need to be historically contextualized. The principal time period considered is the era of global capitalism, which to a large extent is the primary force organizing the social metabolism. Through the application of metabolic and rift analysis, we provide a better understanding of the dynamic relationships involved in global climate change. But to understand global climate change, the accumulation of carbon dioxide in the atmosphere, and the anthropogenic drivers of current patterns in carbon dioxide emissions, we first provide a discussion of the biosphere and the carbon cycle.

The Formation of the Biosphere and the Structure of the Carbon Cycle

The composition of gases in the atmosphere was not always as it is today or even as it has been in previous centuries. Building on the concept of the biosphere introduced by Vladimir Vernadsky in 1926, J. B. S. Haldane in Britain and A. I. Oparin in the Soviet Union independently introduced a materialist hypothesis of the origins of life, which argued that the emergence of life on earth radically transformed the conditions that made this event possible.[25] Life—in interaction with the existing environment—created the atmosphere as we know it. Life exists only in the lower regions of the sky and upper regions of the soil and ocean. An interrelationship between living and nonliving materials within the biosphere produces a cycling of chemical elements. Thus the history of life and the biosphere is a story of coevolution.[26] Vernadsky noted that increasingly human activities acted like a geological force reshaping the planet.[27]

The composition of gases in the atmosphere is the product of biological processes on Earth. Three billion years ago Earth's atmosphere had a dramatically lower concentration of oxygen than it does today. Unsurprisingly, anaerobic bacteria (bacteria that survive in the absence of oxygen) dominated Earth. The long evolutionary history of bacteria led to numerous transformations that greatly affected the composition of gases in the atmosphere. Early bacteria survived by fermentation, breaking down the sugars and chemicals existing in the surrounding environment.[28] Fermenting bacteria metabolizing sugars produced methane and carbon dioxide (key greenhouse gases) as waste products, helping to create the conditions to hold heat in the biosphere. Evolutionary changes led to an early form of photosynthesis, which is quite different from the process found in plants today. These photosynthesizing bacteria used hydrogen sulfide from volcanism (not water) as a source of hydrogen and combined it with energy from sunlight and carbon dioxide from the air "to form organic compounds." Oxygen was not released in this process.

Further evolutionary changes in bacteria led to the emergence of a special type of blue-green bacteria (a distant ancestor of the modern era's blue-green algae) that developed the ability to use sunlight of higher energy to split the stronger bonds of hydrogen and oxygen found in water. The hydrogen was used for building sugars, while the oxygen was released. Over many years, oxygen began to accumulate in the environment, causing toxic reactions with organic matter and the destruction of essential biochemical compounds. Oxygen pollution killed numerous species, creating a punctuated change in the evolutionary development of the bacterial world.[29] Thus blue-green bacteria were able to make use of the waste (oxygen) produced by their photosynthesis process, which, to some degree, regulated the oxygen present in the atmosphere. After several billion years of evolution, life created a mixture of atmospheric gases that provided conditions for the evolution of oxygen-breathing organisms, thus changing the historical conditions of the world.

Life on earth depends upon energy from the sun for its existence. The sun's energy is captured by plants, which store and convert it into chemical energy for their own growth. At the same time, animals eat plants to derive the necessary energy for their lives. Through plants and animals, energy is captured, stored, converted, and deposited throughout the environment, maintaining a viable world for life and its evolutionary processes. Fossil fuels hidden deep within the earth are the remains of past life, especially the first wave of gigantic ferns and giant trees.[30]

This past life captured energy, helping make life possible on the land; at the same time, these plants stored energy in their cells before they were buried deep within Earth. Historic geological processes effectively concentrated energy by removing large quantities of carbon (in the form of hydrocarbons) from the biosphere and burying it deep underground. Otherwise, the energy in the biosphere is primarily stored in plants until they die and release this energy through decay or combustion.

Since the time that oxygen-breathing organisms evolved, the principal gases that envelop the earth have been roughly stable at approximately the current level—nitrogen comprises 78 percent of the atmosphere and oxygen approximately 21 percent. Trace gases, including greenhouse gases such as carbon dioxide, make up the remaining fraction, which regulate the "temperature to life-supporting levels."[31]

Studies of the human transformation of climate date back to Theophrastus, who was one of Aristotle's students. He noted that in draining wetlands, humans influenced the "moderating effects of water and led to greater extremes of cold, while clearing woodlands for agriculture exposed the land to the Sun and resulted in a warmer climate."[32] David Hume in the 1750s contended that advanced cultivation of the land had led to a gradual change in climate. In the 1820s, Jean-Baptiste Fourier studied the heating of Earth, pointing out that "the thickness of the atmosphere and the nature of the surface 'determine' the mean value of the temperature each planet acquires."[33] But he did not know how this happened.[34] He proposed that when the sun heated the earth's surface invisible infrared radiation was emitted back into the atmosphere, but somehow the earth's atmosphere retained some of the infrared radiation, holding the heat, while the rest disappeared into space.[35]

In 1859, John Tyndall sought to understand how the atmosphere regulated Earth's temperature. Part of his interest in this particular issue stemmed from his ongoing investigations regarding the causes of the prehistoric ice age—an issue introduced by Louis Agassiz in 1837.

Scientists assumed that "all gases are transparent to infrared radiation."[36] But Tyndall wanted to confirm such a hypothesis, so he investigated "the radiative properties of various gases."[37] He found that nitrogen and oxygen, the main gases in the atmosphere, are transparent and that heat passes through them. He continued his research, testing the methane gas that was pumped into his lab. He noted that infrared rays confronted this gas in a different way; in fact, "this gas was . . . opaque."[38] He found the same situation with carbon dioxide and water vapor. With further research, he determined that carbon dioxide and other gases, which make up only a small proportion of gases in the atmosphere, absorbed heat via infrared radiation. This heat—transferred to the air and surface—helped warm the earth (the "greenhouse effect") to create a habitable climate.[39]

Tyndall concluded that variations in the atmosphere, whether gas or water vapor, could affect climate conditions. He contended that water vapor was so important to determining the conditions of life on earth that it was "a blanket more necessary to the vegetable life of England than clothing is to man. Remove for a single summer-night the aqueous vapor from the air . . . and the sun would rise upon an island held fast in the iron grip of frost."[40] Thus the relationship between the atmosphere's humidity and temperature could influence changes in climate, leading to such things as an ice age. Marx, who followed closely the work of natural scientists during his day, and attended some of Tyndall's lectures, was extremely impressed by Tyndall's experiments on solar radiation.[41]

Building on the work of Fourier and Tyndall, August Arrhenius, a Swedish scientist, also argued that part of the cause of the ice age was fluctuations in the concentration of carbon dioxide in the atmosphere, which contributed to variations in the amount of water vapor in the air. At the same time he considered the potential for warming beyond average temperatures. From his work on this issue, he then pointed out, in an 1896 paper, "On the Influence of Carbonic Acid in the Air Upon the Temperature of the Ground," that humans were introducing massive quantities of carbon dioxide into the atmosphere via an economy based on the burning of fossil fuels and that this had contributed to an increase in temperature on Earth.[42] Not realizing the systematic expansion of industrial operations, Arrhenius believed that it would take hundreds of years for humans to double the carbon dioxide in the atmosphere, so he did not concern himself with the possibility of the rapid warming of the planet. Instead, Arrhenius believed that the release of carbon dioxide would help slow down the freezing of Earth.[43] But his work, as well as Tyndall's, provided the basis for scientific work on global warming, which did not gain momentum until the 1960s. In the end, it has been recognized that the concentration of greenhouse gases in the atmosphere has substantially changed over geological, and even historic, time. The concern today is that the changes in the concentration of greenhouse gases in the atmosphere are unparalleled and the socioeconomic forces driving the situation show no signs of slowing down.

The carbon cycle involves the whole biosphere, as carbon moves through the air, soil, water, and all living things in a cyclical process. All life is dependent on this process, and carbon serves as "the principal element of which all living beings . . . are made."[44] In part, carbon is absorbed and contained in non-living forms, such as oceans, that serve as sinks, helping to limit the accumulation of carbon dioxide in the atmosphere. Carbon from the atmosphere enters the life cycle through carbon fixation (a process Liebig helped confirm) in which plants' photosynthetic process converts—in conjunction with water, chlorophyll, and the sun's energy—carbon dioxide into carbohydrates and oxygen.[45] From this point, some carbon reenters the atmosphere through the respiration of plants. But much of the carbon is passed on to other species, and onward through the food chain, where carbon enters the soil and water as waste, as dead matter, or as carbon dioxide through the respiration of animals. Thus carbon dioxide is released into the atmosphere only to be recirculated to the earth through a variety of pathways in natural processes. Each part of the carbon cycle (absorption by plants, oceans, and glaciers, the processing of carbon materials by animals, and the circulation in the atmosphere) contributes to the regulatory processes that help govern the complex relationships of interchange of carbon throughout the biosphere. Hence the carbon cycle has a particular carbon metabolism, influenced by interpenetrating relationships throughout the circulation of carbon, which has helped sustain the conditions of life, as we know it, on Earth.

Over the past 400,000 years, the carbon cycle and climate system have operated in a relatively constrained manner, sustaining the temperature of Earth and maintaining the balance of gases in the atmosphere. There have been natural climate variations in global temperatures through the centuries, which contributed to a "Little Ice Age," but the scientific consensus is that the accumulation of carbon in the atmosphere driving global climate change today is the result of human activities.[46] Understanding the basic operations of the carbon cycle is necessary for understanding climate change. In the sections that follow, we focus on the social forces that are driving the historic accumulation of carbon dioxide in the atmosphere and causing a rupture in the carbon cycle.

The Expansion of Capitalist Production and the Accumulation of Carbon Dioxide in the Biosphere

The advent of *Homo sapiens* brought forth unprecedented social interactions with nature, which included the purposeful use of fire. The anthropogenic burning of plants and trees released stored solar energy into the atmosphere. The ability to control fire decreased human vulnerability to nature. Of course, it was not until the rise of capitalism, and especially the development of industrial capital, that anthropogenic carbon dioxide emissions greatly expanded in scale through the burning of coal and petroleum, exploiting the historic stock of energy that was stored deep in the earth and releasing it back into the atmosphere. As a result, the concentration of carbon dioxide in the atmosphere has increased dramatically, overwhelming the ability of natural sinks—which have also been disrupted by anthropogenic forces—to absorb the additional carbon.

To understand the rift in carbon metabolism, one needs to understand the forces that drive carbon dioxide emissions. It is now widely recognized that humans alter the global climate "by interference with the natural flows of energy through changes in atmospheric composition. . . . Global changes in atmospheric composition occur from anthropogenic emissions of greenhouse gases, such as carbon dioxide that results from the burning of fossil fuels and methane and nitrous oxide from multiple human activities."[47] Much worse, "we have driven the Earth system from the tightly bounded domain of glacial-interglacial dynamics," one that defined the Earth system for over 400,000 years.[48] Though not recognizing the potential dangers associated with increasing global temperatures, Arrhenius noted that industrial operations were contributing to an increase of the carbon dioxide in the natural world.[49] We know now that the quantity of carbon dioxide in the atmosphere "has increased 31 percent since preindustrial times" and that "half of the increase has been since 1965."[50]

Often industrialism is identified as the principal factor behind global warming, but this position fails to recognize that industrialism is embedded within a particular global economic system. Understanding the forces and operations of capitalism is necessary for gaining perspective on how industrial social relations function as well as how the human–nature interchange under this system contributes to global climate change. This is not to say that other economic systems do not perpetrate and contribute to environmental degradation. Soviet-type societies caused immense environmental deterioration, but this does not negate the importance and urgency of analyzing the social relations, operations, and development of capitalism since it is the political-economic system that has been and is dominant in the world today.

Environmental crises have existed throughout human history.[51] Indeed, Jason Moore argues that the birth of capitalism was pushed forward, in part, by environmental contradictions and crises in feudalism, namely a metabolic rift particular to the structure of feudal agricultural production.[52] The system found temporary relief through establishing a global economy, which increasingly incorporated the world into a wider metabolic rift of global proportions—as agricultural goods (food and fiber) were transferred from colonies to European nations. Seeking endless accumulation of capital, agricultural practices were intensified, as land was consolidated into fewer hands. Liebig and Marx documented the reemergence of a soil crisis in Europe in the 1800s.[53] This soil crisis led to the global trade of guano to fertilize fields in Europe and eventually to the development of artificial fertilizers, which ever since have been used in larger quantities, despite the associated environmental problems that they create, such as dead zones.[54]

The same logic that dictated the expansion and intensification of agricultural production fueled the drive behind the productive systems in cities.[55] The conditions within cities were in part a consequence of the transformation in the countryside, as people were swept from the land through the concentration of land among fewer landholders. The metabolic rift in the soil nutrient cycle continued to expand with the division between town and country, and new metabolic rifts were created with the ongoing development of capitalism.[56] After being separated from the land, people were forced to seek work in the cities, struggling for survival, under the anarchy of the market. At this point, capitalism's development of technology and its separation from the hands of workers is important. At first, human bodies operated tools, exerting energy for the production of commodities. But the drive to maintain the continuity of production fostered scientific and technological development.[57] Marx discusses how the motive power of production was transformed from humans to machines. "Machine-work" came to dominate "hand-work." The new machines, while being modeled on the craft system of production, forced further technological innovations in production through the

> framework of the machine.... The machine ... is a mechanism that, after being set in motion, performs with its tools the same operations as the worker formerly did with similar tools. Whether the motive power is derived from man, or in turn from a machine, makes no difference here. From the moment that the tool proper is taken from man and fitted into a mechanism, a machine takes the place of a mere implement.[58]

A greater division of labor developed as the mechanization of tools freed capitalism from the limitations of individual workers' labor power and parts of nature were transformed into fuel for the new machines. Tools became embedded within machines that labor operated. Production took place on an enlarged scale, demanding more energy to sustain operations. Marx commented: "An increase in the size of the machine and the number of its working tools calls for a more massive mechanism to drive it; and this mechanism, in order to overcome its own inertia, requires a mightier moving power than that of man, quite apart from the fact that man is a very imperfect instrument for producing uniform and continuous motion."[59]

The movement from human motive power to water and wind to coal-driven steam engines transformed capitalist production, increasing the scale of production by pushing up labor productivity to historically unprecedented levels, and by deepening the exploitation of nature and labor.[60] The social metabolism with nature was intensified to facilitate the accumulation of capital on an ever-larger scale.

Marx outlined how "the growth of machinery and of the division of labor" allowed more commodities to be produced "in a shorter time" and how "the store of raw materials must grow" at the same time.[61] All of this required increases in the quantity of matter-energy throughput, for the expansion of production in the pursuit of the accumulation of capital on a greater scale. Marx explained:

> The material forms of existence of the constant capital, however, the means of production, do not consist only of such means of labour, but also of material for labour at the most varied stages of elaboration, as well as ancillary materials. As the scale of production grows, and the productive power of labour grows through cooperation, division of labour, machinery, etc., so does the mass of raw material, ancillaries, etc. that go into the daily reproduction process. . . . There must always be a greater store of raw material, etc. at the place of production than is used up daily or weekly.[62]

Thus Marx highlights how the drive to accumulate capital fueled the development of industrial productive forces, which at the same time created a growing need for raw materials mined from the earth to power the machines. As capitalism continues to grow, more capital is used to purchase "raw materials and the fuels required to drive the machines."[63] Thus an expansion in productivity and technological development under capitalism increases the quantity of energy throughput that is required to expand the accumulation of capital. The operations of capitalist production became dependent on a constant supply of raw materials that could sustain its operations on an ever-greater scale. Thus capital was pushed "to structure the energy economy around fossil fuels (a reality that is now deeply entrenched)."[64]

Just as the expansion of capitalist agricultural production globalized the metabolic rift in the soil nutrient cycle, capitalist expansion pushed forward technological development that allowed industrial production to take place at ever-greater levels. Previous modes of production primarily operated within the "solar-income constraint," which involves using the immediate energy captured and provided by the sun. By mining the earth to remove stored energy (past plants and animals) to fuel machines of production, capitalist production has "broken the solar-income budget constraint, and this has thrown [society] out of ecological equilibrium with the rest of the biosphere."[65] Herman Daly warns that as a result of these developments natural cycles are overloaded and the "life-support services of nature are impaired" because of "too large a throughput from the human sector."[66] The ability to take coal and petroleum from the earth accelerated the expansion of capital, releasing large quantities of carbon dioxide into the atmosphere. This pattern, just as the rift in the soil nutrient cycle, continues, given the logic of capital.

Ongoing capitalist development continues to dump carbon dioxide into the atmosphere, placing greater demands upon the carbon cycle to metabolize this material. This uneven process only worsens, given the character of capital. To survive, capital must expand. It is engaged in a process of ceaseless expansion and constant motion. The capitalist system is incessantly struggling to transcend existing barriers, both social and natural (such as the regulative laws of natural cycles). At the same time capitalism confronts new barriers (such as natural limits and rifts in metabolic cycles) as the world is reshaped and reorganized in the pursuit of profit. Given that capitalism operates globally, there is no natural confinement or pressure to stop the ruin of ecosystems, short of global collapse.[67]

The basic characteristic of capitalism "is that it is a system of self-expanding value in which accumulation of economic surplus—rooted in exploitation and given the force of law by competition—must occur on an ever-larger scale."[68] The accumulation of capital remains the primary objective in capitalist economies. Sweezy perceptively described the accumulation process and its relationship to nature, in stating:

> A system driven by capital accumulation is one that never stands still, one that is forever changing, adopting new and discarding old methods of production and distribution, opening up new territories, subjecting to its purposes societies too weak to protect themselves. Caught up in this process of restless innovation and expansion, the system rides roughshod over even its own beneficiaries if they get in its way or fall by the roadside. As far as the natural environment is concerned, capitalism perceives it not as something to be cherished and enjoyed but as a means to the paramount ends of profit-making and still more capital accumulation.[69]

In some respects, this is a self-propelling process, as the surplus accumulated at one stage becomes the investment fund for the next. In this, the scale of capitalist operation is ever-increasing, driven by ceaseless economic growth. To sustain this process, capital requires constant access to, and an increasingly large supply of, natural materials (such as petroleum). Capital freely appropriates nature's supplies and leaves wastes behind.[70] As the economic system grows under capitalism, the throughputs of materials and energy increase and capital incorporates ever-larger amounts of natural resources into its operations.

The law of value remains central to understanding capitalism and the ecological crisis.[71] For Marx, "the earth . . . is active as an agent of production in the production of a use-value."[72] But the exchange value (or value) of a particular commodity is determined entirely by the socially necessary labor exercised in its production (while indirectly affected by the capital/technology invested in production). Nature is treated under capitalism as a "free gift" to the property owner. Monopoly rents are imposed for natural resources, based on scarcity, degree of monopoly, and what the market will bear. Value under capitalism therefore systematically excludes nature's contribution to wealth. As Marx put it, land (nature) and labor together constitute "the original sources of wealth."[73] Yet, nature's contribution is absent from capitalist value accounting.

Under capitalism, money serves as a universal equivalent in exchange. It is the reification of universal labor-time, "the product of universal alienation and of the suppression of all individual labour" and "a form of social existence separated from the natural existence of the commodity."[74] Money mystifies labor and nature. In exchange, the qualitative dimensions of social production are erased. "Money 'solves,'" Paul Burkett, a Marxist economist, notes, "the contradiction between the generality of value and the particularity of use values by abstracting from the qualitative differentiation of useful labor as conditioned by the material diversity of human and extra-human nature—the true sources of wealth."[75]

There is no drive to maintain the social metabolism in relation to the natural metabolism (a measure of sustainability) under capital. Capital cannot operate under conditions that require the reinvestment of capital into the maintenance of nature. Short-term profits provide the immediate pulse of capitalism. Capital itself is a manifestation of competition in the accumulation of wealth.[76] Money serves as a universal measure and means for international trade and aids capital in its international expansion, as it incorporates more people and nature into the global system. The monetary process comes to dominate the organization of the material processes of production. In this, capitalism conquers the earth (including the atmosphere), taking its destructive field of operation to the planetary level. The exploitation of nature is universalized, increasingly bringing all of nature within the sphere of the economy, subjecting it to the rationality of profitability.[77]

Capital is the systematic force organizing social production and driving industrialism to intensify the exploitation of nature. Given the logic of capital and its basic operations, the rift in the carbon cycle and global climate change are intrinsically tied to capitalism. In fact, the continued existence of capitalism guarantees the continuation of these events. "Short of human extinction," Burkett stresses, "there is no sense in which capitalism can be relied upon to permanently 'break down' under the weight of its depletion and degradation of natural wealth."[78]

Numerous human activities contribute to the accumulation of carbon dioxide and global climate change, including deforestation, desertification, and expanded agricultural production, but the burning of fossil fuel is the primary source of greenhouse gases. Carbon dioxide is the most abundant greenhouse gas. Society is adding "carbon to the atmosphere at a level that is equal to about 7 percent of the natural carbon exchange of atmosphere and oceans."[79] The increasing concentrations of greenhouse gases have contributed to the warming of the earth.[80]

Capitalism, in organizing the social relations of commodity production, effectively plunders the historical stock of concentrated energy that has been removed from the biosphere only to transform and transfer this stored energy (coal, oil, and natural gas) from the recesses of the earth to the atmosphere in the form of carbon dioxide. In this, capitalism is disrupting the carbon cycle by adding carbon dioxide to the atmosphere at an accelerating rate. At the same time, capital's constant demand for energy necessitates the continual plundering of the earth for new reserves of fossil fuel.[81] With over 23 billion metric tons of carbon dioxide released into the atmosphere per year, current production is creating "waste emissions faster than natural systems can absorb them."[82] As a result, carbon dioxide is accumulating—as atmospheric waste—at alarming rates, warming the earth, and causing dramatic climate change.

The Jevons Paradox: A Dilemma for Ecological Modernization

Scientific recognition of the accumulation of carbon dioxide and climate change has made carbon dioxide emissions a major social concern and culminated in social pressure throughout the world to reduce emissions. Capital and neoclassical economists attempt to assuage fears of environmental deterioration as an inherent part of capitalist economic operations. They typically assert that capitalist development will lead to improved technologies and efficient raw material usage, and that this will decrease emissions and environmental degradation. They argue there is an "environmental Kuznets curve" for many types of environmental impacts. The environmental Kuznets curve suggests that environmental impacts, such as pollution, increase in the early stages of development within nations as an industrial economy is established, but level off and eventually decline as economies "mature," because environmental quality is a luxury good, affordable only by the affluent.[83] Proponents of "green capitalism," such as Paul Hawken, claim that if the value of nature were properly accounted for, capitalism would develop in an ecologically benign direction.[84] In other words, they argue that through innovative technological development and appropriate reformist government policy, the economy can be dematerialized, reducing the throughput of raw materials and energy that the system requires.[85]

Ecological modernization theorists adhere to this technological optimism, claiming that the forces of modernization lead to the dematerialization of society and the decoupling of the economy from energy and material consumption, allowing human society, under capitalism, to transcend the environmental crisis.[86] In particular, ecological modernization theorists argue that rationality, a cornerstone of modernity, percolates into all institutions of "advanced" societies.[87] This process leads to the emergence of "ecological rationality," which focuses on the necessity of maintaining the resources and ecosystem functions upon which societies depend, and shifts the focus away from the pure economic rationality that prevailed in the early stages of modernization.

Hence, ecological modernization theory is at base a functionalist approach in that it does not see the emergence of ecological rationality as coming primarily from social conflict, but rather from ecological enlightenment within the key institutions in societies. Such theorists contend, then, that radical ecological reform does not require radical social reform—the institutions of capitalist modernity can avert a global environmental crisis without a fundamental restructuring of the social order.

In the same vein, Luc Boltanski and Ève Chiapello argue that capitalism is a flexible system that is able to respond to social and natural barriers, social movements, and criticism. It is a system that can incorporate an interest for the common good of society into its operations. In fact, past criticisms of the system have helped direct capitalism in ways that allow it to flourish in order to meet social needs and desires. Expanding knowledge is seen as a force that propels bureaucrats and capitalists to respond readily to social concerns.[88]

Marxist critics, of course, do not deny that capitalism is a dynamic system. However, the likelihood of capital pursuing the common or social good (and by extension environmental sustainability) is contested. Still, ecological modernization theorists and others of this general persuasion believe that capitalism is fully able to respond to climate change through pursuing socio-technical innovation, without challenging the prevailing political-economic structure. Indeed, economic growth has produced new technologies that are more efficient, and some technological improvements have reduced specific types of pollution.

In contrast, we contend that the belief that these changes lead to benign ecological relationships needs further consideration, especially considering that capitalist expansion of commodity production—which includes energy sources as throughputs—has outstripped improvements in the efficiency of energy use. Empirical research suggests that carbon *efficiency* (economic output per unit of carbon emissions) may follow an environmental Kuznets curve, but per capita emissions increase monotonically with economic development.[89] Ironically, the most efficient nations are often the biggest consumers of natural resources.[90]

William Stanley Jevons, in *The Coal Question*, explained that improved efficiency in the use of coal made it more cost effective as an energy source and therefore more desirable to consumers. Thus, he argued, greater efficiency in resource use often leads to *increased* consumption of resources.[91] This relationship has become known as the Jevons Paradox.[92] Jevons pointed to an observed relationship between efficiency and total consumption, but he did not explain why this was the case. He needed to connect this fact—that rising efficiency is associated with *rising* consumption, at least in the case of coal—with the drive for the accumulation of capital, which entails the continued material consumption of transformed nature to fuel its operations. Doug Dowd, a political economist, explains how capitalist production for the accumulation of capital, rather than representing production to meet social needs and the demands of environmental sustainability, generates enormous waste throughout its operations. Given capital's drive to economic expansion, it produces ever-greater amounts of waste, in concentrations that threaten ecosystems.[93] Empirical research and other analyses support the contention that economic growth and expansion typically outstrip gains made in efficiency.[94]

Straightforward calculations based on aggregate data for a selection of "advanced" capitalist nations illustrate the paradoxical relationship between efficiency and resource consumption. Over the period 1975 to 1996, the carbon efficiency of the economy—economic output, measured in terms of gross domestic product (GDP), per metric ton of carbon dioxide emissions—increased dramatically in the United States, the Netherlands, Japan, and Austria (see Table 1). However, over this same period, total carbon dioxide emissions, and even per capita emissions, *increased* in all four of these nations despite the improvements in efficiency.[95] Thus gains in the efficiency of the use of fossil fuel have typically resulted in the expansion of their use in industrialized capitalist nations. As a result, carbon emissions generally increase with modernization and its concomitant "improvements" in technology and gains in efficiency.[96] It is noteworthy that Marx explained that capitalism prevents the truly rational application of new science and technologies because they are simply used to expand the operations of capital.[97]

TABLE 1. Changes in the Economic Carbon Efficiency (GDP Per Unit of Carbon Dioxide Emissions), Total Carbon Dioxide Emissions, and Carbon Dioxide Emissions Per Capita of Four "Advanced" Capitalist Nations between 1975 and 1996.[98]

NATION	CARBON EFFICIENCY (CHANGE)	TOTAL CO_2 EMISSIONS (CHANGE)	CO_2 EMISSIONS PER CAPITA (CHANGE)
United States	+34.0%	+29.7%	+5.9%
Netherlands	+30.0%	+24.3%	+9.1%
Japan	+64.0%	+25.9%	+12.0%
Austria	+50.2%	+11.6%	+4.9%

Capitalism, at this stage of its development, depends upon massive quantities of fossil fuel to continue to operate at the current scale of production, to say nothing of an *increasing* scale of production. State promoted carbon-market policies and carbon-sequestration technologies are ploys to continue capitalist production as is, and they are unlikely to deal directly with global climate change. Carbon-market policies establish a floor under carbon emissions without effectively establishing a ceiling. Sequestration technologies would have their own ecological concerns and are likely too large scale to operate and too expensive to fund in the capitalist economic system.[99]

The recovery of agricultural nutrients has proven to be insurmountable under capitalism for over a hundred years. Thus a massive quantity of artificial fertilizer and oil—which contributes to the accumulation of carbon dioxide—is needed to sustain food production. Recovering carbon waste from the atmosphere will likely prove to be a much more difficult task. The social structure of the capitalist system sets limits and constraints on what mitigating actions will and can be taken.[100] All the while, the rift in the carbon cycle continues to deepen as carbon dioxide continues to accumulate in the atmosphere and capital pursues profit. The social project to mend the carbon rift is not simply a technological one but requires the struggle to establish an entirely new social metabolism with nature. This requires a new social system driven by human development—which by necessity must also be ecological to sustain the conditions of life—not the accumulation of capital.[101]

The Disruption and Flooding of Carbon Sinks

Since modern capitalist societies are emitting carbon dioxide into the atmosphere at an extraordinary and escalating pace, it is important to understand what happens to carbon when it enters the atmosphere. Carbon dioxide has a long atmospheric lifetime, remaining in the atmosphere for up to 120 years.[102] As it accumulates in the atmosphere—as Tyndall discovered—it blocks and absorbs infrared radiation, effectively preventing the radiation from escaping into space. Higher concentrations of carbon dioxide increase the temperature of the atmosphere and the oceans. As described earlier, carbon has an established cycle, where it moves through the biosphere, is absorbed by plants, and is used in the production of carbohydrates before being released back into the atmosphere through a variety of pathways. Oceans and forests serve as natural sinks, absorbing large quantities of carbon dioxide.

The creation of a rift at one point in a cycle (that is, the accumulation of carbon dioxide in the atmosphere) can generate system-wide crises. The metabolic rift in regard to carbon metabolism is caused by the introduction of low entropy fossil fuels. These consist of plant matter that accumulated over geological time and were effectively removed from the environment. The burning of these fuels has allowed economies to break the solar budget, but this has come at the cost of rapid atmospheric change.[103] The gravity of the situation in regard to the carbon cycle in the current historical era is that capitalism is disrupting the carbon cycle at two points, further complicating matters. The circulation of carbon and the stabilization of it within certain parameters depend on the availability of carbon sinks and their ability to absorb carbon dioxide. The oceans and forests are the largest and primary sinks for carbon dioxide, but the amount of carbon dioxide in the atmosphere has exceeded the capacity of nature to absorb these gases. Thus the sinks may be approaching the limits of their capacity to sequester carbon, as environmental destruction throughout the world degrades and depletes them.

"For the globe as a whole," stated Rachel Carson, "the ocean is the great regulator, the great stabilizer of temperatures."[104] The oceans influence the concentration of carbon dioxide in the atmosphere as the gas is continuously exchanged at the surface. Between a third and a half of the carbon dioxide produced is absorbed by the oceans, but its ability to do this is changing due to global warming. In the water, "CO_2 forms a weak acid that reacts with carbonate anions and water to form bicarbonate. The capacity of the oceanic carbonate system to buffer changes in CO_2 concentration is finite and depends on the addition of cations from the relatively slow weathering of rocks."[105]

The scale of anthropogenic carbon dioxide exceeds the supply of cations, and may come to exceed the saturation point in the water. In the long run the capacity of the ocean to absorb carbon dioxide will likely decrease.[106] This rift may only deepen and further limit the sequestration of carbon, leaving more carbon dioxide in the atmosphere. Already the accumulation of carbon dioxide has led to the warming of the earth, which has increased the melting of glaciers (also historic sinks of carbon dioxide), releasing even more carbon dioxide into the atmosphere.

With the ability of the oceans to absorb carbon dioxide potentially in decline, the absorption capacity of terrestrial ecosystems becomes particularly important. Forests are the primary carbon sink on land. Deforestation not only destroys a carbon sink, but leads to the emission of substantial quantities of carbon dioxide into the atmosphere when forests are burned. The removal of forests changes a sink of carbon dioxide into a source of it.[107] Dramatic deforestation, particularly in the tropics (given that many core nations have already destroyed the bulk of their forests and depend on wood imports from less developed nations), continues to decrease the absorption of carbon dioxide by terrestrial ecosystems. Worldwide, there was on average a net loss of over 90,000 square kilometers (an area approximately the size of Portugal) of forest each year during the 1990s.[108] Global deforestation appears to be driven, at least in part, by the increasing globalization of markets and the influence of global capital, where the natural environment of periphery nations is degraded in the extension of trade—especially in agricultural goods and natural resources—and as a result of the expansion of global poverty, as the poor and other landless people clear ground for survival.[109] The reduction in forest area and the potential carbon saturation of available forests only increases the accumulation of carbon dioxide in the atmosphere. Sinks helped slow the accumulation of carbon dioxide, but there is no "natural savior" waiting to absorb and assimilate all the carbon dioxide produced by capitalist processes. "Humans have affected virtually every major biochemical cycle" and sinks "cannot mitigate against continued accumulation of the gas in Earth's atmosphere."[110] The social metabolism in relation to the natural metabolism is causing the sinks to overflow, leading to increases in global temperatures.

Global Inequalities, Per Capita Emission Allowances, and Biospheric Crisis

At the planetary level, ecological imperialism has resulted in the appropriation of the global commons—the atmosphere and oceans, which are used as sinks for waste—and the carbon absorption capacity of the biosphere, primarily to the benefit of a relatively small number of countries at the center of the capitalist world economy. The core nations rose to wealth and power in part through high fossil fuel consumption and exploitation of the global South. Anthropogenic greenhouse gases emissions, while stemming from localized sources, are distributed throughout the atmosphere and accumulate as waste, which degrades the atmosphere and leads to further alteration of the biosphere, creating a global crisis.

Theorists from both the Marxist and world-systems perspectives provide valuable studies regarding the unevenness of carbon dioxide emissions among nations. Non-core nations of the global economy emit significant amounts of carbon dioxide into the atmosphere, use inefficient energy technologies, and, in the case of China, burn coal primarily to meet their energy needs. All of this reflects inequalities in the global economy and the unevenness of capitalist development. But it is core nations that cause disproportionate amounts of emissions due to industries, automobiles, and affluent lifestyles. Timmons Roberts explains, "Overall, the richest 20% of the world's population is responsible for over 60% of its current emissions of greenhouse gases. That figure surpasses 80% if our past contributions to the problem are considered."[111]

Thus the affluent core nations of the global economy are primarily responsible for global climate change, whether it is in regard to emissions, the quantity of carbon dioxide in the atmosphere that floods the sinks, or the hegemonic economic forces that foster the destruction of sinks, such as forests.

The IPCC estimated that at least a 60 percent reduction in carbon emissions from 1990 levels is necessary to reduce the risk of further climate change. The core nations' carbon output alone exceeds the world's total allowable amount. The severity of the situation and the extreme global inequalities are clearly seen when we consider that the per capita emissions of carbon for the United States, in 1999, were over 5.6 metric tons per year.[112] The per capita emissions for the G-7 nations were 3.8 metric tons per year; and the rest of the world had per capita emissions of 0.7 metric tons per year.[113]

Global shifts in production brought an immense amount of capital to peripheral nations, where industrial production increasingly takes place. The profits are then transferred back to the core nations. Nevertheless, this relocation of productive operations to peripheral nations has increased their carbon emissions, despite few immediate economic gains.[114] Marx commented that "the more a country starts its development on the foundation of modern industry, like the United States, for example, the more rapid is this process of destruction."[115] While he was describing the social and environmental exploitation that takes place under capitalist agriculture, Marx's statement captures the social relations under capital in general, especially the human relationship with the biosphere.

Ecological Consequences, Biospheric Crisis, and Redrawing the Maps

It has been recognized that the concentration of carbon dioxide in the atmosphere has varied substantially over geological, and even historic time. But the social metabolism of capitalism has led to an unprecedented change in the atmosphere. Using ice core samples from Greenland and Antarctica, scientists have been able to determine carbon dioxide concentrations in the atmosphere over the last 650,000 years.[116] The concentration of carbon dioxide never exceeded 300 parts per million (ppm) during these years. But the constant drive of capital accumulation via its industrial operations and structured environment (such as the organization of cities) has surpassed 300 ppm even reaching 390 ppm in 2010. The accumulation of carbon dioxide has led to a steady increase in temperatures in the atmosphere and oceans.

As Vernadsky contended, human life is a significant force reshaping the planet.[117] The coevolutionary relation between human society and nature, as organized under capitalist production, is creating an ecological rift that is generating numerous changes in environmental conditions, some of which threaten to hasten the accumulation of carbon dioxide in the atmosphere and wreak social havoc.

Sweezy claimed that the expanding scale of capitalist production created ever greater threats to the natural world, especially to nature's resiliency.[118] Seeking short-term profit, capital continues to push for faster turnover rates to expand the accumulation of capital, as if there were no natural limits to its advance. The social metabolism increasingly comes into greater conflict with the natural metabolism, as metabolic rifts deepen, creating ecological crises in the natural conditions of life. In regard to global climate change, inaction creates a difficult position for the future. If current trends continue, global climate change could spiral out of control, potentially threatening the survival of human beings. An "ecological discontinuity" can occur with few, if any, immediate warning signs.[119] Global climate change is causing extreme weather patterns, such as hurricanes, floods, and droughts. It is affecting glaciers, ice caps, soil nutrients, species extinction, biodiversity, and ocean currents. A metabolic rift in the carbon cycle is creating a biospheric crisis that is radically altering natural processes and cycles and the conditions of nature and life.

Global climate change is setting in motion a series of transformations in the physical world. The rapid melting of sea ice in the Arctic is decreasing the earth's albedo (reflectivity) stepping-up global warming. Glaciers throughout the world are retreating as a result of global warming, threatening much of the world's population with water shortages—in terms of both drinking water and irrigation. The thawing of the permafrost in the tundra, due to global warming, will potentially release 10 times the annual carbon dioxide emissions of humans as well as methane gas, intensifying the process of climate change.[120]

The melting of the ice in Antarctica and Greenland presents another grave danger to human civilization. The ice shelves are breaking up and melting at an astonishing pace. In part this is due to pools of water formed from rising temperatures, which continue to warm. This water heats the ice below, driving deep holes of warm water within an ice shelf. As the ice shelves over water melt, the ice over land begins to shift and break off into the ocean. It is the melting of ice over land that leads to an increase in the sea levels. The melting of ice over land in Antarctica and Greenland creates huge streams of meltwater that pour down crevasses and through tunnels (moulins). When the meltwater reaches the land beneath the ice, it both warms the ice underneath and serves as a lubricant that could lead to the ice shifting and falling into the sea.[121] The melting of Greenland's ice alone could raise the worldwide sea level 24 feet. Hundreds of millions of people living by the sea would be threatened by such a situation, and nations would be disproportionately affected. Many islands, some densely populated, and low-lying countries such as Bangladesh, would be inundated with severe floods and partially submerged. The maps of the world would have to be redrawn and millions of "ecological refugees" would be created.[122] Already the people of Tuvalu, in the South Pacific, are being dislocated as a result of climate change.

Carbon dioxide "will be responsible for more than half of the anticipated global warming over the next century."[123] "The rate of climate change driven by human activity," Hansen states, "is reaching a level that dwarfs natural rates of change."[124] Rapidly changing natural conditions are causing ecosystems to collapse and species extinction. High-mountain ecosystems are already experiencing higher temperatures, causing changes in the types of plants and animals that exist there, as plants from lower elevations move up the mountains.[125] Climate change is leading to a redistribution of plants on earth, as some species respond better to higher carbon dioxide concentration than others and, therefore, displace other plants. Coupled with habitat destruction via human encroachment, anthropogenic climate change is driving a mass extinction of species.[126]

The loss of biodiversity furthers environmental degradation, decreasing the resiliency of ecosystems, which increases the chances of a regime shift in ecosystems to "less desired states in their capacity to generate ecosystem services."[127] This means that continual environmental degradation threatens the very operations of natural processes and conditions that are necessary to sustain the conditions of life. New constraints will be placed on life and its development as the social metabolism intensifies with the growth of capital and expands the metabolic rift. Vitousek warns that the changes that are forced on the world will not be smooth, as humans continue to "alter the structure and function of Earth as a system."[128]

Changes in climate have led to severe storms and droughts. Some areas there have experienced more precipitation, but it is more concentrated, causing floods. In other areas, with the relocation of precipitation, drought conditions persist. Global warming increases the rate of evaporation, sucking the moisture from the soil. Intensive agriculture and water shortages are creating deserts. Dust from dry agricultural lands is being swept up by the wind into the mountains, where it is deposited on the snow. Dirty snow attracts more heat and causes the snow to melt faster, leading to an overflow of water early in the summer and short supplies latter.[129] The warming of the oceans is contributing to more intense tropical storms and hurricanes, because warmer temperatures increase the water content of the air. Global warming is leading to coral bleaching, whereby a healthy coral reef that helps support and sustain a rich biodiversity in oceans is turned into a skeleton. The added carbon dioxide changes the chemistry of the oceans, due to all the carbonic acid that negatively influences "the saturation levels of calcium carbonate in the oceans," altering the natural metabolism of the seas.[130]

The continuation of capitalist operations, which promises to continue to increase carbon emissions, threatens to undermine the capacity of the biosphere to support life on earth. "Humans have, over historical time but with increased intensity after the industrial revolution," note Carl Folke and his associates, "reduced the capacity of ecosystems to cope with change through a combination of top-down . . . and bottom-up impacts . . . as well as through alterations of disturbance regimes including climatic change."[131]

The IPCC expectation of an increase in temperature of 1.5-6.0°C during this century signifies a rift in the carbon metabolism, as carbon waste accumulates in the biosphere. Given the logic of capital, we know that it cannot mend the carbon cycle; capital's productive cycle takes precedence over the maintenance of the natural world. Thus capitalism continuously creates and deepens metabolic rifts, separating the social metabolism from the natural metabolism. The metabolic rift in the carbon cycle is inducing alterations in natural conditions, undermining the regulatory processes that govern the cycle and creating positive feedbacks in the release of even more carbon dioxide into the atmosphere, which may accelerate global climate change via abrupt climate change, forcing a switch in the environmental threshold on which humans depend.

Capital continues to profit while increasing the degradation of nature, and it even finds ways to profit from this process, whether it is increased sales of air conditioners or new power plants to meet increasing energy usage within the megacities. The systematic degradation of natural conditions and the increasing scale of economic operations are undermining the regenerative capabilities of ecosystems and the biosphere in general, increasing the likelihood of ecological overshoot. As a result, we are living in an ecological crisis that threatens large-scale ecocide.[132] It is not nature that is the problem, but society.

6. The Planetary Moment of Truth

It is impossible to exaggerate the environmental problem facing humanity in the twenty-first century. Available evidence now strongly suggests that, under a regime of "business as usual" with no substantial lessening of the drivers of environmental destruction, we could be facing within a decade or so a major "tipping point," leading to irrevocable and catastrophic climate change.[1] Other ecological crises—such as species extinction, the rapid depletion of the oceans' bounty, desertification, deforestation, air pollution, water shortages and pollution, soil degradation, the imminent peaking of world oil production (creating new geopolitical tensions), and a chronic world food crisis—all point to the fact that the planet as we know it and its ecosystems are stretched to the breaking point.[2] The moment of truth for the earth and human civilization has arrived.

To be sure, it is unlikely that the effects of planetary ecological degradation, though extraordinarily rapid, will prove "apocalyptic" for human civilization within a single generation, even under conditions of capitalist business as usual. Measured by normal human life spans, there is considerable time left before the full effect of the current human degradation of the planet comes into play. It is not ourselves but our grandchildren who will be most affected. Yet the period remaining in which we can *avert* future environmental catastrophe, before it is essentially out of our hands, is much shorter. Indeed, the growing sense of urgency of environmentalists has to do with the prospect of various tipping points being reached as critical ecological thresholds are crossed, leading to the possibility of a drastic contraction of life on earth.

Such a tipping point, for example, would be an ice-free Arctic. In summer 2007, the Arctic lost *in a single week* an area of ice almost twice the size of Britain. In 2007 the end-of-summer sea ice was 40 percent below that of the late 1970s. The vanishing Arctic ice cap means an enormous reduction in the earth's reflectivity (albedo), thereby sharply increasing global warming (a positive feedback known as the "albedo flip"). At the same time, the prospective rapid disintegration of the ice sheets in West Antarctica and Greenland, on top of melting mountain glaciers, points to rising world sea levels, threatening coastal regions and islands. As Orin Pilkey and Rob Young write in *The Rising Sea*, "Sea level has clearly been rising at an accelerating rate through the twentieth century and into the twenty-first."[3]

In 2008, James Hansen, director of NASA's Goddard Institute for Space Studies, captured the state of the existing "planetary emergency," with respect to climate change:

> Our home planet is dangerously near a tipping point at which human-made greenhouse gases reach a level where major climate changes can proceed mostly under their own momentum. Warming will shift climatic zones by intensifying the hydrologic cycle, affecting freshwater availability and human health. We will see repeated coastal tragedies associated with storms and continuously rising sea levels. The implications are profound, and the only resolution is for humans to move to a fundamentally different energy pathway within a decade. Otherwise, it will be too late for one-third of the world's animal and plant species and millions of the most vulnerable members of our own species.[4]

According to environmentalist Lester Brown in his *Plan B 3.0*: "We are crossing natural thresholds that we cannot see and violating deadlines that we do not recognize. Nature is the time keeper, but we cannot see the clock. . . . We are in a race between tipping points in the earth's natural systems and those in the world's political systems. Which will tip first?"[5] As the clock continues to tick and little is accomplished, it is obvious that decisive and far-reaching changes are required to stave off ultimate disaster. This raises the question of more revolutionary social change as an ecological as well as a social necessity.

Yet if revolutionary solutions are increasingly required to address the ecological problem, this is precisely what the existing social system is guaranteed *not* to deliver. Today's environmentalism is aimed principally at those measures necessary to lessen the impact of the economy on the planet's ecology *without* challenging the economic system that in its very workings produces the immense environmental problems we now face. What we call "*the* environmental problem" is in the end primarily a problem of political economy. Even the boldest establishment economic attempts to address climate change fall far short of what is required to protect the earth—since the "bottom line" that constrains all such plans under capitalism is the necessity of continued, rapid growth in production and profits.

The Dominant Economics of Climate Change

The economic constraint on environmental action can easily be seen by looking at what is widely regarded as the most far-reaching establishment attempt to date to deal with *The Economics of Climate Change* in the form of a massive study issued in 2007 under that title, commissioned by the UK Treasury Office.[6] Subtitled the *Stern Review*, after the report's principal author Nicholas Stern, former chief economist of the World Bank, it is widely viewed as the most important and most progressive mainstream treatment of the economics of global warming.[7] The *Stern Review* focuses on the target level of carbon dioxide equivalent (CO_2e) concentration in the atmosphere necessary to stabilize global average temperature at no more than 3°C (5.4°F) over preindustrial levels. CO_2e refers to the six Kyoto greenhouse gases—carbon dioxide, methane, nitrous oxide, hydrofluorocarbons, perfluorocarbons, and sulfur hexafluoride—all expressed in terms of the equivalent amount of carbon dioxide. Whereas carbon dioxide concentration in the atmosphere is 390 parts per million (ppm), CO_2e is around 450 ppm.

James Hansen and other climatologists at NASA's Goddard Institute for Space Studies have recently argued: "If humanity wishes to preserve a planet similar to that on which civilization developed and to which life on Earth is adapted, paleoclimate evidence and ongoing climate change suggest that CO_2 will need to be reduced . . . to at most 350 ppm"—around 400 ppm CO_2e.[8]

In contrast, the *Stern Review* proposes that the target for CO_2e in the atmosphere should be 550 ppm (around 450 ppm carbon dioxide). In doing so, it targets an average global temperature increase of least a 3°C—well beyond what climate science consider dangerous. A 3°C increase would bring the earth's average global temperature to a height last seen in the "middle Pliocene around 3 million years ago." Furthermore, such an increase might be enough, the *Stern Review* explains, to trigger a shutdown of the ocean's thermohaline circulation that warms Western Europe, creating abrupt climate change, and thereby plunging Western Europe into Siberian-like conditions. Other research suggests that water flow in the Indus may drop by 90 percent by 2100 if global average temperatures rise by 3°C, potentially affecting hundreds of millions of people.

To make matters worse, the *Stern Review* admits that "for stabilisation at 550 ppm CO_2e, the chance of exceeding 3°C" is "30–70%." Or, as stated further on, a 550 ppm CO_2e stabilization level suggests "a 50:50 chance of a temperature increase above or below 3°C, and the Hadley Centre model predicts a 10% chance of exceeding 5°C [9°F] even at this level." Indeed, studies by climatologists indicate that at 550 ppm CO_2e there is more than a 5 percent chance that global average temperature could rise in excess of 8°C (14.4°F). All of this suggests that a stabilization target of 550 ppm CO_2e could be catastrophic for the earth, as we know it, as well as for its people.

Why then, if the risks to the planet and civilization are so enormous, does the *Stern Review* focus on a CO_2e atmospheric concentration target of 550 ppm (what it describes at one point as "the upper limit to the stabilisation range")? Given the enormous dangers, why not aim at deeper cuts in greenhouse gas emissions, a lower level of atmospheric CO_2e, and a smaller increase in global average temperature?

The reason is economics, pure and simple. The *Stern Review* is explicit that a radical mitigation of the problem *should not be attempted.* The costs to the world economy of ensuring that atmospheric CO_2e stabilized at present levels or below would be prohibitive, destabilizing capitalism itself. "Paths requiring very rapid emissions cuts," we are told, "are unlikely to be economically viable." If global greenhouse gas emissions peaked in 2010, the annual emissions reduction rate necessary to stabilize CO_2e at 450 ppm, the *Stern Review* suggests, would be 7 percent, with emissions dropping by about 70 percent below 2005 levels by 2050. This scenario is viewed as economically insupportable.

The *Stern Review*'s own preferred scenario, as indicated above, is a 550 ppm (CO_2e) target that would see global greenhouse gas emissions peak in 2015, with the emission cuts that followed at a rate of 1 percent per year. By 2050 the reduction in the overall level of emissions (from 2005 levels) in this scenario would only be 25 percent. (The report also considers, with less enthusiasm, an in-between 500 ppm CO_2e target, peaking in 2010 and requiring a 3 percent annual drop in global emissions.) Only the 550 ppm target, the *Stern Review* suggests, is truly economically viable because it is practically impossible to accomplish emission reductions of as much as 1 percent per year except in the case of a major economic downturn or a social collapse (such as the fall of the Soviet Union).

The only example the *Stern Review* can find of a sustained annual cut in greenhouse gas emissions of 1 percent or more, coupled with economic growth, among leading capitalist states is the United Kingdom in 1990–2000. Due to the discovery of North Sea oil and natural gas, the United Kingdom was able to switch massively from coal to gas in power generation, resulting in a 1 percent average annual drop in its greenhouse gas emissions during that decade. France came close to such a 1 percent annual drop in 1977–2003, reducing its greenhouse gas emissions by 0.6 percent per year due to a massive switch to nuclear power. By far the biggest drop for a major state was the 5.2 percent per year reduction in greenhouse gas emissions in the former Soviet Union in 1989–98. This, however, went hand in hand with a social-system breakdown and a drastic shrinking of the economy. All of this signals that any reduction in CO_2e emissions beyond around 1 percent per year would make it virtually impossible to maintain strong economic growth—the bottom line of the capitalist economy. Consequently, in order to keep the treadmill of accumulation going the world needs to risk environmental Armageddon.[9]

Although the foregoing was the position of the 2006 *Stern Review* commissioned by the British government, Stern himself has, it should be noted, subsequently adopted a slightly more aggressive position. In his 2009 book *The Global Deal* he argued that the goal should be to reduce CO_2e to less than 500 ppm (rather than 550 ppm as emphasized in the original report) by 2050. Stern acknowledges that Hansen has "raised strong and serious arguments" for targeting 350 ppm carbon dioxide (or 400 CO_2e) if we are to avoid climate catastrophe. But he argues that the "magnitude of the risks" is outweighed by economic "costs and opportunities"—so that the higher target should be adopted. The only thing that can be said "in favor" of Stern's position is that most climate economists—William Nordhaus, for example—are even worse.[10]

Accumulation and the Planet

None of this should surprise us. Capitalism since its birth, as Paul Sweezy wrote in 1989 in his article "Capitalism and the Environment," has been "a juggernaut driven by the concentrated energy of individuals and small groups single-mindedly pursuing their own interests, checked only by their mutual competition, and controlled in the short run by the impersonal forces of the market and in the longer run, when the market fails, by devastating crises." The inner logic of such a system manifests itself in the form of an incessant drive for economic expansion for the sake of class-based profits and accumulation. Nature and human labor are exploited to the fullest to fuel this juggernaut, while the destruction wrought on each is externalized so as to not fall on the system's own accounts.

"Implicit in the very concept of this system," Sweezy continued, "are interlocked and enormously powerful drives to both creation and destruction. On the plus side, the creative drive relates to what humankind can get out of nature for its own uses; on the negative side, the destructive drive bears most heavily on nature's capacity to respond to the demands placed on it. Sooner or later, of course, these two drives are contradictory and incompatible." Capitalism's overexploitation of nature's resource taps and waste sinks eventually produces the negative result of undermining both, first on a merely regional, but later on a world and even planetary basis (affecting the climate itself). Seriously addressing environmental crises requires "a reversal, not merely a slowing down, of the underlying trends of the last few centuries." This, however, cannot be accomplished without economic regime change.[11]

Naturally, most mainstream environmentalists have been averse to such conclusions. The more accumulation or economic growth seems to be the source of the problem, the more they declare that technology by itself can work magic, providing a perpetual free lunch in which economic expansion occurs without cost to the environment—allowing for open-ended economic growth together with environmental sustainability. Nature thus becomes the proverbial "free gift" of classical and neoclassical economics. Yet such is the depth of the environmental predicament today that some representatives of mainstream environmentalism seem to be breaking away—arriving at the more realistic and radical conclusion that the root problem is capitalism itself.[12] A case in point is James Gustave Speth, who has been called the "ultimate insider" within the environmental movement. Speth served as chairman of the Council on Environmental Quality under President Jimmy Carter, founded the World Resources Institute, co-founded the Natural Resources Defense Council, was a senior adviser in Bill Clinton's transition team, and administered the United Nations Development Programme from 1993 to 1999. At present, he is dean of the prestigious Yale School of Forestry and Environmental Studies. He is a winner of Japan's Blue Planet Prize.

However, with the publication in 2008 of his book *Bridge at the Edge of the World*, Speth has emerged as a trenchant critic of modern capitalism's destruction of the environment.[13] In this radical rethinking, he has sought to confront the perils brought on by the present economic regime, with its pursuit of growth and accumulation at any cost.

Speth has long emphasized that exponential economic growth rather than population growth represents the greatest threat to the planet. As early as 1991 he stressed that while population had increased three-fold in the twentieth century, economic output had increased twenty-fold.[14] Yet anyone reading Speth's earlier books would have had little reason to suspect that he would eventually turn into a strong environmental critic of capitalism. His 2004 book *Red Sky at Morning: America and the Crisis of Global Environmentalism* was noteworthy for its ecological-modernizing notion that technology and markets would save the day, and ensure that the business-as-usual capitalist economy could continue virtually unchanged. As he wrote with respect to the technological factor:

> We urgently need a worldwide environmental revolution in technology—a rapid ecological modernization of industry and agriculture. The prescription is straightforward but challenging: the principal way to reduce pollution and resource consumption while achieving expected economic growth is to bring about a wholesale transformation in the technologies that today dominate manufacturing, energy, transportation, and agriculture.... The focus should be on "dematerializing" the economy through a new generation of environmentally benign technologies that sharply reduce the consumption of natural resources and the generation of residual products per unit of economic output. Capital investment will shape the future, and investment is all about technology choice.[15]

With regard to markets, *Red Sky at Morning* argued: "We seek a market transition to a world in which market forces are harnessed to environmental ends, particularly by making prices reflect the full environmental costs. . . . Full-cost pricing is everywhere thwarted today by the failure of governments to eliminate environmentally perverse subsidies and to ensure that external environmental costs—including damages to public health, natural resources, and ecosystem services—are captured in market prices. The corrective most needed now is environmentally honest prices."[16]

Speth felt good enough about these arguments two years later in his book *Global Environmental Governance*, written together with Peter Haas, to suggest that they constituted the principal contributions of *Red Sky at Morning*. This position placed his analysis squarely in the ruling environmental camp. Indeed, capitalism is conspicuous in its absence in both books.[17]

In contrast, Speth's *The Bridge at the Edge of the World* represents a radical departure, reflecting a growing conviction on his part—fed it seems by the failure of the World Summit on Sustainable Development in Johannesburg and by the acceleration of climate change—that current environmental policies have failed to address a system that is heading off the cliff.[18] "Most of us with environmental concerns," he writes, "have worked within the system, but the system has not delivered. The mainstream environmental community as a whole has been the 'ultimate insider.' But it is time for the environmental community—indeed everyone—to step outside the system and develop a deeper critique of what is going on."

"Capitalism as we know it today," Speth declares in the same book, "is incapable of sustaining the environment." The crucial problem from an environmental perspective is exponential economic growth, which is the driving element of capitalism. Little hope can be provided in this respect by so-called dematerialization since it can be shown that the expansion of output overwhelms all increases in efficiency in throughput of materials and energy. Hence, one can only conclude that "right now . . . growth is the enemy of [the] environment. Economy and environment remain in collision." Here, the issue of capitalism becomes unavoidable. "Economic growth is modern capitalism's principal and most prized product." Speth favorably quotes Samuel Bowles and Richard Edwards's *Understanding Capitalism*, which bluntly stated: "Capitalism is differentiated from other economic systems by its drive to accumulate, its predisposition toward change, and its built-in tendency to expand."

The principal environmental problem for Speth, then, is capitalism as the "operating system" of the modern economy. "Today's corporations have been called 'externalizing machines.'" Indeed, "there are fundamental biases in capitalism that favor the present over the future and the private over the public." Quoting system defenders Paul Samuelson and William Nordhaus in the seventeenth edition of their textbook *Macroeconomics*, Speth points out that capitalism is the quintessential "Ruthless Economy," engaged "in the relentless pursuit of profits."

Building on this critique, Speth goes on to conclude in his book that: (1) "today's system of political economy, referred to here as modern capitalism, is destructive of the environment, and not in a minor way but in a way that profoundly threatens the planet"; (2) "the affluent societies have reached or soon will reach the point where, as Keynes put it, the economic problem has been solved . . . there is enough to go around"; (3) "in the more affluent societies, modern capitalism is no longer enhancing human well-being"; (4) "the international social movement for change—which refers to itself as 'the irresistible rise of global anti-capitalism'—is stronger than many imagine and will grow stronger; there is a coalescing of forces: peace, social justice, community, ecology, feminism—a movement of movements"; (5) "people and groups are busily planting the seeds of change through a host of alternative arrangements, and still other attractive directions for upgrading to a new operating system have been identified"; and (6) "the end of the Cold War . . . opens the door . . . for the questioning of today's capitalism."

Speth does not embrace socialism as part of the solution, which he associates, in the Cold War manner, primarily with Soviet-type societies in their most regressive form. Thus he argues explicitly for a "nonsocialist" alternative to capitalism. Such a system would make use of markets (but not the self-regulating market society of traditional capitalism) and would promote a "New Sustainability World" or a "Social Greens World" (also called "Eco-Communalism") as depicted by the Global Scenario Group. The latter scenario has been identified with socialist thinkers like William Morris (who was inspired by both Karl Marx and John Ruskin).[19]

A Bridge Too Far?

Speth is an important figure because he is the "ultimate insider" who has been compelled by the force of events to come face to face with the "moment of truth" and recognize that the primary ecological problem is derived from modern capitalism. He can also be seen as illustrating just how far it is possible to go—and no further—in the environmental critique of capitalism while not being dismissed as a dangerous radical. But despite his clear-sighted and often courageous vision, he has a tendency to pull back at the brink when faced with the more revolutionary implications of his own argument. If the *bridge at the edge of the world*, as he sometimes seems to suggest, requires a complete departure from capitalism and a shift to a socialist ecology, then Speth makes it abundantly clear that for him it is a *bridge too far*.

Speth thus appears to avoid a complete break with the system, despite his often trenchant criticisms of capitalism, leaving himself an escape hatch whereby an ecological reconciliation with the system can be envisioned. The problem of the rapid destruction of the environment is rationalized, in the end, as residing in "the operating system" of "modern capitalism," rather than in capital's inherent laws of motion, that is, its drive to accumulation on an ever-expanding scale. Despite the fact that his book is from beginning to end an environmental critique of modern capitalism, Speth chooses to reject not capitalism categorically, but rather socialism. "The important question," he writes is "no longer the future of socialism, rather it is to identify the contours of a new *nonsocialist* operating system that can transform capitalism as we know it." What we should fight for, he tells us, is "a nonsocialist alternative to today's capitalism," which defends private property while making room for additional public property. Important initiatives have been proposed "to transform the market and consumerism, redesign corporations, and focus growth on high-priority human and environmental needs. If pursued, they would change modern-day capitalism in fundamental ways. We would no longer have capitalism as we know it. The question whether this something new is beyond capitalism or is a reinvented capitalism is largely definitional."[20]

We are informed that "the growth fetish" characteristic of modern capitalism is not at all intrinsic to capitalism, but can be dispensed with in Speth's proposed non-socialist but not necessarily non-capitalist alternative. A kind of capitalism identified simply in terms of private property and competitive markets is seen as still feasible—strangely divorced from profits and accumulation (indeed from its own expanded reproduction). Here Speth praises Peter Barnes's notion of a "capitalism 3.0" that redirects investment in the public interest through a variety of essentially philanthropic schemes. The basic class dynamic of the capitalist system and the entrenched vested interests that this creates are largely ignored in this analysis, and seen as nonessential to the system. Corporations and markets can simply be redesigned in the new operating system to fit nonprofit objectives without affecting the underlying hard drive.

It is perhaps not surprising, given this slippery logic, that Speth has in the space of two years fallen back to an analysis that refrains altogether from naming—or even speciously alluding to—the capitalist economy. In a widely circulated manifesto, titled "Towards a New Economy and a New Politics," in the journal *Solutions*, he addresses the combined economic and environmental crisis by questioning what he vaguely calls "today's system of political economy"—a system that remains unnamed and undefined. Speth's intention in this article is to counter what he calls no less vaguely the present "growth economy" with an equally abstract new "post-growth society."

"Challenging the current order" in this context means: (1) questioning the notion that "GDP growth is an unalloyed good"; (2) insisting that markets are to be increasingly regulated in the public interest; (3) making sure that the corporate model is shifted from "one ownership and motivation model to many"; and (4) ensuring that consumption is transformed from consumerism to "mindful consumption." In none of this, however, does Speth address accumulation as the driving force of the economy—or class, profit, or property relations. If the object is to move the wealthy societies away from the "growth fetish," all the genuine historical issues associated with actually accomplishing this have nonetheless been set aside.[21]

In a capitalist society any critique of capital accumulation is difficult to maintain within a reformist framework, since this constitutes the main condition of existence of the system. In 2010, in the midst of the greatest economic crisis since the Great Depression, it is clear to all that capitalism is a grow-or-die system.[22] To raise the question of a no-growth economy and capitalism in the same breath in this context would thus mean a clear rejection of the latter. The notion of a stationary-state capitalism as a sustainable society would at this time fool no one. Rather than promote such views and put himself perhaps in the untenable political position of an out-and-out rejection of the system, Speth has to resort to vague generalities. The critique of capitalism has been removed from his analysis precisely when the question becomes most serious, that is, during a major crisis of accumulation.

It is in its critique of accumulation and its emphasis on qualitative human development, rather than endless quantitative economic expansion, that socialism offers a powerful alternative to capitalism, and the possibility of a sustainable society. As early as the 1950s Paul Baran argued in *The Political Economy of Growth* for the institution in a socialist economy and society of a "planned surplus," one that would represent the "optimum" use of society's overall economic surplus generating potential. As Baran himself put it, such an optimum does not

> presuppose the maximization of output that might be attainable in a country at a given time. It may well be associated with a less than maximum output in view of a voluntarily shortened labor day, of an increase in the amount of time devoted to education, or of conscious discarding of

> certain noxious types of production (coal mining for example). What is crucial is that the volume of output would not be determined by the fortuitous outcome of a number of uncoordinated decisions on the part of individual businessmen and corporations, but by a rational plan expressing what society would wish to produce, to consume, to save, and to invest at any given time.[23]

Indeed, a sustainable or steady-state economy (in countries that have reached a sufficient level of development) is only possible, as Baran's argument suggests, under conditions of rational socialist planning.[24]

Overcoming the ecological rift (and the social rift that lies beneath it) thus demands the transcendence of capitalism and the development of a genuine socialist alternative associated with substantive equality and socioeconomic-ecological planning. Not surprisingly, some progressive establishment figures, like Speth and Stern, have been known to hesitate, once they realize that crossing the ecological *bridge at the edge of the world* means leaving capitalism behind and entering a socialist-oriented, because radically ecological, landscape on the other side. For Speth (at least at the present time) this is clearly *a bridge too far*. Hence, he has retreated and now offers only vague depictions of the path ahead.[25]

But if some are hesitating, afraid of what lies beyond, others can already be seen crossing the bridge that lies before us in search for a more egalitarian society and *a world that is our friend*.[26] Given the limitless ecological crisis emanating from today's business as usual, all hope for the future of humanity and the earth must lie in this direction.

PART TWO

Ecological Paradoxes

7. The Return of the Jevons Paradox

The nineteenth century was the century of coal. It was coal above all else that powered British industry, and thus the British Empire. But in 1863 the question was raised by industrialist Sir William George Armstrong, in his presidential address to the British Association for the Advancement of Science, as to whether Britain's world supremacy in industrial production could be threatened in the long run by the exhaustion of readily available coal reserves.[1] At that time, no extensive economic study had been conducted on coal consumption and its impact on industrial growth.

In response, William Stanley Jevons, who would become one of the founders of neoclassical economics, wrote, in only three months, a book entitled *The Coal Question: An Inquiry Concerning the Progress of the Nation, and the Probable Exhaustion of Our Coal-Mines* (1865). Jevons argued that British industrial growth relied on cheap coal, and that the increasing cost of coal, as deeper seams were mined, would lead to the loss of "commercial and manufacturing supremacy," possibly "within a lifetime," and a check to economic growth, generating a "stationary condition" of industry "within a century."[2] Neither technology nor substitution of other energy sources for coal, he argued, could alter this.

Jevons's book had an enormous impact. John Herschel, one of the great figures in British science, wrote in support of Jevons's thesis that "we are using up our resources and expending our national life at an enormous and increasing rate and thus a very ugly day of reckoning is impending sooner or later."[3] In April 1866, John Stuart Mill praised *The Coal Question* in the House of Commons, arguing in support of Jevons's proposal of compensating for the depletion of this critical natural resource by cutting the national debt. This cause was taken up by William Gladstone, Chancellor of the Exchequer, who urged Parliament to act on debt reduction, based on the uncertain prospects for national development in the future, due to the anticipated rapid exhaustion of coal reserves. As a result, Jevons's book quickly became a bestseller.[4]

Yet Jevons was stunningly wrong in his calculations. It is true that British coal production, in response to increasing demand, more than doubled in the thirty years following the publication of his book. During the same period in the United States, coal production, starting from a much lower level, increased ten times, though still remaining below the British level.[5] Yet no "coal panic," due to exhaustion of available coal supplies, ensued in the late nineteenth and early twentieth centuries. Jevons's chief mistake had been to equate the energy for industry with coal itself, failing to foresee the development of substitutes for coal, such as petroleum and hydroelectric power.[6] In 1936, seventy years after the parliamentary furor generated by Jevons's book, John Maynard Keynes commented on Jevons's projection of a decline in the availability of coal, observing that it was "overstrained and exaggerated." One might add that it was quite narrow in scope.[7]

The Jevons Paradox

But there is one aspect of Jevons's argument—associated with what is now known as the Jevons Paradox—that continues to be considered one of the pioneering insights into ecological economics.[8] In chapter 7 of *The Coal Question*, entitled "Of the Economy of Fuel," Jevons responded to the common notion that, since "the falling supply of coal will be met by new modes of using it efficiently and economically," there was no problem of supply, and that, indeed, "the amount of useful work got out of coal may be made to increase manifold, while the amount of coal consumed is stationary or diminishing." In sharp opposition to this, Jevons contended that increased efficiency in the use of coal as an energy source only generated increased demand for that resource, not decreased demand, as one might expect. This was because improvement in efficiency led to further economic expansion. "*It is wholly a confusion of ideas*," he wrote, "*to suppose that the economical use of fuel is equivalent to a diminished consumption. The very contrary is the truth.* As a rule, new modes of economy will lead to an increase of consumption according to a principle recognised in many parallel instances. . . . The same principles apply, with even greater force and distinctness, to the use of such a general agent as coal. It is the very economy of its use which leads to its extensive consumption."[9]

"Nor is it difficult," Jevons wrote, "to see how this paradox [of increased efficiency leading to increased consumption] arises." Every new technological innovation in the production of steam engines, he pointed out in a detailed description of the steam engine's evolution, had resulted in a more thermodynamically efficient engine. And each new, improved engine had resulted in an increased use of coal. The Savery engine, one of the earlier steam engines, he pointed out, was so inefficient that "practically, the cost of working kept it from coming into use; *it consumed no coal, because its rate of consumption was too high*."[10] Succeeding models that were more efficient, such as Watt's famous engine, led to higher and higher demand for coal with each successive improvement. "Every such improvement of the engine, when effected, does but accelerate anew the consumption of coal. Every branch of manufacture receives a fresh impulse—hand labour is still further replaced by mechanical labour, and greatly extended works can be undertaken which were not commercially possible by the use of the more costly steam-power."[11]

Although Jevons thought that this paradox was one that applied to numerous cases, his focus in *The Coal Question* was entirely on coal as a "general agent" of industrialization and a spur to investment goods industries. The power of coal to stimulate economic advance, its accelerated use, despite advances in efficiency, and the severity of the effects to be expected from the decline in its availability, were all due to its dual role as the necessary fuel for the modern steam engine and as the basis for blast furnace technology.

In the mid-nineteenth century, coal was seen not simply as the fuel for the steam engine, but also as the key material input for blast furnaces in the production of iron, the crucial industrial product and the foundation of industrial dominance.[12] It was by virtue of its greater development in this area, as "the workshop of the world," that Britain accounted for about half of world output of iron in 1870.[13] Greater efficiency in the use of coal thus translated into a greater capacity to produce iron and an expansion of industry in general, leading to spiraling demand for coal. As Jevons put it:

> If the quantity of coal used in a blast-furnace, for instance, be diminished in comparison with the yield, the profits of the trade will increase, new capital will be attracted, the price of pig-iron will fall, but the demand for it [will] increase; and eventually the greater number of furnaces will more than make up for the diminished consumption of each. And if such is not always the result within a single branch, it must be remembered that the progress of any branch of manufacture excites a new activity in most other branches, and leads indirectly, if not directly, to increased inroads upon our seams of coal.[14]

What made this argument so powerful at the time was that it seemed immediately obvious to everyone in Jevons's day that industrial development depended on the capacity to expand iron production cheaply. This meant that a reduction in the quantity of coal needed in a blast furnace would immediately translate into an expansion of industrial production, industrial capacity, and the ability to capture more of the world market—hence more demand for coal. The tonnage of coal consumption by the iron and steel industries of Britain in 1869, 32 million tons, exceeded the combined amount used in both general manufactures, 28 million tons, and railroads, 2 million tons.[15]

This was the age of capital and the age of industry, in which industrial power was measured in terms of coal and pig iron production. Output of coal and iron increased basically in tandem in this period, both tripling between 1830 and 1860.[16] As Jevons himself put it: "Next after coal . . . iron is the material basis of our power. It is the bone and sinews of our laboring system. Political writers have correctly treated the invention of the coal-blast furnace as that which has most contributed to our material wealth. . . . The production of iron, the material of all our machinery, is the best measure of our wealth and power."[17]

Hence none of Jevons's readers could fail to perceive the multiplier effects on industry of an improvement in efficiency in the use of coal, or the "increased inroads" upon "seams of coal" that this would tend to generate. "Economy," he concluded, "multiplies the value and efficiency of our chief material; it indefinitely increases our wealth and means of subsistence, and leads to an extension of our population, works, and commerce, which is gratifying to the present, but must lead to an earlier end."[18]

A Natural Law

In treating coal as the "chief material" of British industry, Jevons emphasized what he saw as a shift in industrial development over time from what he referred to as one "staple produce of the country" to another. The great battle over the Corn Laws had already pointed to the fact—noted by his father, Thomas Jevons, among others—that a lower price for a staple product would greatly expand demand and ultimately scarcity (which, in the case of corn, was to be satisfied by imports).[19] But by the late nineteenth century, it was coal, not corn, that was the focus of a kind of Malthusian scarcity.[20]

"It was Jevons's thesis in this book," Keynes noted, "that the maintenance of Great Britain's prosperity and industrial leadership required a continuous growth of her heavy industries on a scale which would mean a demand for coal increasing in a geometrical progression. Jevons advanced this principle as an extension of Malthus's law of population, and he designated it the *Natural Law of Social Growth*. . . . From this it is a short step to put *coal* into the position occupied in Malthus's theory by *corn*."[21]

Extending Malthus's theory to coal, Jevons wrote: "Our subsistence no longer depends upon our produce of corn. The momentous repeal of the Corn Laws throws us from corn upon coal. It marks, at any rate, the epoch when coal was finally recognised as the staple produce of the country;—it marks the ascendancy of the manufacturing interest, which is only another name for the development of the use of coal." Jevons contended that although population had "quadrupled since the beginning of the nineteenth century," the consumption of coal had increased by "sixteenfold," and that this growth of coal production "per head" was a necessity of rapid industrial development, which must come to an end.[22]

Yet the chief contradiction behind the paradox that Jevons raised—the whole dynamic of accumulation or expanded reproduction intrinsic to capitalism—was not analyzed in *The Coal Question*. As one of the early neoclassical economists, Jevons had abandoned the central emphasis on class and accumulation that distinguished the work of the classical economists. His economic analysis took the form of static equilibrium theory. There is nothing in his argument resembling Karl Marx's notion of capital as self-expanding value, and the consequent need for continual expansion.

Jevons's economic framework was thus ill equipped to deal concretely with issues of accumulation and economic growth. The expansion of population, industry, and the demand for coal (as the "central material" of industrial life) was, in his view, simply the product of an abstract Natural Law of Social Growth, building on Malthus. Viewing capitalism more as a natural phenomenon than a socially constructed reality, he could find no explanation for continuously increasing economic demand, other than to point to individual behavior, Malthusian demographics, and the price mechanism. Rather than emphasizing the profit motive itself, he drew on Justus von Liebig's abstract law of power: "Civilisation, says Baron Liebig, is *the economy of power*, and our power is coal."[23] The forces driving economic expansion, feeding industrialization, and resulting in the growing demand for coal, were thus strangely weak and undeveloped in *The Coal Question*, reflecting the fact that Jevons lacked a realistic conception of a capitalist economy and society.

Industrial Hegemony, Not Ecological Sustainability

British hegemony, rather than ecology, lay at the bottom of Jevons's concerns. Despite the emphasis he placed on resource scarcity and its importance for ecological economics, it would be a mistake to see *The Coal Question* as predominantly ecological in character. Jevons was unconcerned with the environmental problems associated with the exhaustion of energy reserves in Great Britain or the rest of the world. He even failed to address the air, land, and water pollution that accompanied coal production. Charles Dickens, decades before, had described the industrial towns, with their concentrated coal burning, as characterized by a "plague of smoke, [which] obscured the light, and made foul the melancholy air" in a ceaseless progression of "black vomit, blasting all things living or inanimate, shutting out the face of day, and closing in on all these horrors with a dense dark cloud."[24] Of this, there is not a trace in Jevons. Similarly, the occupational illnesses and hazards confronting workers in the coal mines and coal-fed factories did not enter his analysis, though such concerns were evident in the work of other nineteenth-century analysts, as witnessed by Frederick Engels's *The Condition of the Working Class in England.*[25]

Indeed, there was in Jevons no concern for nature as such. He simply assumed that the mass disruption and degradation of the earth was a natural process. Although the shortage of coal, as an energy source, generated questions in his analysis about whether growth could be sustained, the issue of ecological sustainability itself was never raised. Because the economy must remain in continual motion, Jevons disregarded sustainable sources of energy, such as water and wind, as unreliable, limited to a particular time and location.[26] Coal offered capital a universal energy source to operate production, without disruption of business patterns.

Jevons therefore had no real answer to the paradox he raised. Britain could either rapidly use up its cheap source of fuel—the coal on which its industrialization rested—or it could use it up more slowly. In the end, he chose to use it up rapidly: "If we lavishly and boldly push forward in the creation of our riches, both material and intellectual, it is hard to over-estimate the pitch of beneficial influence to which we may attain in the present. *But the maintenance of such a position is physically impossible. We have to make the momentous choice between brief but true greatness and longer continued mediocrity*."[27]

Expressed in these terms, the path to be taken was clear: to pursue glory in the present and accept the prospect of a drastically degraded position for future generations. Since Jevons had no answer to what he saw as the inevitable and rapid depletion of Britain's coal stocks—and British capital and the British government saw no other conceivable course than "business as usual"—the response to Jevons's book largely took the form, oddly enough, of an added justification for reduction of the national debt. This was presented as a precautionary measure in the face of the eventual slowdown of industry. As Keynes wrote, "The proposition that we were living on our natural capital" gave rise to the irrational response that it was necessary to effect "a rapid reduction of the dead-weight debt."[28]

Indeed, nearly the entire political impact of Jevons's book was confined, ironically, to its penultimate chapter, "Taxes and the National Debt." Jevons and other figures, such as Mill and Gladstone, who took up his argument, never seriously raised the idea of the conservation of coal. There was no mention anywhere in Jevons's analysis of the point raised by Engels in a letter to Marx, in which industrial capitalism was characterized as a "squanderer of *past* solar heat" as evidenced by its "squandering [of] our reserves of energy, our coal, ore, forests, etc."[29] For Jevons, the idea of an alternative to business as usual was never discussed, and doubtless never entered his mind. Nothing was further from his general economic outlook than the transformation of the social relations of production in the direction of a society governed, not by the search for profit, but by people's genuine needs and the requirements of socio-ecological sustainability. In the end, the problems he foresaw were delayed in the actual historical course of events by the expansion in the use of other fossil fuels—oil and natural gas—as well as hydroelectric power, and by the ongoing exploitation of the resources of the entire globe. All of this, however, has prepared the way for our current planetary dilemma and the return of the Jevons Paradox.

The Rediscovery of the Jevons Paradox

The Jevons Paradox was forgotten in the heyday of the age of petroleum during the first three-quarters of the twentieth century, but reappeared in the 1970s due to increasing concerns over resource scarcity associated with the Club of Rome's *Limits to Growth* analysis, heightened by the oil-energy crisis of 1973–74. As energy efficiency measures were introduced, economists became concerned with their effectiveness. This led to the resurrection, at the end of the 1970s and the beginning of the 1980s, of the general question posed by the Jevons Paradox, in the form of what was called the "rebound effect." This was the fairly straightforward notion that engineering efficiency gains normally led to a decrease in the effective price of a commodity, thereby generating increased demand, so that the gains in efficiency did not produce a decrease in consumption to an equal extent. The Jevons Paradox has often been relegated to a more extreme version of the rebound effect, in which there is a *backfire*, or a rebound of more than 100 percent of "engineering savings," resulting in an *increase* rather than decrease in the consumption of a given resource.[30]

Technological optimists have tried to argue that the rebound effect is small, and therefore environmental problems can be solved largely by technological innovation alone, with the efficiency gains translating into lower throughput of energy and materials (dematerialization). Empirical evidence of a substantial rebound effect is, however, strong.

For example, technological advancements in motor vehicles, which have increased the average miles per gallon of vehicles by 30 percent in the United States since 1980, have not reduced the overall energy used by motor vehicles. Fuel consumption per vehicle stayed constant while the efficiency gains led to the augmentation, not only of the numbers of cars and trucks on the roads (and the miles driven), but also their size and "performance" (acceleration rate, cruising speed, etc.)—so that SUVs and minivans now dot U.S. highways. At the macro level, the Jevons Paradox can be seen in the fact that, even though the United States has managed to double its energy efficiency since 1975, its energy consumption has risen dramatically. Over the last thirty-five years, Juliet Schor notes

> energy expended per dollar of GDP has been cut in half. But rather than falling, energy demand has increased, by roughly 40 percent. Moreover, demand is rising fastest in those sectors that have had the biggest efficiency gains—transport and residential energy use. Refrigerator efficiency improved by 10 percent, but the number of refrigerators in use rose by 20 percent. In aviation, fuel consumption per mile fell by more than 40 percent, but total fuel use grew by 150 percent because passenger miles rose. Vehicles are a similar story. And with soaring demand, we've had soaring emissions. Carbon dioxide from these two sectors has risen 40 percent, twice the rate of the larger economy.

Economists and environmentalists who try to measure the direct effects of efficiency on the lowering of price and the immediate rebound effect generally tend to see the rebound effect as relatively small, in the range of 10 to 30 percent in high-energy consumption areas such as home heating and cooling and cars. But once the indirect effects, apparent at the macro level, are incorporated, the Jevons Paradox remains extremely significant. It is here at the macro level that scale effects come to bear: improvements in energy efficiency can lower the effective cost of various products, propelling the overall economy and expanding overall energy use.[31] Ecological economists Mario Giampietro and Kozo Uno argue that the Jevons Paradox can only be understood in a macro-evolutionary model, where improvements in efficiency result in changes in the matrices of the economy, such that the overall effect is to increase scale and tempo of the system as a whole.[32]

Most analyses of the Jevons Paradox remain abstract, based on isolated technological effects, and removed from the historical process. They fail to examine, as Jevons himself did, the character of industrialization. Moreover, they are still further removed from a realistic understanding of the accumulation-driven character of capitalist development. An economic system devoted to profits, accumulation, and economic expansion without end will tend to use any efficiency gains or cost reductions to expand the overall scale of production. Technological innovation will therefore be heavily geared to these same expansive ends. It is no mere coincidence that each of the epoch-making innovations (namely, the steam engine, the railroad, and the automobile) that dominated the eighteenth, nineteenth, and twentieth centuries were characterized by their importance in driving capital accumulation and the positive feedback they generated with respect to economic growth as a whole—so that the scale effects on the economy arising from their development necessarily overshot improvements in technological efficiency.[33] Conservation in the aggregate is impossible for capitalism, however much the output/input ratio may be increased in the engineering of a given product. This is because all savings tend to spur further capital formation (provided that investment outlets are available). This is especially the case where core industrial resources—what Jevons called "central materials" or "staple products"—are concerned.

The Fallacy of Dematerialization

The Jevons Paradox is the product of a capitalist economic system that is unable to conserve on a macro scale, geared, as it is, to maximizing the throughput of energy and materials from resource tap to final waste sink. Energy savings in such a system tend to be used as a means for further development of the economic order, generating what Alfred Lotka called the "maximum energy flux," rather than minimum energy production.[34] The deemphasis on absolute (as opposed to relative) energy conservation is built into the nature and logic of capitalism as a system unreservedly devoted to the gods of production and profit. As Marx put it: "Accumulate, accumulate! That is Moses and the prophets!"[35]

Seen in the context of a capitalist society, the Jevons Paradox therefore demonstrates the fallacy of current notions that the environmental problems facing society can be solved by purely technological means. Mainstream environmental economists often refer to "dematerialization," or the "decoupling" of economic growth, from consumption of greater energy and resources. Growth in energy efficiency is often taken as a concrete indication that the environmental problem is being solved. Yet savings in materials and energy, in the context of a given process of production, as we have seen, are nothing new; they are part of the everyday history of capitalist development.[36] Each new steam engine, as Jevons emphasized, was more efficient than the one before. "Raw materials-savings processes," environmental sociologist Stephen Bunker noted, "are older than the Industrial Revolution, and they have been dynamic throughout the history of capitalism." Any notion that reduction in material throughput, per unit of national income, is a new phenomenon is therefore "profoundly ahistorical."[37]

What is neglected, then, in simplistic notions that increased energy efficiency normally lead to increased energy savings overall, is the reality of the Jevons Paradox relationship—through which energy savings are used to promote new capital formation and the proliferation of commodities, demanding ever greater resources. Rather than an anomaly, the rule that efficiency increases energy and material use is integral to the "regime of capital" itself.[38] As stated in *The Weight of Nations*, an important empirical study of material outflows in recent decades in five industrial nations (Austria, Germany, the Netherlands, the United States, and Japan): "Efficiency gains brought by technology and new management practices have been offset by [increases in] the scale of economic growth."[39]

The result is the production of mountains upon mountains of commodities, cheapening unit costs and leading to *greater squandering of material resources*. Under monopoly capitalism, moreover, such commodities increasingly take the form of artificial use values, promoted by a vast marketing system and designed to instill ever more demand for commodities and the exchange values they represent—as a substitute for the fulfillment of genuine human needs. Unnecessary, wasteful goods are produced by useless toil to enhance purely economic values at the expense of the environment. Any slowdown in this process of ecological destruction, under the present system, spells economic disaster.

In Jevons's eyes, the "momentous choice" raised by a continuation of business as usual was simply "*between brief but true* [national] *greatness and longer continued mediocrity*." He opted for the former—the maximum energy flux. A century and a half later, in our much bigger, more global—but no less expansive—economy, it is no longer simply national supremacy that is at stake, but the fate of the planet itself. To be sure, there are those who maintain that we should "live high now and let the future take care of itself." To choose this course, though, is to court planetary disaster. The only real answer for humanity (including future generations) and the earth as a whole is to alter the social relations of production, to create a system in which efficiency is no longer a curse—a higher system in which equality, human development, community, and sustainability are the explicit goals.

8. The Paperless Office and Other Ecological Paradoxes

At the core of the broad program aimed at achieving environmental sustainability is a concern with how the dynamics of economic systems can be brought into harmony with ecosystems. One major challenge for this program is to understand the dynamics of market economies with respect to natural resource consumption. In particular, it is important for environmental social scientists to assess whether some modern economies are dematerializing—reducing the absolute quantity of natural resources they consume—and, if so, why.[1] Here we discuss two ecological paradoxes in economics that call into question whether the dematerialization of economies can be achieved through either of two routes that are commonly suggested: (1) improvements in the efficiency of resource use in the production process; and (2) the development of substitutes for some types of natural resources.[2] The first paradox we discuss is a classical one, the Jevons Paradox, which suggests that improvements in efficiency do not necessarily lead to a reduction in resource consumption; in fact they may lead to an increase in resource consumption. The second paradox is one that has not previously been explicitly identified as a paradox, the Paperless Office Paradox, which suggests that the development of substitutes for some resources may not lead to a reduction in consumption of those resources and in some cases may actually lead to an increase in consumption.

The Jevons Paradox

William Stanley Jevons, one of the foundational writers in ecological economics, in his famous book *The Coal Question* (1865) identified what is perhaps the most widely known paradox in ecological economics, which has subsequently become known as the Jevons Paradox.[3] Jevons observed that as the efficiency of coal use by industry improved, thereby allowing for the production of more goods per unit of coal, total coal consumption *increased*. At least two potentially complementary explanations for this paradox stand out. First, following classical economic reasoning, as the efficiency of coal use increases, the cost of coal per unit of goods produced decreases. This reduction in cost makes coal more desirable to producers as an energy source, thus leading producers to invest in technologies that utilize coal. Second, following political-economic reasoning, the drive to increase profits inherent in capitalist modes of production leads producers to try to reduce costs by reducing resource inputs per unit of production (improving efficiency) as well as increase revenues by expanding the quantity of goods and services produced and sold, thus necessitating the expansion of resource consumption.[4] The political-economic explanation of the Jevons Paradox suggests that the association between efficiency and total consumption is primarily due to a third factor that drives both, that is, profit-seeking behavior by capitalists, although it recognizes a potentially direct link in that profits stemming from improvements in efficiency can be invested in expanding production. The classical economic explanation sees efficiency and total consumption as causally linked through the cost of coal per unit of production. Of course, both processes are potentially complementary and may operate together or alternately in different historical moments, and other processes may well be at work, too.

Determining the extent to which the Jevons Paradox does indeed exist and how generally applicable it is—how commonly is rising efficiency in the use of a resource associated with an escalation of consumption of that resource?—is an important task for environmental social scientists, since arguments that more efficient production technologies will help solve environmental crises are a staple in public policy discussions in most developed nations and are at least implicit, and frequently explicit, in various research programs, such as ecological modernization.[5] Environmental social scientists can provide a great service by assessing the contexts in which a paradoxical association exists between the efficiency of use and the total consumption of natural resources, and the reasons for this association.[6] Although Jevons focused on the association at a specific level (industry) between a specific type of efficiency (output per unit of resource use) and a specific natural resource (coal), it is important to establish how generalized the association between efficiency and total resource consumption is. After all, if rising efficiency is frequently associated with escalating resource consumption, then a focus on improving efficiency may be both misguided and misleading. We present two examples that suggest that the Jevons Paradox, as a factual proposition about the association between efficiency of production and the consumption of resources, may have broad applicability and characterize a variety of situations and types of efficiency.

Eco-Efficiency of National Economies

Stephen Bunker, an environmental sociologist, found that over a long stretch of recent history, the world economy as a whole showed substantial improvements in resource efficiency (economic output per unit of natural resource), but that the total resource consumption of the global economy continually escalated.[7] Similarly, recent research has shown that at the national level, high levels of affluence are, counter intuitively, associated with both greater eco-efficiency—GDP output per unit of ecological footprint—of the economy as a whole and with a higher per capita ecological footprint, suggesting that empirical conditions characteristic of the Jevons Paradox often may be applicable to the generalized aggregate level.[8] Indeed, this type of pattern appears to be quite common. Statistical analyses using elasticity models of the effect of economic development (GDP per capita) on environmental impacts, such as carbon dioxide emissions, have shed light on the relationship between efficiency and total environmental impact.[9] With such a model, an elasticity coefficient for GDP per capita (which indicates the percentage increase in the environmental impact of nations for a 1 percent increase in GDP per capita) of between 0 and 1 (indicating a positive inelastic relationship) implies a condition where the aggregate eco-efficiency of the economy improves with development but the expansion of the economy exceeds improvements in efficiency, leading to a net increase in environmental impact. This type of research does not establish a causal link between efficiency and total environmental impact or resource consumption, but it does empirically demonstrate that an association between rising efficiency and rising environmental impacts may be common, at least at the national level.[10] These findings also suggest that improving eco-efficiency in a nation is not necessarily, or even typically, indicative of a decline in resource consumption.[11]

Fuel Efficiency of Automobiles

The fuel efficiency of automobiles is obviously an issue of substantial importance, since motor vehicles consume a large share of the world's oil. It would seem reasonable to expect that improvements in the efficiency of engines and refinements in the aerodynamics of automobiles would help to curb motor fuel consumption. However, an examination of recent trends in the fuel consumption of motor vehicles suggests a paradoxical situation where improvements in efficiency are associated with increases in fuel consumption. For example, in the United States an examination of a reasonable indicator of fuel efficiency of automobiles stemming from overall engineering techniques, pound-miles per gallon (or kilogram-kilometers per liter) of fuel, supports the contention that the efficiency of the light-duty fleet (which includes passenger cars and light trucks) improved substantially between 1984 and 2001, whereas the total and average fuel consumption of the fleet *increased*.

For the purposes of calculating CAFE (corporate average fuel economy) performance of the nation's automobile fleet, the light-duty fleet is divided into two categories, passenger cars and light trucks (which includes sports utility vehicles), each of which has a different legally enforced CAFE standard.[12] In 1984 the total light-truck fleet CAFE miles per gallon (MPG) was 20.6 (~8.8 kilometers per liter; KPL) and the average equivalent test weight was 3,804 pounds (~1,725 kilograms), indicating that the average pound-miles per gallon was 78,362 (20.6 • 3,804) (~15,100 kilogram-KPL). By 2001, the total light truck fleet CAFE MPG had improved slightly to 21.0 (~8.9 KPL), while the average vehicle weight had increased substantially, to 4,501 pounds (~2,040 kilograms). Therefore the pound-miles per gallon had increased to 94,521 (21.0 • 4,501) (~18,200 kilogram-KPL), a 20.6 percent improvement in efficiency from 1984. A similar trend happened in passenger cars over this same period. In 1984 the total passenger car fleet CAFE was 26.9 MPG (~11.4 KPL) and the average equivalent test weight was 3,170 pounds (~1,440 kilograms), indicating that the pound-miles per gallon was 85,273 (26.9 • 3,170) (~16,400 kilogram-KPL). By 2001, the total passenger car fleet CAFE MPG had improved to 28.7 (~12.2 KPL) while the average vehicle weight had increased to 3,446 pounds (~1,560 kilograms), making the average fleet pound-miles per gallon 98,900 (28.7 • 3,446) (~19,070 kilogram-KPL)—a 16 percent improvement since 1984.

Clearly engineering advances had substantially improved the efficiency of both light trucks and passenger cars in terms of pound-MPG (or kilogram-KPL) between 1984 and 2001. The observation of this fact in isolation might lead one to expect that these improvements in efficiency were associated with a reduction in the fuel consumption of the total light-duty fleet. However, this is not what happened. Over this period, light trucks, which on average are heavier and consume more fuel than passenger cars, grew from 24.4 percent of the light duty fleet to 46.6 percent. Because of this shift in composition, the CAFE MPG for the combined light-duty fleet declined from 25.0 to 24.5 (from ~10.6 to ~10.4 KPL), a 2 percent decrease. Clearly, engineering advances had improved the efficiency of engines and other aspects of automobiles, but this did not lead to a less fuel-thirsty fleet since the size of vehicles increased substantially, particularly due to a shift from passenger cars to light trucks among a large segment of drivers.[13] It is worth noting that even if the total fleet MPG had improved, a reduction in fuel consumption would have been unlikely to follow, since over this period the distance traveled by drivers per year increased from little more than 15,000 km (~9,300 miles) per car, on average, to over 19,000 km (~11,800 miles).[14] And, finally, an increase in the number of drivers and cars on the road drove up fuel consumption even further. For example, between 1990 and 1999, the number of motor vehicles in the United States increased from 189 million to 217 million due to both population growth and a 2.8 percent increase in the number of motor vehicles per 1,000 people (from 758 to 779).[15]

It appears that technological advances that improved the engineering of cars were in large part implemented, at least in the United States, in expanding the size of vehicles, rather than reducing the fuel the average vehicle consumed. The causal explanations for this are likely complex, but the fact that, despite engineering improvements, the U.S. light-duty fleet increased its total and average fuel consumption over the past two decades does suggest that technological refinements are unlikely in and of themselves to lead to the conservation of natural resources. Furthermore, it is possible that improvements in efficiency may actually contribute to the expansion of resource consumption, since it is at least plausible that success at improving the MPG/KPL of a nation's automobile fleet may encourage drivers to travel more frequently by car, due to the reduction in fuel consumption per mile/kilometer—a situation directly analogous to the one Jevons observed regarding coal use by industry.

The Paperless Office Paradox

Paper is typically made from wood fiber, so paper consumption puts substantial pressure on the world's forest ecosystems. It would seem on the face of it that the rise of the computer and the capacity for the storage of documents in electronic form would lead to a decline in paper consumption and, eventually, the emergence of the "paperless office"—which would be decidedly good news for forests. This, however, has not been the case, as Abigail J. Sellen and Richard H. R. Harper clearly document in their aptly titled book *The Myth of the Paperless Office*.[16] Contrary to the expectations of some, computers, e-mail, and the World Wide Web are associated with an increase in paper consumption. For example, consumption of the most common type of office paper (uncoated free-sheet) increased by 14.7 percent in the United States between the years 1995 and 2000, embarrassing those who predicted the emergence of the paperless office.[17] Sellen and Harper also point to research indicating that "the introduction of e-mail into an organization caused, on average, a 40% increase in paper consumption."[18] This observation suggests that there may be a direct causal link between the rise of electronic mediums of data storage and paper consumption, although further research is necessary to firmly establish the validity of this possible causal link.

The failure of computers and electronic storage mediums to bring about the paperless office points to an interesting paradox, which we label the Paperless Office Paradox: the development of a substitute for a natural resource is sometimes associated with an increase in consumption of that resource. This paradox has potentially profound implications for efforts to conserve natural resources. One prominent method advocated for reducing consumption of a particular resource is to develop substitutes for it. For example, the development of renewable energy resources, such as wind and solar power, are commonly identified as a way to reduce dependence on fossil fuel, based on the assumption that the development of alternative sources of energy will displace, at least to some extent, fossil fuel consumption. However, just as the Jevons Paradox points to the fact that efficiency may not lead to a reduction in resource consumption, the Paperless Office Paradox points to the fact that the development of substitutes may not lead to a reduction in resource consumption.

The reasons that computers led to a rise in paper consumption are not particularly surprising. Although computers allow for the electronic storage of documents, they also allow for ready access to innumerable documents that can be easily printed using increasingly ubiquitous printers, which explains in large part the reason for escalating office paper consumption.[19] Due to the particularistic reasons for the association between electronic storage mediums and paper consumption, the Paperless Office Paradox may not represent a generality about the development of substitutes and resource consumption. However, this paradox does emphasize the point that one should *not* assume that the development of substitutes for a natural resource will lead to a reduction in consumption of that resource.

For example, over the past two centuries we have seen the rise of fossil fuel technologies and the development of nuclear power, so that whereas in the eighteenth century biomass was the principal source of energy in the world, biomass now only provides a small proportion of global energy production. However, it is worth noting that even though substitutes for biomass—such as fossil fuel and nuclear power—have expanded dramatically, the *absolute* quantity of biomass consumed for energy in the world has *increased* since the nineteenth century.[20] This is likely due, at least in part, to the fact that new energy sources fostered economic and population growth, which in turn expanded the demand for energy sources of all types, including biomass. This observation raises the prospect that the expansion of renewable energy production technologies, such as wind turbines and photovoltaic cells, may not displace fossil fuel or other energy sources, but merely add a new source on top of them, and potentially foster conditions that expand the demand for energy. Clearly, further theoretical development and empirical research aimed at assessing the extent to which substitutes actually lead to reductions in resource consumption is called for, and faith that technological developments will solve our natural resource challenges should at least be called into question.

Coda

Here, we have drawn attention to two ecological paradoxes in economics, the Jevons Paradox and the Paperless Office Paradox. The Jevons Paradox is a classical one, based on the Jevons observation that rising efficiency in the utilization of coal led to an escalation of coal consumption. We presented two examples, which suggest that the Jevons Paradox may have general applicability to a variety of circumstances. The Paperless Office Paradox is a new one, and draws attention to the fact that the development of computers and electronic storage mediums has not led to a decline in paper consumption, as some predicted, but rather to more paper consumption. It is important to note that these are empirically established paradoxes—they point to the correlation between efficiency or substitutes and resource consumption. Each paradox may actually house phenomena that have a diversity of theoretical explanations. Therefore, underlying these two paradoxes may be many forces that need to be theorized.

Together, these paradoxes suggest that improvements in the efficiency of use of a natural resource and the development of substitutes for a natural resource may not lead to reductions in consumption of that resource—in some circumstances they may even lead to an escalation of consumption of that resource. Although improvements in efficiency and utilization of substitutes will reduce consumption of a resource *all else being equal* (if the scale of production remains constant), economies are complex and dynamic systems with innumerable interactions among factors.

Changes in the type and efficiency of resource utilization will likely influence many other conditions, thus ensuring that all else will rarely be equal. Relying on technological advances alone to solve our environmental problems may have disastrous consequences. The two paradoxes we present here suggest that social and economic systems need to be modified if technological advances are to be translated into natural resource conservation.

9. The Treadmill of Accumulation

In 1994 one of us (John Bellamy Foster) was invited to give a keynote luncheon address to "Watersheds '94," a conference organized by the Environmental Protection Agency (EPA), Region 10, to be held in September of that year in Bellevue, Washington. The invitation was to provide a full analysis of the planetary ecological crisis and its social causes, but there was one catch: it was crucial, the organizers made clear, not to name the system; all explicit references to capitalism needed to be left out.

This restriction created a dilemma. A serious ecological critique necessarily involves a critique of the capitalist system. How could one present such a critique of the system without naming it? The solution adopted on that occasion was to call capitalism the "treadmill of production" and use that as a device to bring out its most essential dynamics. Environmental sociologists Allan Schnaiberg and Kenneth Alan Gould had just published *Environment and Society: The Enduring Conflict*.[1] In many ways, this book was a step back theoretically from Schnaiberg's classic *The Environment: From Surplus to Scarcity*.[2] Nevertheless, Schnaiberg and Gould had, in one succinct formulation in their new book, developed the concept of the "treadmill of production," introduced by Schnaiberg in his earlier book, to the point that it was then almost the functional equivalent of capitalism.

The resulting address for the EPA, Region 10, was titled "Global Ecology and the Common Good" and was published as the Review of the Month in the February 1995 issue of *Monthly Review*. It argued that calls for moral transformation of our society to deal with ecological degradation typically ignored

> the central institution of our society, what might be called the global "treadmill of production." The logic of this treadmill can be broken down into six elements. First, built into this global system, and constituting its central rationale, is the increasing accumulation of wealth by a relatively small section of the population at the top of the social pyramid. Second, there is a long-term movement of workers away from self-employment and into wage jobs that are contingent on the continual expansion of production. Third, the competitive struggle between businesses necessitates on pain of extinction the allocation of accumulated wealth to new, revolutionary technologies that serve to expand production. Fourth, wants are manufactured in a manner that creates an insatiable hunger for more. Fifth, government becomes increasingly responsible for promoting national economic development while ensuring some degree of "social security" for at least a portion of its citizens. Sixth, the dominant means of communication and education are part of the treadmill, serving to reinforce its priorities and values.
>
> A defining trait of the system is that it is a kind of giant squirrel cage. Everyone, or nearly everyone, is part of this treadmill and is unable or unwilling to get off. Investors and managers are driven by the need to accumulate wealth and to expand the scale of their operations to prosper within a globally competitive milieu. For the vast majority, the commitment to the treadmill is more limited and indirect: they simply need to obtain jobs at livable wages. But to retain those jobs and to maintain a given standard of living in these circumstances it is necessary, like the Red Queen in *Through the Looking Glass*, to run faster and faster in order to stay in the same place.[3]

Most of this drew directly on the succinct description of the treadmill of production to be found in Schnaiberg and Gould's *Environment and Society*.[4] This was acknowledged in a footnote in the published version of the talk. It was also noted that in Schnaiberg's earlier work, *The Environment*, the treadmill was "situated in the historical context of monopoly capitalism as described in [Paul] Baran and [Paul] Sweezy's *Monopoly Capital* and James O'Connor's *Fiscal Crisis of the State*."[5]

The use of the treadmill of production concept proved highly successful on the occasion of the EPA luncheon address. It facilitated a powerful criticism of capitalism for its ecological shortcomings without ever mentioning the system by name. Nor, indeed, was it necessary to mention the existence of a system as such at all. The treadmill metaphor had such a concrete, pragmatic character that it was greeted as a mere description of reality with none of the usual political and ideological baggage. At the same time, it allowed for a treatment of both the micro and macro aspects of the system. The talk could then be published word for word in *Monthly Review*, where it was understood to be a critique of capitalism.

Still, the very use of the treadmill of production concept in this way involved the adoption of a kind of Aesopian language. The concept in itself did not add anything indispensable analytically—not to be found in the more general Marxian (or neo-Marxian) ecological critique of capitalism. All of this raises the question of what the significance of the famous treadmill analysis within environmental sociology really is. Is it simply a Trojan horse for getting a radical ecological critique of capitalism inside the gates? What is the relation of this theoretical perspective to Marxian political economy? Does the treadmill of production perspective go beyond the critique of capitalism to a critique of Soviet-type societies as well? How has the treadmill perspective evolved? Is the "treadmill of *production*" even the right treadmill—should it be called "the treadmill of *accumulation*" instead?

Marx and the Treadmill

It is useful to comment first on the treadmill metaphor and the historical reality that lay behind it. Although the term is a familiar one, probably none of us have seen a literal treadmill apart from the ubiquitous exercise machine, and few of us have any clear sense of the historical meaning and the significance of such a treadmill of production. Of course, we all have a general sense of what it is, and it certainly does not seem complex. Nevertheless, our commonplace image falls far short of the horrific reality as it was experienced by English workers in the nineteenth century. This can be seen in an engraving of workers on a treadmill at the House of Correction at Petworth in England early in the nineteenth century, which was published by the Select Committee on Gaols and Houses of Correction of the House of Commons in 1835, and was reproduced in the October 1971 *Scientific American*. It shows a row of fifteen workers forced to climb in unison in a machine-like motion. As Eugene S. Ferguson explained in the article in *Scientific American*, Sir William Cubbit in 1818 reintroduced English prisoners to the treadmill, which employed men in "grinding grain or in providing power for other machines. Each prisoner had to climb the treadmill a total vertical distance of 8,640 feet (2,630 meters) in six hours. The feat was the equivalent of climbing the stairs of the Washington Monument 16 times, allowing about 20 minutes for each trip."[6]

Among modern thinkers, the one who gave the greatest attention to the historical significance of the treadmill as a relation of work and exploitation was Karl Marx. Marx pointed out that under capitalism, "the crudest *modes* (and *instruments*) of human labor reappear; for example, the *tread-mill* used by Roman slaves has become the mode of production and mode of existence of many English workers."[7] For Marx, the reintroduction of "the treadmill within civilisation" meant that "barbarism reappears, but created within the lap of civilisation and belonging to it; hence leprous barbarism, barbarism as leprosy of civilization."[8] The imposition of the treadmill on English workers symbolized the tendency of the capitalist mode of production to degrade the work and, hence, the worker in mind and body: "Not only do the poor devils receive the most wretched and meagre means of subsistence, hardly sufficient for the propagation of the species," he wrote of the factory conditions of the time, "their activity, too, is restricted to revolting, unproductive, meaningless, drudgery, such as work at the treadmill, which deadens both body and mind."[9]

It is not clear to what extent Schnaiberg himself was aware of this treatment of the treadmill in Marx. There are only a couple of direct references to Marx in *The Environment*, the most important one in a footnote. That footnote is devoted to the Marxian conception of technology as Janus-faced: constituting both a key ingredient of development and also a means of the degradation of the worker under capitalism.[10]

The Radical Origins of the Treadmill of Production Concept

There is no doubt that in organizing his critique of environmental degradation around the concept of the treadmill of production, Schnaiberg was taking up a primarily Marxist theme (although one that overlapped with some non-Marxist critiques, such as those of Joan Robinson, John Kenneth Galbraith, and C. Wright Mills). The treadmill theory grew out of a dialogue on the theory of monopoly capital, associated with Paul Baran, Paul Sweezy, and Harry Magdoff and what has sometimes been known as the "Monthly Review School" within Marxian political economy. Although some interpretations of Schnaiberg's work have emphasized his references to James O'Connor's *Fiscal Crisis of the State* (1973), references to Baran, Sweezy, and Magdoff in the central chapter on production in *The Environment* exceeded those to O'Connor by a factor of five, whereas mentions of the revisionist historian Gabriel Kolko, whose approach to the state was adopted by Schnaiberg, outnumbered mentions of O'Connor three to one.[11] Most references in the chapter were to authors who were writing for, or had a close association with and would later write for *Monthly Review*, including—beyond Baran, Sweezy, and Magdoff—Giovanni Arrighi, Harry Braverman, Raford Boddy, Samuel Bowles, Richard Cloward, James Crotty, Richard Du Boff, Herb Gintis, John Gurley, Stephen Hymer, Jacob

Morris, James O'Connor, Frances Fox Piven, John Saul, Howard Sherman, and Immanuel Wallerstein.[12]

The treadmill of production concept is introduced with direct reference to Galbraith's *The Affluent Society*. There, Galbraith had written, "Production only fills the void it has created . . . the individual who urges the importance of production is precisely in the position of the onlooker who applauds the efforts of the squirrel to keep abreast of the wheel that is propelled by its own effort."[13] Inspired by this notion, Schnaiberg writes, "Paralleling his [Galbraith's] concept of the squirrel cage, we can trace out a 'treadmill' of production."[14]

This concept, Schnaiberg claims, could in a loose way be applied to both capitalism and Soviet-style socialism, but it is more specifically associated with capitalism, and in particular monopoly capitalism. As he himself puts it, "While production expansion has occurred in socialist as well as capitalist societies, the particular form of the treadmill is more evident in the latter. The basic social force driving the treadmill is the inherent nature of competition and concentration of capital."[15] Here, he quoted from an article by Harry Magdoff, in which the latter had pointed to the tendency of capitalism to pursue accumulation above all else.[16] This accumulation tendency was rooted in class relations, activated by competition, and led to the concentration and centralization of capital.[17] Schnaiberg referred specifically to the "monopoly capital treadmill," growing out of the analysis of Baran and Sweezy's *Monopoly Capital*.[18] "Both the volume and source of a treadmill of production," he wrote, "is high-energy monopoly capital industry."[19]

In terms of labor, Schnaiberg's original analysis relied heavily on the two-sector model of the competitive and monopolistic areas of the economy as proposed by Baran and Sweezy and O'Connor.[20] In this conception, workers in the monopoly sector benefited from some part of the surplus and thus tended to provide some degree of support to the system. Schnaiberg also drew heavily on Magdoff and Sweezy's analysis of the buildup of the credit-debt system (or the tendency toward financial explosion and periodic meltdowns in a system normally mired in stagnation) as a principal force in the expansion of the treadmill.[21]

With respect to the state, Schnaiberg's *The Environment* relied most heavily on Kolko's analysis of political capitalism, which sees the state as providing a key accumulation function; as well as on Kolko's notion of how the corporations (the regulated) had captured the regulatory system. This was coupled with Baran and Sweezy's explanation of how the state, by promoting military spending and an economy of waste, sought to expedite accumulation (the treadmill) despite growing social and environmental irrationalities. As Schnaiberg summed it up, "The bloom is partly off the monopoly capital rose."[22]

In *The Environment*, Schnaiberg made it clear that the ecological problem consisted mainly of "increased environmental withdrawals and additions" as the treadmill sped up. He did not expect much from the gains in efficiency lauded by economists and today by ecological modernizationists within sociology. Such efficiencies tended to be economic, not ecological. As he stated, "While some capital intensification of production may lead to more efficient production techniques, these often involve substitutions of energy for older materials. The case of plastics is a prototype. Plastics are high-energy products that serve to substitute for larger volumes of wood and metals."[23] This is a theme taken from Barry Commoner, but Schnaiberg was also well aware of Nicholas Georgescu-Roegen's critique of the economic process from the standpoint of entropic degradation.

The most radical element in Schnaiberg's analysis was the recognition that the treadmill was a system, monopoly capitalism, and that the system, understood in these terms, could not be reversed short of a major revolt from below. Schnaiberg underscored the social and environmental implications of Sweezy's dramatic conclusion: "Capitalism's utopia in a sense is a situation in which workers live on air, allowing their entire product to take the form of surplus value, and in which the capitalists accumulate all their surplus value."[24]

The main hope, Schnaiberg stressed in his classic study, drawing on Magdoff and Sweezy, was that the economic and ecological contradictions of the monopoly capital treadmill would so destabilize the system as to create room for substantial change from below. He concluded his chapter on production by emphasizing the "education of labor" as the basis for change. This meant: "(1) educating labor to the discomforts—the environmental hazards—of the treadmill; (2) educating labor to the socially inefficient role of the state in the allocation of surplus to the treadmill; and (3) educating labor to the alternatives to the present state-supported treadmill system."[25] Without such education and the practical agency for change arising from it, he argued, it would be impossible for the treadmill "to be slowed and reversed."[26] This was a class struggle perspective linked to a conception of environmental necessity.

The Environment was published in 1980 and had a vast impact on environmental sociology in the United States, emerging as the most influential theoretical perspective. Nevertheless, environmental sociology, which had been enormously creative and growing in the late 1970s, went into decline in the conservative climate of the early Reagan era and did not begin to recover until almost a decade later.[27] When it did, it was a less radical environmentalism overall (though the same period saw the growth of Marxist ecology) with the emphasis of the environmental movement no longer the deep green critique of capitalism but rather sustainable development (often interpreted as sustaining the economy) and, increasingly, ecological modernization. Schnaiberg and his younger associates tried to adjust to this changing reality in subsequent works, toning down (without substantially altering) his original critique, but at considerable cost to its overall coherence.

Environmental sociologists since the 1990s have became more professionalist, hence intellectually insular—more involved in the internal discussions of their own field and less involved in questions of imperialism, war, and economic crisis. Environmental sociology has thus become increasingly disconnected from the root critique of the system emanating from Marxism. One can easily see the discursive differences between Schnaiberg and Gould's 1994 *Environment and Society* and Schnaiberg's earlier *The Environment*. Direct references to capitalism had all but disappeared in the later work, displaced almost totally by the metaphor of the treadmill. Magdoff and Sweezy—the thinkers most frequently referred to in Schnaiberg's *magnum opus*—vanished completely from this second book, as did nearly all the writers associated with *Monthly Review*—apart from Braverman, Kolko, and O'Connor, each of whom appeared only once.[28] The discussion of labor was sharply curtailed, and the hope for "educating labor" so evident in the first book was conspicuous in its absence in the second. The historical specificity of the argument, which had been rooted in the analysis of the monopoly stage of production, was gone as well. Although *The Environment* was closely integrated with a theory of economic crisis—namely, Magdoff and Sweezy's analysis of stagnation and the debt explosion—this was no longer the case in any significant sense in *Environment and Society*. The same disappearances are evident in the more recent book, *The Treadmill of Production*, authored by Gould, David Pellow, and Schnaiberg. Here the concept of capitalism scarcely makes an appearance, and Braverman, Kolko, and O'Connor have now also disappeared from the analysis. Recognition of these changes in the basic perspective caused Frederick Buttel—the late environmental sociologist—to describe the treadmill model in its later development as "extra-Marxist."[29]

Treadmill of Production or Treadmill of Accumulation?

The first and biggest weakness of the treadmill of production theory from a historical materialist perspective is that it concentrated on the wrong treadmill. To understand the major thrust and inherent dangers of capitalism, it is necessary to see the problem as one of a treadmill of *accumulation* much more than *production*. Of course, the two are not separate. In the Marxist perspective, all is traceable to the relations of production and to the social formation arising out of the mode of production at a historically specific period. But the core issue where capitalism is concerned is accumulation. It is this which accounts for the dynamism and the contradictions of the capitalist mode. The best way to describe this is in terms of Marx's general formula for capital—M-C-M′. In this formula, money capital is transformed into a commodity (via production), which then has to be sold for more money, realizing the original value plus an added or surplus value, distinguishing M′ (or M + Δ m). In the next period of production M′ is reinvested with the aim of obtaining M″, and so on. In other words, capital, by its nature, is self-expanding value. This accumulation dynamic is enforced by the competitive tendencies of the system and is at one with the concentration and centralization of production. It is rooted in a system of class exploitation. As Sweezy put it in "Capitalism and the Environment":

> The purpose of capitalist enterprise has always been to maximize profit, never to serve social ends. Mainstream economic theory since Adam Smith has insisted that by *directly* maximizing profit the capitalist (or entrepreneur) is *indirectly* serving the community. All the capitalists together, maximizing their individual profits, produce what the community needs while keeping each other in check by their mutual competition. All this is true, but it is far from being the whole story. Capitalists do not confine their activities to producing the food, clothing, shelter, and amenities society needs for its existence and reproduction. In their single-minded pursuit of profit, in which none can refuse to join on pain of elimination, capitalists are driven to accumulate ever more capital, and this becomes both their subjective goal and the motor force of the entire economic system.
>
> It is this obsession with capital accumulation that distinguishes capitalism from the simple system for satisfying human needs it is portrayed as in mainstream economic theory. And a system driven by capital accumulation is one that never stands still, one that is forever changing, adopting new and discarding old methods of production and distribution, opening up new territories, subjecting to its purposes societies too weak to protect themselves. Caught up in this process of relentless innovation and expansion, the system runs roughshod over even its own beneficiaries if they get in its way or fall by the roadside. As far as the natural environment is concerned, capitalism perceives it not as something to be cherished and enjoyed but as a means to the paramount ends of profit-making and still more capital accumulation.[30]

The treadmill of production model, particularly in Schnaiberg's earliest version, certainly encompasses the accumulation dynamic of capitalism to some extent. But the emphasis of the model is not on accumulation and the social relations of accumulation, but rather on production and technology. Consequently, there is a significant tendency to underestimate the role of accumulation as the "juggernaut" of capital, as Marx termed it, along with the crisis tendencies it generates. Indeed, many readers will doubtless not see the relation of the treadmill of production to accumulation at all. The treadmill of production metaphor, so useful in some ways, feeds into the abstract notion of *growth* divorced from the specific form that this takes under the regime of capital—as a *system of accumulation*. It is the accumulation drive, according to Marx, that "gives capital no rest and continually whispers in its ear: 'Go on! Go on!'"[31]

The second weakness (not an inherent one, but rather one of emphasis rooted in the central metaphor) is that the treadmill of production framework is focused almost exclusively on *scale* and relatively little on *system*—except insofar as this is related to scale. The problem becomes the uni-directionality and the speed of the treadmill, which means increasing scale. This fits the argument on additions and withdrawals and conforms to the dominant emphasis within the environmental movement on the problem of carrying capacity. It captures the quantitative aspect of the confrontation between economy and ecology. But the more qualitative dimensions of the environmental problem frequently get lost. It is not simply a question of scale but of dislocations or rifts in the environment. Capitalism seeks to reduce and *simplify* human labor to exploit it more effectively. Similarly, with the environment, capital seeks, for example, to replace an old-growth forest with all of its natural complexity with a simplified industrial tree plantation that is ecologically sterile, dominated by a single species, and "harvested" at accelerated rates. A detailed "division of nature" thus accompanies the detailed division of labor under capitalism, often with disastrous results.[32]

To highlight scale almost exclusively is often to lose sight of this complex, coevolutionary dialectic. It also detracts from knowledge of the micro-toxicity that is a major part of today's environmental problem. After all, the level of production can remain the same while the level of toxicity goes up, a reality not normally captured by scale or carrying capacity concepts. Hence the dialectical complexity and historicity of nature, as well as the contradictory nature of the human exchange with the environment, lie outside of the treadmill of production analysis. An exclusive emphasis on scale as opposed to system can blind one to what Marx called the "metabolic rift" in the organic human relation to nature. The naturally given human relation to nature is torn asunder through the polarization of town and country and the extreme division of nature with horrendous results.

Quantitative scale problems have to be seen in relation to the qualitative aspects transforming the nature of the environmental problem. Capitalism is certainly governed by the drive toward growth and, more concretely, accumulation. But its own distorted existence means that this has to be done in certain ways. It is easier for the system to grow by producing depleted uranium shells to be used in imperialist wars or by expanding agribusiness devoted to producing luxury crops to be consumed by the relatively well-to-do in the rich countries than it is to protect the integrity of the environment or to provide food for those actually in need. To reduce the whole environmental problem to the issue of scale—however much that constitutes the first step in addressing the problem—is to underestimate the systematic obstacles like the conflict between use value and exchange value built into the structure of the existing system. At the same time, it downplays the full range of possibilities that might be opened up if system change allowed qualitative, not simply quantitative, transformations in human and human-natural relations. Here what Joan Robinson called "the second crisis of economic theory"—the question of what (as opposed to how much) is produced and the systematic waste built into current production for the sake of profits—remains central.[33]

All of this raises serious questions about the treadmill of production model. Nevertheless, the power of Schnaiberg's original vision and its close relation to a more general Marxian political economic critique allows for a renewed synthesis. In some ways, environmental sociology, in adopting Schnaiberg's analysis, is in an analogous position to institutionalist economics in the United States, which originated with Thorstein Veblen. A radical critique, as represented by Veblen's work (particularly *Absentee Ownership*), was watered down by the instituitonalists who followed. But Veblen's original foundations remain forever intact, providing the possibility of rebuilding on this more solid basis.[34]

The comparison between the treadmill of production perspective and institutional economics is useful in other respects as well. In both cases radical traditions of political-economic critique were promoted—ones that had close relations to Marxist thought—while also reflecting a distinctly American pragmatism. Although often weak theoretically (descending into a narrow empiricism), the commitment to the concrete aspects of reality also makes such pragmatic thinking relatively immune to mystifications. In the widening debate within environmental sociology between ecological modernization analysis and the treadmill of production theory, the latter has had the benefit of both a solid empiricism and a strong (albeit weakened) connection to the Marxian critique. The result has been one devastating refutation after another of ecological modernization analysis.[35] The treadmill of production model has provided an elegant, empirically testable set of hypotheses at the level of middle-range theory. Lacking a fully developed ecological critique of the kind that Marxian theory as a whole might have provided—but which the treadmill theory largely abandoned in its current extra-Marxist form—it nonetheless has had the advantage over ecological modernization analysis of a greater degree of realism arising from its radical roots.

There is no doubt, then, that the treadmill of production model raises key institutional questions that have to be addressed if the environmental crisis is to be recognized for what it is. Perhaps the greatest virtue of this perspective, as Buttel observed, is that it describes "how the relentless reinforcing processes of capital-intensive economic expansion create [environmental] additions and withdrawals, and at the same time, highly constrain the movements that mobilize to redress the processes of environmental degradation."[36]

In describing the environmental problem as arising from a treadmill of production, Schnaiberg captures the futility and irrationality of a system of production that frequently degrades the minds and bodies of workers while pursuing an endless Sisyphean labor. It is well to remember that for Marx the treadmill stood for the barbarism within civilization. The treadmill, which was long utilized in English houses of correction, earned the hatred of workers, who feared being consigned to it. By choosing this particular metaphor, therefore, Schnaiberg—whether fully cognizant of its historical significance or not—highlighted a major contradiction of capitalism. Despite its profession of civilization and modernization, capitalism never truly surmounted a brutal, barbaric relation to human beings and nature; indeed, it has robbed both on an ever-increasing scale. The dire global implications of this predatory as well as pecuniary system (to borrow Veblenian language) are only now becoming fully apparent.

Hence the treadmill of production perspective through its central metaphor serves to remind us of the barbaric, unsustainable character of capitalism's relation to humanity and nature, even as the system seemingly expands and prospers. It remains in this respect a revolutionary ecological perspective.

10. The Absolute General Law of Environmental Degradation under Capitalism

Capitalism is a system of contradictions. Here we briefly reflect on what has been termed, by James O'Connor, the "first and second contradictions" of capitalism. The first contradiction, following Marx, can be referred to as "the absolute general law of capitalist accumulation."[1] The second contradiction may then be designated as "the absolute general law of environmental degradation under capitalism." It is characteristic of capitalism that the second of these "absolute general laws" derives its momentum from the first; hence it is impossible to overthrow the second without overthrowing the first. Nevertheless, it is the second contradiction rather than the first that increasingly constitutes the most obvious threat not only to capitalism's existence but to the life of the planet as a whole.

The first contradiction, O'Connor tells us, "expresses capital's social and political power over labor, and also capitalism's inherent tendency toward a realization crisis, or crisis of capital over-production."[2] It finds its expression in the limitless drive to increase the rate of exploitation. This "absolute general law of capitalist accumulation" results in the amassing of wealth at one pole and relative human misery and degradation at the other. It reflects an "oscillation of wages" that is "kept penned within limits satisfactory to capitalist exploitation" by the continual reproduction of a relative surplus population of the unemployed/underemployed. Today the "field of operation" of this law is the entire world.[3]

The second contradiction of capitalism, or the "absolute general law of environmental degradation," is more difficult to characterize since bourgeois political economy (together with its classical Marxist critique) has—for reasons related to the functioning of capitalism itself—never incorporated what Marx termed the "conditions of production" (natural, personal, and communal) into its internal logic.[4] Nevertheless, this contradiction can be expressed as a tendency toward the amassing of wealth at one pole and the accumulation of conditions of resource-depletion, pollution, species and habitat destruction, urban congestion, overpopulation, and a deteriorating sociological life-environment (in short, degraded "conditions of production") at the other.

Under capitalism "the greater the social wealth, the functioning capital, the extent and energy of its growth," the greater are capital's ecological demands and the level of environmental degradation. Although the second law of thermodynamics guarantees that there will be an increase in "entropic degradation" with the advance of production, the existence of a capitalist mode of appropriation, with its goal of promoting private profits with little regard for social or environmental costs, guarantees that this entropic degradation globally will tend toward maximum economically feasible levels at any given historical phase of development. Worse still, the contemporary structure of commodity production, with its built-in dependence on pesticides, petrochemicals, fossil fuels, and nuclear power generation, and its treatment of external habitats as a vast commons to be freely exploited by capital, tends to maximize the overall toxicity of production and to promote accelerated habitat destruction, creating problems of ecological sustainability that far outweigh the general entropic effect.[5]

Although the "absolute general law of environmental degradation" in this sense relates primarily to the realm of natural-material processes and use value rather than exchange value, the costs borne by the environment rebound on the economic realm in multiple unforeseen ways, reflecting what Frederick Engels called "the revenge" of nature that follows every human "conquest over nature." "Labor," Marx observed, "is not the only source of material wealth, of use-values produced by labor. As William Petty puts it, labour is its father and the earth its mother." Capitalism grows, Marx contends, by exploiting the former and "robbing" the latter.[6]

It stands to reason that such a freebooting relation to ecological systems cannot long persist without disastrous consequences for the conditions of life, which, obviously, includes the economy itself. Thus we have witnessed the emergence of what has come to be known globally as "the environmental crisis" (beginning with the onset of the nuclear age in the second half of the twentieth century)—at a point in the development of the system when the scale and extent of its operation is in danger of overwhelming the major ecological cycles of the planet. This new awareness of environmental degradation, moreover, has forced itself on the consciousness of society primarily through its economic effects. For it is only at this stage in the system's development that general physical barriers increasingly translate into specific economic barriers to capital's advance.[7]

The reordering of capitalism that occurred with the rise of its monopoly stage in the twentieth century resulted in the enlargement of the first contradiction, making it more and more essential for capital to expand the circle of consumption while keeping the basic relation between capital and labor intact.[8]

Thus the penetration of the sales effort into production, already perceived by Veblen, has become increasingly evident, undermining capitalism's claim to conform to the necessary conditions of production in general.[9] An ordinary English muffin, for example, has been shown to pass through seventeen "energy steps" following the growth and harvesting of the wheat, with the result that nearly twice as much energy is now utilized to process the muffin as to grow it.[10] Supply-price therefore no longer conforms to rational principles of cost-containment. Instead, ever more baroque "commodity chains" are emerging, with each link in the chain deriving its justification from the increment of profit it provides together with its contribution to the salability of the final commodity.[11] Synthetic products, poisonous to natural and human environments, have become intrinsic to the development of the system.[12] It was an understanding of this problem (together with the expansion of armaments) that led Joan Robinson to insist that "the second crisis of economic theory" (the question of the content as opposed to the level of production) is now paramount.[13]

Since the early 1970s, the world economy has been suffering from relative stagnation (or a decline in the secular growth trend) accompanied by rising unemployment and excess capacity. Capital has responded to this crisis in its usual fashion through supply-side "restructuring," or the opening up of the system to a more intensive exploitation (and super-exploitation) of labor and the environment. Many regulations previously put in place to protect the conditions of production were cast aside—as Karl Polanyi leads us to expect—under the ideological mantle of the "self-regulating market."[14] At the same time, the system's core has been shifting away from the production of the goods and services that constitute GNP and toward the speculative proliferation of financial assets.[15] One result of both of these processes has been an acceleration of the pace of environmental degradation. Hence it is no accident that there has been a speedup in the destruction of the remaining natural forest ecosystems throughout the world, which by Wall Street criteria are viewed as non-performing assets to be liquidated as quickly as possible.

The second contradiction of capitalism therefore is rapidly gaining on the first—partly due to measures taken to compensate for the first—without the first ever abating. The result is a "hyper-capitalist" disorder in which the system is obsessed with both enlarging markets and finding ways around rising environmental costs.[16] Since only a tiny proportion of environmental costs have thus far been internalized by capital and the state, it is a foregone conclusion that the economic repercussions of the second contradiction will grow by leaps and bounds—partly under the pressure of social movements—marking nature's ultimate "revenge" on the accumulation process.

From a movement perspective the implications seem clear. Any struggle that attempts to combat only one of capitalism's "absolute general laws" while perpetuating the other will prove ineffectual. The future of humanity and the earth therefore lies with the formation of a labor-environmentalist alliance capable of confronting both of capitalism's absolute general laws. The forging of such an alliance would mark the rise of socialist ecology as a world-historical force—and the onset of the struggle that is likely, more than any other, to define the course of the twenty-first century.

PART THREE

Dialectical Ecology

11. The Dialectics of Nature and Marxist Ecology

For the philosophical tradition of "Western Marxism" no concept internal to Marxism has been more antithetical to the genuine development of historical materialism than the "dialectics of nature."[1] Commonly attributed to Frederick Engels rather than Karl Marx, this concept is often seen as the *differentia specifica* that beginning in the 1920s separated the official Marxism of the Soviet Union from Western Marxism. Yet as Georg Lukács, who played the leading role in questioning the concept of the dialectic of nature in his *History and Class Consciousness*, was later to admit, Western Marxism's rejection of it struck at the very heart of the classical Marxist ontology—that of Marx no less than Engels.[2]

The question of the dialectics of nature has therefore constituted a major contradiction within Marxist thought, dividing its traditions. On the one hand, the powerful dialectical imagination that characterized Western Marxism rested on a historical-cultural frame of analysis focusing on human praxis that excluded non-human nature. On the other hand, Marx's own dialectical and materialist ontology was predicated on the ultimate unity between nature and society, constituting a single reality and requiring a single science. Marx's original method had pointed to the complex interconnections between society and nature, utilizing a dialectical frame in analyzing both—although the nature dialectic was much less explicitly developed within his thought than the social dialectic. The unbridgeable chasm between nature and society that was to arise with Western Marxism was entirely absent in his work (as were the positivistic tendencies of what became official Marxism in the Soviet Union).

In recent decades, the larger consequences associated with the Western Marxist repudiation of the dialectics of nature have been highlighted by the growth of Marxist ecology, which is concerned with the complex coevolutionary relations between society and nature. Yet even as it has brought this weakness of the Western Marxist tradition to the fore, the growth of Marxist ecology has provided new ways of transcending the contradiction, building both on classical Marxist thought and new understandings of the material relations between humanity and nature emerging in the context of a planetary ecological crisis.

Lukács and the Dialectics of Nature

The birth of "Western Marxism" as a distinct philosophical tradition has commonly been traced to Georg Lukács's famous footnote 6 in chapter 1 of *History and Class Consciousness*, in which he rejected any extension of the dialectical method from society to nature:

> It is of the first importance to realize that the method is limited here to the realms of history and society. The misunderstandings that arise from Engels' account of dialectics can in the main be put down to the fact that Engels—following Hegel's mistaken lead—extended the method to apply also to nature. However, the crucial determinants of dialectics—the interaction of subject and object, the unity of theory and practice, the historical changes in the reality underlying the categories as the root cause of changes in thought, etc.—are absent from our knowledge of nature.[3]

Lukács suggests here that the dialectical method in its full sense necessarily involves reflexivity, the identical subject-object of history.

The subject (the human being) recognizes in the object of his/her activity the results of humanity's own historical self-creation. We can understand history, as Vico said, because we have "made" it. The dialectic thus becomes a powerful theoretical means of discovery rooted in the reality of human praxis itself, which allows us to uncover the totality of social mediations. Yet such inner, reflexive knowledge arising from human practice, he insists, is not available where external nature is concerned. There one is faced with the inescapable Kantian thing-in-itself. Hence the "crucial determinants of dialectics" are inapplicable to the natural realm; there can be no dialectics of nature—as a method—equivalent to the dialectics of history and society. "Engels—following Hegel's mistaken lead—" was therefore wrong in extending "the method to apply also to nature."

As Lukács observed a few years later in a 1925 review of Nikolai Bukharin's *Historical Materialism*, "Engels reduced the dialectic to 'the science of the general laws of motion, both of the external world and of human thought.'"[4] By applying the dialectical method to nature, Engels had overstepped its proper realm of application.

This prohibition against the extending of the dialectic to nature, which was to become a distinguishing feature of Western Marxism following Lukács's lead, had its counterpart in a prohibition against the undialectical introduction of the methods of natural science into the realm of the social sciences and humanities, that is, against positivism. Engels's mistake, as Lukács suggested, was to have argued in *Anti-Dühring* that "nature is the proof of dialectics," and that the dialectic could be grasped by studying the development of natural science—thus the method of the latter could be a key to the method for the analysis of society itself.[5] In his critique of Bukharin's *Historical Materialism*, Lukács pointed to the way that Bukharin, following Engels's lead, had embraced a "contemplative materialism" that drew largely on the external, objective view of nature—and then attempted to apply this to human society. "Instead of making a historical-materialist critique of natural sciences and their methods," Bukharin, Lukács wrote, extended "these methods to the study of society without hesitation, uncritically, unhistorically, and undialectically"—thus falling prey to positivism and the reification of both nature and society.[6]

Antonio Gramsci, who, along with Lukács and Karl Korsch, helped found Western Marxism as a philosophical tendency in the 1920s, was likewise critical of Bukharin's *Historical Materialism* for its tendency to impose natural-scientific views on society. But Gramsci, though skeptical about the dialectics of nature, was nonetheless disturbed by the potential implications of what seemed to be Lukács's outright rejection of the concept in *History and Class Consciousness*:

> It would appear that Lukács maintains that one can speak of the dialectic only for the history of men and not for nature. He might be right and he might be wrong. If his assertion presupposes a dualism between nature and man he is wrong because he is falling into a conception of nature proper to religion and to Graeco-Christian philosophy and also to idealism which does not in reality succeed in unifying and relating man and nature to each other except verbally. But if human history should be conceived also as the history of nature (also by means of the history of science) how can the dialectic be separated from science? Lukács, in reaction to the baroque theories of the *Popular Manual* [Bukharin's *Historical Materialism*], has fallen into the opposite error, into a form of idealism.[7]

Yet for Gramsci, too, the dialectic of nature remained outside his analysis. He rejected any tendency to "make science the base of life" or to suggest that the philosophy of praxis "needs philosophical supports outside of itself."[8] In his philosophical practice the question of the dialectics of nature and the materialist conception of nature remain unexplored.

The seriousness of this contradiction for Marxist theory as a whole cannot be overstated. As Lucio Colletti observed in *Marxism and Hegel*, a vast literature "has always agreed" that differences over (1) the existence of an objective world independent of consciousness (philosophical materialism or realism), and (2) the existence of a dialectic of matter (or of nature) constituted "the two main distinguishing features between 'Western Marxism' and 'dialectical materialism.'"[9] Lukács launched "Western Marxism" in *History and Class Consciousness* by calling into question both of these epistemological propositions (the first through his critical identification of reification and objectification and hence his Hegelian-Marxist critique of the subject-object distinction; the second through his reservations about the dialectics of nature). Yet in a dramatic turnaround in his later years, Lukács reinstated the very principles he had earlier rejected, on the grounds that he had violated Marx's own materialist ontology. Therefore Lukács was both the founder of Western Marxism and its most potent critic. Both through the contradictions of his thought and through his repudiation of his earlier views he guaranteed that the problem of the dialectics of nature in Western Marxist thought would be raised primarily as a problem of his own philosophy.

Even in *History and Class Consciousness* there were signs that Lukács's rejection of the dialectic of nature was not absolute. At the end of the most important essay in his book, "Reification and the Consciousness of the Proletariat," Lukács wrote:

> Hegel does perceive clearly at times that the dialectics of nature can never become anything more exalted than a dialectics of movement witnessed by the detached observer, as the subject cannot be integrated into the dialectical process, at least not at the stage reached hitherto.... From this we deduce the necessity of separating the merely objective dialectics of nature from those of society. For in the dialectics of society the subject is included in the reciprocal relation in which theory and practice become dialectical with reference to one another. (It goes without saying that the growth of *knowledge* about nature is a social phenomenon and therefore is to be included in the second dialectical type.) Moreover, if the dialectical method is to be consolidated concretely it is essential that the different types of dialectics should be set out in concrete fashion.... However, even to outline a typology of these dialectical forms would be well beyond the scope of this study.[10]

From this it is clear that Lukács that did not entirely abandon the notion of the dialectics of nature even at the time of *History and Class*

Consciousness but rather saw it as limited in the sense that it could never be "more exalted than a dialectics of movement witnessed by the detached observer" and was "merely objective dialectics," lacking an internal subject. Moreover, his criticism of this "merely objective dialectics of nature" excluded the knowledge of nature, which was a social phenomenon and so fell within the dialectics of society. The larger "typology" of "dialectical forms" that Lukács referred to here was never concretely taken up in his analysis. To make matters even more complicated, in a review of the work of Karl Wittfogel published two years after *History and Class Consciousness*, Lukács stated: "For the Marxist as a historical dialectician both *nature* and all the forms in which it is mastered in theory and practice are *social categories*; and to believe that one can detect anything supra-historical or supra-social in this context is to disqualify oneself as a Marxist."[11] But this would in itself seem to deny the possibility of an "objective dialectics of nature"—or the Marxist nature of such an inquiry.

Recently, with the discovery of Lukács's *Tailism and the Dialectic*, written two or three years after the publication of *History and Class Consciousness*, and presenting a defense of that work in the face of the harsh criticisms to which it was subjected by Soviet Marxists associated with dialectical materialism, a more detailed and nuanced look at the early Lukács's position on the dialectics of nature has become available. By quoting extensively from the section on "The Dialectics of Nature" we can see the full complexity of Lukács's position:

> Self-evidently society arose *from* nature. Self-evidently nature and its laws existed *before* society (that is to say before humans). Self-evidently the dialectic *could* not possibly be effective as an *objective principle of development* of society, if it were not already effective as a principle of development of nature before society, if it did not already *objectively exist*. From that, however, follows neither that social development could produce no new, equally objective forms of movement, dialectical moments, nor that the dialectical moments in the development of nature would be *knowable* without the mediation of these new social dialectical forms. . . .

> This [metabolic] *exchange* of matter with nature [that is, production] cannot possibly be achieved—even on the most primitive level—without possessing a certain degree of objectively correct knowledge about the processes of nature (which exist prior to people and function independently of them).... The type and degree of this knowledge depends on the economic structure of society....
>
> I am of the opinion that our knowledge of nature is socially mediated, because its material foundation is socially mediated; and so I remain true to the Marxian formulation of the method of historical materialism: "it is social being that determines consciousness." ...
>
> The sentence [in footnote 6 of chapter 1 of *History and Class Consciousness* where changes in concepts accompanying changes in reality are referred to] means that a change in material (the reality that underlies thought) must take place, in order that a change in thought may follow. . . . That objective dialectics are in reality independent of humans and were there before the emergence of people, is precisely what was *asserted* in this passage; but . . . for the dialectic as knowledge . . . thinking people are necessary.[12]

Here Lukács contends that even in his controversial criticism of Engels in his footnote to chapter 1 of *History and Class Consciousness* he assumed the existence of an objective dialectic of material change (both natural and social) that formed the condition for the change in concepts—the rise of *dialectical knowledge*. The issue for him, then, is not whether an objective dialectic exists as a process independent of human beings and containing within it matter and motion, contradiction, the interdependence of opposites, the transformation of quality and quantity, the mediation of totality, and so forth. Rather he contends that this objective dialectic as such is inaccessible apart from the working out of the metabolic relation between human beings and nature as evident in the development of human social production. Dialectical knowledge is necessarily socially mediated, and does not constitute an immediate relation to nature. It is a product of human praxis. Moreover, the implication of Lukács's whole argument up to this point is that insofar as the dialectic of nature does not arise directly out of the transformations resulting from the metabolic exchange with nature (that is, insofar as it is not a social dialectic in which nature is simply a part) it can take no form "more exalted" than a contemplative materialism or contemplative dialectic—"merely objective dialectics." Thus the dialectical method in its full sense can never be applied to nature except as mediated by social production, that is, praxis.[13]

Marx, Lukács observed in *Tailism and the Dialectic*, had suggested in letters to Ferdinand Lassalle (December 21, 1857, and February 22, 1858) that the ancient materialists Heraclitus and Epicurus created dialectical systems, but that these lacked self-conscious awareness of themselves as such. It is only with the emergence of historical materialism (with Marx himself), Lukács contends, that there arises a self-conscious dialectical conception resting on materialist foundations. And it is this materialistically apprehended dialectic, which has now become "for-us" and not "only-in-itself," that reveals for the first time the real basis of natural science and of knowledge in general. Such a self-conscious materialist dialectic is knowledge that is aware it is socially mediated by historical production or the "*capitalist* exchange of matter with nature." It therefore becomes possible on this basis to apprehend how other societies with very different productive relations (different exchanges of matter with nature) would generate very different conceptual understandings. The historical recognition of the material evolution of society and of human consciousness thus becomes possible.[14]

In his 1967 "Preface to the New Edition" of *History and Class Consciousness*, in which he repudiated many of his earlier views, Lukács is less explicit on the question of the dialectic of nature, but condemns his early work for arguing "in a number of cases that nature is a societal category." He strongly criticizes *History and Class Consciousness* for narrowing down the economic/materialist problem "because its basic Marxist category, labour as the mediator of the metabolic interaction between society and nature, is missing. . . . It is self-evident that this means the disappearance of the ontological objectivity of nature" upon which the process of historical production/change is rooted. From this, he insisted, other mistakes arose, including (1) the confusion of objectification with alienation (causing a lapse into Hegelianism), and (2) the tendency to "view Marxism exclusively as a theory of society, as social philosophy, and hence to ignore or repudiate it as a theory of nature." These errors, he wrote, "strike at the very roots of Marxian ontology." Lukács suggested that what he had called the objective dialectics of nature existed. Moreover, he concedes that he was wrong in characterizing as a stance of "pure contemplation" Engels's attempt to transcend the problem of subjectivity with respect to natural science by arguing that experimentation allowed a reconciliation of theory and practice. But Lukács points nevertheless to the continuing weaknesses of this argument of Engels, and leaves the question of the application of the dialectical method to nature hanging—implying that this is an insurmountable problem, that could not be solved as in his early work by "an overextension of the concept of praxis" to all of reality.[15]

There is in Lukács a different conception that recognizes that through the development of human labor (the developing metabolic exchange between nature and society), a historical transformation in the human consciousness of nature occurs that allows the expansion of the knowledge of nature—the genuine progress of the science of nature as dialectic. As he declared in his famous *Conversations* of 1967—taking place the same year he wrote his new preface to *History and Class Consciousness*—"You will remember how enthusiastically Marx greeted Darwin, despite many methodological reservations, for discovering the fundamentally historical character of being in organic nature. As for inorganic nature, it is naturally extremely difficult to establish its historicity. . . . The problem . . . is whether present-day physics is to be based on a so to speak obsolete standpoint—either that of vulgar materialism, or the purely manipulative conception of neo-positivism—or whether we are moving towards a historic and genetic conception of inorganic nature."[16] For Lukács such a historic and genetic understanding of nature (he gave the specific example of the breakthrough in the understanding of the origins of life introduced by J. B. S. Haldane and Alexander Oparin) clearly approached a dialectical conception (falling within his typology of "dialectical forms") and captured in some way nature's own objective dialectic—without, however, representing the actual application of the full dialectical method in the sense that this pertained to human history and society. The problem of objectification remained and limited the pretensions of a dialectic of praxis. But "since human life is based on a metabolism with nature, it goes without saying that certain truths which we acquire in the process of carrying out this metabolism have a general validity—for example the truths of mathematics, geometry, physics and so on."[17]

In summation, the "Lukács Problem" as we have presented it here can be seen as consisting of: (1) the rejection of the notion that the dialectical method is applicable to nature; and (2) the assertion that a "merely objective dialectic of nature" nonetheless exists and that this is essential to Marxist theory. Lukács, as we have seen, ended up by emphasizing this disjuncture within Marxist thought without being able to resolve it in any way, leaving behind a kind of Kantian dualism. Within Western Marxism generally the Lukács Problem is recognized only one-sidedly. The repudiation of the dialectics of nature and the rejection of any positivistic intrusion of natural science into social science is accepted. Yet the problem of the knowledge of nature and its relation to totality that persists in Lukács through his acknowledgment of an "objective dialectics of nature" is largely ignored. Although for Lukács the dualism that seemed to emanate from his thought was a source of concern, Western Marxism in general has been more content to accept it—or else to adopt a more explicitly idealist way of resolving the contradiction through the subsumption of nature entirely under society.[18]

According to Herbert Marcuse, the dialectic of nature was part of the Hegelian dialectic of totality but was missing from Marx's own dialectical conception of social ontology rooted in labor—except insofar as nature entered into the social realm. Thus as he wrote in *Reason and Revolution*:

> The dialectical totality...includes nature, but only in so far as the latter enters and conditions the historical process of social reproduction.... The

> dialectical method has thus of its very nature become a historical method. The dialectical principle is not a general principle equally applicable to any subject matter. To be sure, every fact whatever can be subjected to a dialectical analysis, for example, a glass of water, as in Lenin's famous discussion. But all such analyses would lead into the structure of the socio-historical process and show it to be constitutive in the facts under analysis. . . . Every fact can be subjected to dialectical analysis only in so far as every fact is influenced by the antagonisms of the social process.[19]

Marcuse's view did allow for a complex social-natural dialectic, if not a dialectics of nature separate from society. Yet the rejection of the dialectics of nature characteristic of Western Marxism was so thoroughgoing in general as to leave little room for the consideration of nature outside of human nature. In *From Hegel to Marx*, Sidney Hook wrote: "Galileo's laws of motion and the life history of an insect have nothing to do with dialectic except on the assumption that all nature is spirit. . . . Whether natural phenomena are continuous at all points or discontinuous at some is an empirical question. It is strictly irrelevant to the solution of any *social* problem. . . . The natural objective order is relevant to dialectic only when there is an implied reference to the way in which it conditions historical and social activity."[20] For Jean-Paul Sartre in the *Critique of Dialectical Reason*: "In the historical and social world . . . there *really* is dialectical reason; by transferring it into the 'natural' world, and forcibly inscribing it there, Engels stripped it of its rationality: there was no longer a dialectic which man produced by producing himself, and which, in turn, produced man; there was only a contingent law, of which nothing could be said except *it is so* and not otherwise."[21]

Pathway I: Classical Marxism and the Dialectics of Nature

"Western Marxism," in the sense referred to above, therefore provided little or no answer other than the dismissal of the question itself to the larger Lukács problem on the status of the dialectic of nature within Marxist thought. It is therefore necessary to turn to pathways out of this dilemma offered by those whose position on the dialectic of nature and society is more complex. Here we will briefly and schematically consider mainly two such pathways, those offered by classical Marxism (mainly Marx and Engels themselves) and the development of Marxist ecology.

Within classical Marxism the concept of the dialectic of nature stemmed largely from Engels's heroic, if not always successful, attempt, first in *Anti-Dühring* and then in *The Dialectics of Nature*, to extend the dialectical method beyond society to nature—in ways consistent with a materialist outlook and developments in nineteenth-century science. The extant evidence suggests that Marx was broadly supportive of the efforts on Engels's part and regarded it as part of their overall collaboration. Yet Marx's philosophical background was far deeper than that of Engels, and his treatment of nature and the dialectic in many ways more philosophically complex. This became evident with the publication of Marx's early philosophical writings.

Still, rather than searching in Marx's expanded corpus for solutions to the problem of a materialist dialectical method extending to both society and nature, Western Marxists insisted instead that Marx's greater philosophical sophistication had led him to adopt a social ontology cordoned off from nature and natural science—and all questions of philosophical materialism. A wedge was driven between Marx and Engels and between Marx and nature. Marx's early naturalism and humanism became humanism alone—and if Marx was faulted it was for giving way unduly to naturalism in his later writings. Materialism, like the dialectic, related only to society and was narrowed down to an abstract concept of economic production—abstract since removed from all natural conditions. Sebastiano Timpanaro, a lone voice against this trend, was thus inspired to open his book *On Materialism* with the ironic statement: "Perhaps the sole characteristic common to virtually all contemporary varieties of Western Marxism is their concern to defend themselves against the accusation of materialism."[22] In rejecting dialectical materialism, Western

Marxism rejected materialism (and with it nature) rather than the dialectic, attempting to find a way to define Marxism exclusively as a dialectic of social praxis.

Marx's materialism, as he developed it first in his earlier writings—in particular his dissertation on Epicurus, the *Economic and Philosophical Manuscripts* of 1844, the *Theses on Feuerbach*, *The Holy Family*, and *The German Ideology*—had the distinction that it sought to transcend the division between materialism and idealism by creating a new materialism embodying the active principle previously developed best by idealism, while retaining a materialist starting point and emphasis. From the materialist side he drew upon Epicurus and Feuerbach; on the idealist side from Hegel. It was the immanent dialectic he found in Epicurus that first allowed Marx to envision a materialism rooted in human sensuous activity. Epicurus, in insisting on the truth of the senses, while also emphasizing the role of the human mind in assessing the data of sense experience, laid the grounds for a sophisticated materialism that rejected God, teleology, determinism, and skepticism, all at the same time. Epicurus, Marx wrote, always sought to break through "the bonds of fate."[23] As Jean-Paul Sartre wrote in "Materialism and Revolution," "The first man who made a deliberate attempt to rid men of their fears and bonds, the first man who tried to abolish slavery within his domain, Epicurus, was a materialist."[24]

This gave rise in Marx's thought to what might be called a "natural praxis" underlying his social praxis. For Marx the materialist method was rooted in the senses, but the senses played an active, constitutive role, and were not simply passive instruments, reflective of a passive nature. "Human sensuousness," Marx argued in his dissertation on Epicurus (introducing a philosophical viewpoint that although emanating from Epicurus's physics was to be fundamental to his own materialist dialectic throughout his life), "is . . . embodied time, the existing reflection of the sensuous world in itself." Mere perception through the senses is only possible because it expresses an active and therefore changing relation to nature—and indeed of nature to itself. "In hearing nature hears itself, in smelling it smells itself, in seeing it sees itself."[25]

Epicurus had stated that "death is nothing to us"—referring to the fact that once there is no sensation there is nothing for us. Marx was to echo this in the *Economic and Philosophic Manuscripts* by saying: "*Nature*, taken abstractly, for itself, and rigidly separated from man, is *nothing* for man. . . . Nature as nature," as a mere abstraction devoid of sense, is "nothing proving itself to be nothing."[26]

For Marx all human activity has a basis in nature, is sensuous activity, which does not prevent the development of distinctly human species characteristics, that is, human social activity. The senses are nature touching, tasting, seeing, hearing, and smelling. The tools with which human beings seek to transform the natural world around them constitute the "inorganic body of man"—the social technology that extends the "natural technology" of the human organs and capacities. (Here Marx followed the ancient Greek notion of the bodily organs as tools or natural technology, reflected in the dual meaning of the Greek term "organon," which referred to both the organs and to tools.)[27] Labor and production constituted the active human transformation of nature, but also of human nature, the human relation to nature and of human beings themselves. The alienation of human beings from themselves and their production is also the alienation of human beings from nature, and the alienation of nature, since human beings are "a part of nature."[28]

The "first fact" to be established in analyzing human beings, Marx argued in *The German Ideology*, was their dependence on nature to meet their physical needs—and hence the necessity of production for human subsistence.[29] Human labor (and production), according to Marx in *Capital*, was a metabolic exchange between nature and society without which human beings could not exist and history could not develop. "The fact that man is an *embodied*, living, real, sentient, objective being with natural powers," Marx wrote, "means that he has *real, sensuous objects* as the objects of his being, or that he can only *express* his life in real, sensuous objects. . . . Hunger is a natural *need*; it requires, therefore, a *nature* outside itself, an *object* outside itself, in order to be satisfied and stilled. . . . A being which does not have its nature outside itself is not a *natural* being and does not share in the being of nature."[30]

At every point in his analysis, therefore, Marx insisted on the complex material relation between human beings and nature. The relation was a dialectical one in that it was an internal relation within a single totality. Rather than positing a dualistic relation between human beings and nature, he suggested that the two opposing poles existed radically separated from one another only insofar as alienation in the realm of appearance separated human beings from their essential human capacities as both natural and social beings—beings actively constituting nature's relation to itself through natural and social praxis. As Alfred Schmidt wrote, "The hidden nature speculation in Marx [holds that] . . . the different economic formations of society which have succeeded each other historically have been so many modes of nature's self-mediation. Sundered into two parts, man and material to be worked on, nature is always present to itself in this division. . . . Only in this way [as the self-mediation of nature through human activity] can we speak meaningfully of a 'dialectic of nature.'"[31] In his references to Marx's argument that labor-production was the metabolic relation between society and nature, Lukács had implicitly recognized this while never elaborating it, and nevertheless retaining the notion of nature as an "exalted" sphere not entirely subject to the dialectical method.

Andrew Feenberg has argued that "Marx's theory suggests a solution to some of the problems in Lukács. . . . Specifically, Marx's way of conceiving the relation of man to nature promises to overcome the split between history and nature which mars Lukács' theory."[32] Lukács had pointed to labor and production or the metabolic exchange between human beings and nature as the means of overcoming the subject-object duality. Human beings in Vico's terms can understand nature as history because they have made it. As more and more of nature is transformed by human history the externality of this part of nature disappears and nature becomes subsumed within production. Objectification becomes for human beings the objectification of themselves. Yet this requires an enormous extension of what constitutes the realm of production, of social praxis. Seen merely in terms of production, moreover, such a conception seems absurd. As Feenberg, quoting Marx, asks: "Under what conditions can 'man himself become *the* object?' Will not the realm of independent nature always transcend society, hence the human subject?"[33]

Marx's theory of social praxis rooted in production thus requires what will be referred to here as *natural praxis*—a much larger concept of human praxis that encompasses human activity as a whole, that is, the life of the senses. Here Marx's new materialism rests much more on the old materialism extending back to Epicurus. Feenberg suggests that Marx attempted to overcome these difficulties "by elaborating a remarkable new theory of sensation in which the senses 'become directly theoreticians in practice,' acting on their objects as does the worker on his raw materials. The senses, unlike labor, have traditionally been conceived by philosophy as a potentially universal mode of reception, relating to all possible (real) objects. The senses can therefore take over where actual labor leaves off, supporting the assertion of a universal identity of subject and object in nature."[34] This was the argument that Marx had found in Epicurus's "immanent dialectics" and that had been developed (not always as radically) up through Feuerbach.[35] Thus it was the latter who said: "Only sense and only sense perception give me something as *subject*." And "only sensuous beings act upon one another."[36]

This was a point of view—both materialistic and holistic—aimed at a complex totality. Indeed, in commenting on the materialist tradition in philosophy, which he associated in particular with Epicurus and his modern adherents, Hegel had gone so far as to concede that the aim of materialism was in his terms a dialectical one: "We must recognize in materialism the enthusiastic effort to transcend the dualism which postulates two different worlds as equally substantial and true, to nullify this tearing asunder of what is originally One."[37]

Marx's theory, according to Feenberg, provides a "meta-theoretical reconstruction of sense knowledge as a historically evolving dimension of human being. Marx argues that the object of sensation contains a wealth of meaning available only to the trained and socially developed sense organ."[39] Indeed, Marx insists, in Epicurean-materialist terms, but in ways that display a more active principle or a natural praxis, that "the *distinctive character* of each faculty is precisely its *characteristic* essence and thus also the characteristic mode of its objectification, of its *objectively real*, living *being*. It is therefore not only in thought, but through *all* the senses that man is affirmed in the objective world. . . . The cultivation of the five senses is the work of all previous history." Emancipation from the alienation of private property is for Marx also "the complete emancipation of the human qualities and senses."[39]

Underlying Marx's argument is the proposition, previously advocated most radically by Epicurus, that the senses (if rationally trained and developed) are the source of enlightenment and that the object of sense perception is real, and not a product of human consciousness. Treating John Locke as belonging to the materialist tradition in the line of descent from Epicurus, Marx stated that Locke in his *Essay on Human Understanding* had suggested "indirectly that there cannot be any philosophy at variance with the healthy human senses and reason based on them."[40] There is a necessary relation between the senses and the object of sense perception, which forms the basis for human knowledge. "Sense experience (*see* Feuerbach)," Marx wrote,

> must be the basis of all science. Science is only genuine science when it proceeds from sense experience, in the two forms of *sense perception* and *sensuous* need; i.e. only when it proceeds from nature. . . . Natural science will one day incorporate the science of man, just as the science of man will incorporate natural science; there will be a *single* science. . . . The first object for man—man himself—is nature, sense experience; and the particular sensuous human faculties, which can only find objective realization in *natural* objects, can only attain self-knowledge in the science of natural being.[41]

In Marx's notion of natural praxis, sense perception develops in history in accord with the development of human production. By actively and rationally sensing the world in wider and wider dimensions, human beings are able historically to experience the world as the objectification of their praxis—but only insofar as they are able to

jectively. Hence, in contradistinction to materialism, the *active* side was developed abstractly by idealism—which, of course, does not know real, sensuous activity as such." What Marx called "contemplative materialism," that is, materialism as it existed from Epicurus to Feuerbach, has the failing that it "does not comprehend sensuousness as practical human-sensuous activity."[47]

Marx never abandoned materialism or realism. Nature always existed to some extent independent of human beings and prior to human beings—though human beings and their relations were ultimately conceived as a part of nature within a complex set of internal relations. The senses were limited in the extent to which they could apprehend the world. But this constituted no insuperable obstacle to the understanding of nature since human sense perception and scientific inference based on it was capable of historical development. Moreover, materialism could be combined with the dialectical method—the way in which basic laws of nature (of matter and motion) were handled—in order to develop reasoned science based on sense experience. In a June 27, 1870, letter to his friend Dr. L. Kugelmann, Marx observed that Frederick "Lange [addressing Marx's *Capital*] is naïve enough to say that I 'move with rare freedom' in empirical matter. He hasn't the least idea that this 'free movement in matter' is nothing but a paraphrase for the *method* of dealing with matter—that is, the *dialectic method*."[48]

For Marx, like Hegel, the "true is the whole," and could therefore not be understood apart from its development, making *dialectical reason* necessary.[49] All was transitory, a passing away. Time itself was, as Epicurus first stated and Marx later repeated, the "accident of accidents." The only immutable reality, according to Epicurus (and later Marx), was "death the immortal"—all reality was time and process—there were no set positions in the world, nothing was static.[50] Contingency (symbolized by Epicurus's famous atomic swerve) was everywhere.

It is no mere coincidence that this notion of "death the immortal" that Marx took from Epicurus as a fundamental description of his own philosophy has recently been characterized by Ann Fairchild

Pomeroy in her *Marx and Whitehead* as "the only general statement made possible by Marx's dialectics" and as the point of convergence of Alfred North Whitehead's process philosophy with Marx.[51] It constitutes the recognition that reality is to be conceived as process or as internal relations—and for Marx especially, conceived *historically*—since "death the immortal" (what we now would call the arrow of time) is built into the "nature of things." As Marx wrote in relation to Epicurus: "The temporal character of things and their appearance to the senses are posited as intrinsically one."[52]

Marx's basic ontological scheme for understanding the world, as with Hegel, was one of internal relations. This is far removed from what Engels following Hegel had termed the "metaphysical" view, in which the world is a collection of things, isolated from one another, creating an understanding that is necessarily limited and partial—reductionist, determinist, and dualist.[53] According to the philosophy of internal relations, as Bertell Ollman put it in his *Dialectical Investigations*, "each part is viewed as incorporating in what it is all its relations with other parts up to and including everything that comes into the whole."[54] In Marx's case this was not a question simply of thought relating to itself as in the case of idealism, but represented the complex, changing, contingent, contradictory, and coevolutionary nature of the world itself—that is, the world and each "totality" within it was characterized by internal relations. The common "metaphysical" mistake of translating it into a world of separate things rigidly cut off from each other is therefore wrong. For Marx, each thing consists of the totality of its relations. Faced with a world of reality of this kind, the only rational way of "dealing with matter" was the dialectic method. As Ollman stated:

> Dialectics is an attempt to resolve this difficulty [of comprehending a world that is ever changing and interacting] by expanding our notion of anything to include, as aspects of what it is, both the process by which it has become that and the broader interactive context in which it is found.... Dialectics restructures our thinking about reality by replacing the common sense notion of "thing," as something that *has* a history and *has* external connec-

> tions with other things, with notions of "process," which *contains* its history and possible futures, and "relation," which *contains* as part of what it is its ties with other relations.[55]

Characteristic of Marx's thought was the tendency to see all of reality as historical—not just human society but the natural world itself. Natural history had to be studied along with social history—neither was to be viewed as passive; both were characterized by complex laws of change and contradiction. How Marx viewed natural history is most concretely evident in his broader ecological discussions—that is, in his considerations of evolution and of what he called the "metabolic relation" between human beings and nature embodied in production. In the 1850s and '60s, European and North American agriculture was threatened by the impoverishment of the soil due to the failure to replace the nutrients (nitrogen, phosphorous, and potassium) removed from it and shipped to the cities in the form of food and fiber, where these nutrients ended up as waste polluting the cities. Following Justus von Liebig, the great German chemist, Marx argued that this robbing of the soil constituted a crisis for agriculture. But Marx went further than any other thinker of his time in terms of treating this as a metabolic rift in the human relation to nature based on capitalist relations of production. This theory emphasized the "nature-imposed" conditions of human production outside of society itself, the historical (geological) evolution of the soil, and its complex interaction with the evolution of human society and production.[56]

Marx specifically employed the word *metabolism*, as he stated in his *Notes on Adolph Wagner*, to capture "the 'natural' process of production as the material exchange between man and nature"—in accordance with a complex, dialectical, coevolutionary scheme.[57] Human labor was defined as the metabolic relation exchange between human society and nature, within a context of material permanence. Marx quoted Epicurus (via Lucretius) to emphasize the materialist axiom that nothing in the material world comes from nothing or is in the end reduced to nothing.[58] Rather than standing in a pure theoretical relation to the world, human beings, Marx stressed in *Notes on Adolph Wagner*, apprehended the world through their senses activated in the context of their real practical activity in meeting their essential needs. Communist society was seen as the historical resolution of the contradiction between nature and society, as well as of the class contradictions internal to society, "by organizing the human metabolism with nature in a rational way" via unalienated human production.[59]

At every point Marx's dialectic sought to break the bonds of fate associated with the notion of an objective world separate from and dominating human beings—along with the alienation of the external world and attempts to dominate it on the part of human beings. His materialist dialectic sought to embody both genuine humanism and genuine naturalism, while also adopting a realistic conception of the world beyond humanity—in relation to which human beings evolved in complex, coevolutionary ways. The material world was experienced by human beings as a sensuous reality and hence the liberation of the world required the emancipation of the senses and of the sensuous relation to nature along with the emancipation of society. But this in turn required as its basis the rational regulation of the metabolic relation between nature and society.

Marx's explicit treatment of nature in terms of the dialectical method focused on those realms in which nature was objectified through actual human praxis—both natural and social. It was Engels who tried to provide the philosophical extension of this in terms of science and nature more broadly—treating nature as a system of internal relations characterized by its own dialectical "laws." Engels famously wrote in *Anti-Dühring*: "Nature is the proof of dialectics."[60] This deceptively simple statement, the truth of which was taken for granted within Marxism for the next few decades—but later emerged as perhaps the single greatest point of contention within Marxist theory—stood for the thesis that there existed a dialectic of nature side by side with (or even anterior to) the dialectic of society.

Just as Marx and Engels viewed the materialist conception of history as inseparably bound to the materialist conception of nature, they viewed the dialectics of society as inseparably bound to the dialectics of nature. Both society and nature had to be viewed materialistically and dialectically, since they embodied a dialectical motion in their very being. Without materialism dialectics led to dialectical idealism, of which Hegelianism was the highest form. Without dialectics, materialism led to abstract empiricism and mechanism, as embodied in British political economy and in the so-called scientific materialism then prevalent in Germany.

The first systematic case for a distinct dialectics of nature was made by Engels while Marx was still alive in the late 1870s in *Anti-Dühring*—a work he read to Marx in full prior to its publication and to which Marx contributed a chapter. This was followed by Engels's *Dialectics of Nature*—a voluminous, incomplete manuscript, consisting mainly of fragments, parts of which were written between 1873 and 1882 (thus overlapping in time with *Anti-Dühring*). *The Dialectics of Nature* was not published in its entirety until 1925, three decades after Engels's death. His analysis is often seen in terms of the three laws of dialectics he outlined in these works: (1) "the transformation of quantity into quality and vice versa"; (2) "the law of the interpenetration of opposites"; and (3) "the law of the negation of the negation."[61] Although these "laws," taken from Hegel, constituted pathways to dialectical thinking, their formalization as abstract laws tended to lead to an approach, when turned into a rigid schema by later thinkers, that militated against dialectic. Hence Engels's analysis has been widely criticized on this basis.

Rejection of Engels's position on the nature dialectics was, as we have seen, crucial to the genesis of Lukács's dialectic and "Western Marxism." There is no doubt that "dialectical materialism" as it developed in the Soviet Union in the 1930s frequently turned into a mechanical, and extremely undialectical, way of thinking that all too often gave way to crude positivism. Nevertheless, it is much harder to criticize Engels of such a failing—nor, as we shall see, is it possible to place such criticisms at the door of all those who, inspired by his example, tried to extend the materialist dialectic, applying it in the natural and physical sciences. Indeed, Marx himself, as we have noted, was supportive of Engels's efforts.

Engels's outlook has been criticized most extensively in terms of his rendition of physics, which was still dominated by the mechanistic Newtonian worldview. By the time *Dialectics of Nature* was actually published, several decades after being written, physics had gone through several revolutions making his work seem dated—although his dialectical view had caused him to question mechanism at every point. Yet it was in the biological sphere that Engels's dialectical naturalism was most developed. Here he drew heavily upon the Darwinian revolution and explored lines of coevolution. Applying a dialectical materialist method to the theory of evolution, Engels wrote:

> *Hard and fast lines* are incompatible with the theory of evolution. Even the border-line between vertebrates and invertebrates is now no longer rigid, just as little is that between fishes and amphibians, while that between birds and reptiles dwindles more and more every day.... Dialectics, which likewise knows no *hard and fast lines*, no unconditional, universally valid "either-or" and which bridges the fixed metaphysical differences, and besides "either-or" recognizes also in the right place "both this—and that" and reconciles the opposites, is the sole method of thought appropriate to the highest degree to this stage [in the development of science].[62]

In this view, then, the fluid, changing, interpenetrating forms that clearly characterize nature—as it is given to us as natural beings through our sense experience, both sense perception and sensuous need—demand a method of reasoning (the dialectic) capable of dealing with this. Under the influence of Hegel, Darwin, and the work of the German scientist (electrophysiologist) Emil Du Bois-Reymond, Engels developed a view of the evolution and the origins of life as a *process of emergence*. The idea of emergence has been traced back to Epicurus and was clearly a part of Marx's conception of nature as well. As Z. A. Jordan explained in *The Evolution of Dialectical Materialism*:

> The doctrine of emergent evolution is a hypothesis concerned with novelties which are not mere combinations of their component elements and which are supposed to occur in the course of natural development, such as the emergence of life or of mind.... The fact cannot be denied that the central idea of emergent evolution is to be found in *Anti-Dühring* and *Dialectics of Nature*. . . . In the case of Engels, the emphasis is clearly upon the ontological conception, upon the gradual emergence of the atomic, chemical, and biological level, the latter with its numerous emergent transitions to higher and higher forms of life.[63]

Engels's dialectics of nature thus can be seen as a genetic-historical approach rooted in a philosophy of emergence. The natural world is conceived as an interconnected fabric with an arrow of time (since involving irreversible transformations) that can only be envisioned in terms of qualitative (beyond merely quantitative) transformations, the interpenetration of opposites (that is, internal relations that encompass mutually determining processes), and the negation of the negation (emergence). Our understanding of this world for Engels is dialectical and made possible by human praxis, including the methods of experimentation introduced by science.

The brilliance of Engels's dialectical conception of nature, when he turns specifically to nature-society relations, can be seen in his understanding of the rupture in the evolutionary process and the ecological disaster resulting from the alienated society of capitalism, which goes against all "laws" of natural reproduction and sustainability. Human history, according to Engels, continually comes up against ecological problems that represent contradictions in the human relation to nature—contradictions in nature introduced by society and undermining its own natural conditions. These contradictions can only be transcended by relating to nature rationally through nature's laws, and thus organizing production accordingly—a possibility not open to capitalist society. Warning of the ecological consequences of the alienation of nature in history, Engels writes in *The Dialectics of Nature*:

> Let us not, however, flatter ourselves overmuch on account of our human victories over nature. For each such victory nature takes its revenge on us. Each victory, it is true, in the first place brings about the results we expected, but in the second and third places it has quite different, unforeseen effects which only too often cancel the first. . . . Thus at every step we are reminded that we by no means rule over nature like a conqueror over a foreign people, like someone standing outside nature—but that we, with flesh, blood and brain, belong to nature, and exist in its midst, and that all our mastery of it consists in the fact that we have the advantage over all other creatures of being able to learn its laws and apply them correctly.[64]

Marx and Engels strove to comprehend the dialectic of nature and society, in which natural relations—not simply social relations—were taken seriously as embodying necessity and contradiction. As a result, they were able to perceive in ways far more penetrating than their contemporaries a dialectic of ecology involving new historical contradictions. This is no less true of Marx than Engels. As Marx, analyzing the alienation of nature, was to observe in the *Grundrisse:*

> It is not the *unity* of living and active humanity with the natural, inorganic conditions of their metabolic exchange with nature, and hence their appropriation of nature, which requires explanation or is the result of a historic process, but rather the *separation* between these inorganic conditions of human existence and this active existence, a separation which is completely posited only in the relation of wage labor and capital.[65]

This separation both in material reality and human consciousness was, as Marx and Engels argued, to have disastrous ecological consequences manifested in what Marx called the "irreparable rift" between nature and society.[66]

Pathway II: The Development of Marxist Ecology

Given the direction of Marx and Engels's own thought, which pointed to the disastrous material consequences arising from the alienation of nature along with the alienation of humanity under capitalism, it is not surprising that later advances in the development of the conception of the dialectic of nature were inseparable from the emergence of a Marxist ecological worldview. As recent investigations both into Marx's ecology and into subsequent Marxist and materialist thought have shown, ecological insights of the kind depicted here were not only intrinsic to Marx and Engels's worldview but were deeply rooted in the materialist tradition going back to the ancient Greeks. Moreover, they extended beyond Marx and Engels to some of their early followers. These included figures such as August Bebel, William Morris, Karl Kautsky, Rosa Luxemburg, V. I. Lenin, and Nikolai Bukharin. In the Soviet Union in the 1920s wider ecological analyses blossomed in the work of such scientists as V. I. Vernadsky, N. I. Vavilov, Alexander Oparin, and Boris Hessen.[67]

In contrast, Western Marxism beginning with Lukács's *History and Class Consciousness*, as we have seen, became extremely critical of application of the methods of natural science to the realm of human history and society, while also rejecting the application of the dialectical method to science and nature. Hence an enormous theoretical firewall was erected separating the two spheres and limiting any ecological insights. The Frankfurt School's considerations of nature and the domination of nature in the work of Max Horkheimer and Theodore Adorno tended to attribute these problems to the Enlightenment and its modern positivistic heir—and to be concerned much more with the domination of human nature than nature as a whole.[68] Its critique was thus carried out without any genuine ecological knowledge. In the Soviet Union, too, the pioneering role of scientists in this area largely vanished in the 1930s with the purges, which took the lives of such important figures as Bukharin, Vavilov, and Hessen.

Nevertheless, a powerful engagement with materialist science and the dialectic of nature developed in Britain in the 1930s and '40s in the work of such important figures, most of them scientists, as: J. B. S. Haldane, Hyman Levy, Lancelot Hogben, J. D. Bernal, Joseph Needham, Benjamin Farrington, and Christopher Caudwell. These thinkers self-consciously united a materialist philosophy with roots extending as far back as Epicurus with modern science, dialectical conceptions, and Marxist revolutionary praxis. They all struggled to overcome conceptions of an unbridgeable gulf between nature and society and furthered ecological notions. One of the most important figures in this period was the mathematician Levy, whose early system theory as developed in his 1933 *The Universe of Science* helped inspire British ecologist Arthur Tansley (a Fabian-style socialist and former student of Marx's friend, the biologist E. Ray Lankester) to introduce the concept of "ecosystem" in 1935.[69]

Levy's 1938 *A Philosophy for a Modern Man* advanced a philosophy of internal relations intended to explain the significance of dialectical conceptions for science. In analyzing changes in forest succession as illustrative of dialectical processes, Levy wrote: "Vegetation . . . transforms the environment, and the environment in its turn the vegetation. It is almost like a society of human beings. We may expect, therefore, to find in the growth of vegetation dialectical changes manifesting themselves over and over again, whereby the whole isolate passes from phase to phase."[70] Levy's analysis of dialectical interrelations of nature and society led him to point to ecological rifts resulting from capitalism's intrinsic disregard for this connectedness. Thus the largely unforeseen effects of the steam engine on the coal region of South Wales; the consequences of the unthinking introduction of a pair of rabbits into Australia without any recognition of the ecological consequences; and how red deer could in one context be diminished by the destruction of their forest environment and in another context proliferate and destroy their environment (with both cases attributable to human interference related to profit-making) were all points emphasized in his analysis. He also alluded to the unequal exchange affecting poor countries as the rich countries rapaciously extracted their limited raw materials.[71]

A materialist dialectic therefore could be applied to elucidate both nature-nature and people-nature (or people-nature-people) relations/metabolisms.[72] All material relations of social change between people necessarily involved the changing of natural relations. A fundamental "phase change" in nature or society is a manifestation of "irreversible historical change" involving coevolutionary processes and the changing of conditions. Basic to a materialist dialectical conception, according to Levy, is the notion that the universe not only exists but is constantly changing. And with this arises changes in language (thinking, science) and consciousness. All history was interactive in complex sequences: "We change and alter under the impact of the environment; the environment is changed and altered under our impact. We are its environment." It is characterized by emergence, beginning with "the emergence of living properties in matter" as "a very distinctive step in the natural process." There is an arrow of time at work in the world: "The traffic of the universe flows in one direction only." Yet the underlying contingency of relations prevents a deterministic outlook.[73]

The ecological conceptions emanating from dialectical Marxist scientists of this period resulted in what has been the leading materialist theory of the origins of life (overcoming earlier laboratory-based refutations of spontaneous generation). Thus the dialectical-materialist Haldane-Oparin hypothesis on the origins of life (based on Vernadsky's earlier work on the biosphere) argued that the conditions that had allowed for the origins of life were altered by life itself, which created an atmosphere rich in reactive oxygen.[74] This act of life itself so altered the base conditions of the biosphere that, in the words of Rachel Carson, "this single extraordinary act of spontaneous generation could not be repeated."[75] Commenting on this, J. D. Bernal was to observe in his *Origins of Life*: "The great liberation of the human mind, of the realization, first stressed by Vico and then put into practice by Marx and his followers that *man makes himself*, will now be enlarged with the essential philosophical content of the new knowledge of the origin of life and the realization of its self-creative character."[76]

This ecological conception of the world of life was itself in many ways reflexive and self-creating, arising out of materialist dialectics, and would give birth to many of the most powerful insights associated with the development of modern ecology. Moreover, it can now be argued more generally, extending Engels's early proposition with respect to nature, that "ecology is the proof of dialectics." No other form of thinking about nature and society has so conclusively shown the importance of irreversible change, contingency, coevolution, and contradiction.

As two noted ecological scientists, David Keller and Frank Golley, have observed: "Ecology is captivating due to the sheer comprehensiveness of its scope and complexity of its subject matter; ecology addresses everything from the genetics, physiology and ethology of animals (including humans) to watersheds, the atmosphere, geologic processes, and influences of solar radiation and meteor impacts—in short, the totality of nature. . . . An ecological worldview emphasizes interaction and connectedness."[77] With the growth of ecology dialectical forms of thinking have necessarily advanced, giving new force to a dialectics of nature that at first was limited and at times distorted by the dominant mechanistic conceptions of science. Mechanism, as Whitehead had said, made nature "a dull affair, soundless, scentless, colourless; merely the hurrying of material, endlessly, meaninglessly."[78] In contrast, the dialectical view emerging from ecology is anything but lifeless or mechanical; it has generated a view of nature no longer shorn of life, interconnection, and sensuous reality—no longer deterministic—but a world of coevolution, contradiction, and crisis.

Dialectical thinking in ecology was more a question of necessity than choice. Neither mechanism nor vitalism, neither determinism nor teleology, were adequate in the ecological realm—a realm that demanded an understanding that was at once genetic and relational. The inability to conceive of an ecology that left humans out meant that ecology was from the first at once natural and social.

In the post-Second World War era, the center for the study of the evolutionary sciences shifted from Britain to the United States, which also became the center for work in dialectical biology for those scientists influenced by dialectical materialism. In particular the work of Richard Levins, Richard Lewontin, and Stephen Jay Gould, along with many other scientific colleagues with materialist and dialectical orientations, has shown the power of such forms of thinking when applied to fields such as genetics, paleontology, evolutionary biology, and ecology itself.[79] Lacking the epistemological surety of the dialectic of social praxis, ecological science is nonetheless able to make up for this in part through a genetic or historical form of analysis, together with experiments that in isolating phenomena for study are nonetheless non-reductionistically designed to understand the larger processes of which they are a part. In this analysis scientific inference is directed at uncovering complexity. Change is to be considered constant, opposing forces create change, and life rather than simply responding to its environment also generates its environment—the organism is conceived as both subject and object of evolution.

Ecological dialectics in this complex variegated sense seem a far cry from Lukács's notion of a "merely objective dialectics of nature" in which human beings, entirely removed from praxis in this realm, simply look on from the outside at a distance—able only to derive very general external "exalted" laws regarding a passive nature. Instead the changing conditions of human existence and of natural life as a whole are teaching us that human beings as living, sensuous beings are part of the ecological world—that the biosphere is constitutive of our own existence even as we transform it through our actions. Scientific inference rooted in developing praxis—both natural and social—is providing the intellectual basis for a reflexive ecological science that constantly seeks to transcend the boundaries between natural and social science—and in ways that are dialectical rather than reductionist. This is now altering in many ways our view of the social world—expanding its scope in a universe of internal relations, forcing broader dialectical conceptions on society in relation to an evolving ecological crisis. We can see a new revolutionizing of Marxist materialism so as to take into account these wider ecological conceptions—as well as an attempt to reconnect this to the deep materialist roots of historical materialism.[80]

The historical basis of this new dialectics of ecology has been the development of human production, of what Marx called the metabolic relation between human beings and nature, which has made ecology the most vital of the sciences since our changing relation to the environment (which humankind has too long viewed as a mere external relation rather than an internal relation) is threatening to undermine both the conditions of production and of those of life itself. Hence, it is an increasingly dialectical conception forced on humanity as a result of its own myopic intrusion into the universe of nature—a result of alienated social praxis. To recapture the necessary metabolic conditions of the society-nature interaction what is needed is not simply a new *social praxis*, but a revived *natural praxis*—a reappropriation and emancipation of the human senses and human sensuousness in relation to nature.

This is what revolutionary materialist dialectics has always been about. From the beginning materialism drew its impetus from its revolutionary character—revolutionary in relation to all aspects of human existence. "Epicurus," Sartre wrote, "reduced death to a fact by removing the moral aspect it acquired from the fiction of seats of judgment in the nether world. . . . He did not dare do away with the gods, but reduced them to a mere divine *species*, unrelated to us; he removed their power of self-creation and showed that they were the products of the play of atoms, just as we were."[81] But a merely contemplative materialism, which relies on mechanism, can be as destructive as idealism. A materialism unrelated to praxis and divorced from dialectical conceptions is, as Sartre emphasized, a mere mechanical myth and can itself be a tool of domination. What is needed is an expansion of our knowledge of the universe of praxis—or, to adopt Sartre's term, the universe of concrete "totalizations." Hence, what is required is a more unified understanding of the dialectics of nature and society—recognizing that the dialectical method when applied to nature is our way of handling the complexity of a constantly changing nature. The development of ecology as a unifying science is pointing irrefutably to the validity of Marx's original hypothesis that in the end there will only be "a single science" covering a complex reality in which the dialectic of change subverts all reductionisms.

12. Dialectical Materialism and Nature

What is the nature of Nature? Although for the most part scholars in the environmental social sciences do not directly examine the natural environment or explicitly struggle with this question, their (often implicit) assumptions about the natural world can have a substantial influence on their analyses of human-environment interactions. There are two common conceptualizations of the natural world. One, especially prominent among economists, views nature as fundamentally mechanical and maintains an optimism about the ability of human societies to tinker with its machinery so as to improve its utility (for those in power, at least). Another, common in environmentalist circles, sees nature, when unmolested by industrial society, as existing in a grand harmonious order that human beings must be in sync with if we are to overcome environmental crises. Here our aim is to present a different conception of nature, developed largely by scientists in the Marxist tradition, which is fundamentally materialist, though not mechanical, and concerned with interconnections and emergent order in nature, though not functionalist. Our main concern is with the emergence of apparent order and the nature of change and how these relate to the human-environment relationship, particularly in the current era of global environmental crisis.

Over the past half-century, the human relationship with the environment has become an ever-more prominent topic in public discourse. The first Earth Day in 1970 helped to make environmental degradation and pollution major concerns. The Club of Rome and the 1970s oil crisis placed scarcity and natural limits at the forefront of social concerns. Public debates raged over logging, mining, and drilling on public land. Social movements pushed forward concerns with environmental racism and environmental justice, nuclear energy, and the poisoning of ecosystems. For some, direct action—through various forms of sabotage, such as spiking trees, tree sitting, and road blockades—became the primary means to confront the powerful forces that organize social production.

Corporations shifted their marketing campaigns to present their products as healthy, environmentally friendly items for eco-conscious consumers. Social science scholars slowly came to focus on the environment as an important realm of study, noting human dependence on nature. Sociologists such as William R. Catton Jr. and Riley Dunlap, as well as Allan Schnaiberg, helped raise awareness in the social sciences of the role the natural environment plays in maintaining societies.[1] However, for the most part, the environment remains peripheral to the thinking of most social scientists, and many, particularly in economics, are actively hostile to the notion that environmental crises threaten the sustainability of societies. In social sciences, as well as in various other intellectual and popular communities, nature takes on either an ideal form, existing as a harmonious order separate from society, or a realm that provides resources, waiting to be molded and operated at human convenience. The tension between idealized and mechanistic conceptions of nature has persisted for thousands of years, shaping philosophical discourse and social understandings of the world.[2]

Appropriately, today much of the social science focus—among those concerned with ecological issues—is on the intersection of human society and nature, especially in regard to issues of production. Too often, however, nature remains in the background, as either a passive, harmonious realm "out there" beyond the bounds of urban society or as the source of "free goods" that fuels the engines of industrial society. Little time is spent understanding natural processes and patterns: how they operate on their own, how historical social systems interact with nature, how nature influences social conditions, and how natural processes are transformed by social interactions. The measure of nature remains bound by our assumptions about how it operates and what purpose (if any) it serves. Our understanding of the human-environment relationship, the conditions of nature, and the direction of society is affected by these conceptualizations. In contrast to the economistic and the idealized approaches to nature, a dialectical materialist position offers a dynamic position for grappling with the complexities of the natural world and for assessing the environmental conditions on which we depend.

Economism and "Green" Capitalism: Nature as an Input

In an era when capitalism dominates the world economy and is assumed to be the only political-economic option available, it should come as no surprise that an economistic understanding of human-environment interactions is highly prevalent. Of course, economistic approaches are not unified, given the wide range of economists' interests and variation in the extent that they directly address the environment. But economists are connected by both their mechanical view of nature and optimism that human society can surmount any natural barriers that exist through technological innovation. For them, economics is the measure of the world, in all of its aspects. Nature, if it is considered at all, is seen as a problem, an obstacle to overcome. It remains a world of Newton's clock, mechanical in its organization, malleable before our ingenuity. Proponents of economic modernization, ecological modernization, and green capitalism adhere to the position that the ongoing development of the capitalist economy, often simply referred to as "modernization," will provide the means for addressing and correcting environmental problems.

Although the Club of Rome's report was not without shortcomings, it did highlight that an economy driven by the ceaseless accumulation of capital, through the endless expansion of production and consumption, exists in conflict with a finite world.[3] Furthermore, scientists noted that an economic system based on constant growth generated ecological scarcities and environmental degradation that could not be reversed within human time frames.[4] The short-term focus of economists on profits conflicted with the long-term health of the environment. In response to the concerns being raised by the environmental movement in the 1970s, orthodox economists denied "limits to growth" by arguing that so long as technological innovation continues and substitutes exist for natural resources, no immediate ecological concern existed.[5] In opposition to the findings of environmental science, economists assumed that the conditions of the environment were effectively irrelevant to society. Characteristically, nature was seen as simply a reserve of resources, waiting to be used in the production of commodities for the market.

Although the degree to which environmental concerns occupy public debates and interests has varied, often related to historical events and economic fluctuations, the issue persists as a central concern. The broadening and diversification of the environmental movement to include concerns from the preservation of wilderness to urban pollution and public health has helped to make it an ongoing part of social discourse.[6] At the same time, the range and scale of problems—global climate change, loss of biodiversity, deforestation, the accumulation of radioactive wastes, increasing levels of toxins throughout ecosystems and in our food, the contamination of water, overfishing, and desertification—continue to expand, making the ecological crisis more than a threat that exists in the distant future.[7] All these events have forced economists, corporations, and social scientists to address the environment.

Environmental economists from the neoclassical tradition acknowledge that economic development has generated environmental problems but argue that further economic development can solve these problems rather than add to them. The environment is seen as a luxury good, subject to public demand through the market. Gene Grossman and Alan Krueger, both economists, contend that during the early stages of capitalist development environmental impacts increase, but as the affluence within these societies rises the value the public places on the environment—including wildlife, wilderness, clean air, and clean water—will increase.[8] The public desire for environmental quality, in large part expressed as consumer demand for "green" products and services, will, economists expect, place pressure on the government and businesses to invest in eco-friendly technologies and commodities. Businesses and citizens will be able to afford these "green" commodities due to the wealth generated by economic expansion. Thus environmental economists tend to argue that if the market is allowed to operate without dramatic interference, ongoing economic development will lead to a leveling and eventual decline in the environmental impact of societies. This inverted U-shaped curve, representing the relationship between economic development and environmental impacts, is known as the environmental Kuznets curve and follows the same formulation as Simon Kuznets's discussion of the relationship between economic growth and income inequality.[9] While materialistic, in an economic reductionist sense, nature remains in the background as an entity taken for granted, as a realm to be manipulated as needed. The thinking in economic circles too often goes as follows: The market determines any importance the material world has, so that a "problem" that has no substantial and immediate consequences for economic development is no problem at all. The processes and cycles of nature are not a concern, so long as the environment remains as a resource for production.

Ecological modernization, drawing in part on the work of environmental economists, proposes that the only "possible way *out* of the ecological crisis is by going further *into* the process of modernization."[10] The particular form of modernization embraced is not a radical break with the current economic system and institutions. Rather, the forces of modernization that are believed to lead human society from its past of environmental degradation and exploitation to environmental sustainability are the institutions of modernity, including the market, industrialism, and technology.[11] Ecological modernization theorists, such as Arthur Mol, do not view environmental degradation as an inherent characteristic of capitalist development. They remain zealous socio-techno-optimists, believing that the forces of modernization will lead to the dematerialization of society and the decoupling of the economy from energy and material consumption, allowing human society, under capitalism, to transcend the environmental crisis.[12] Some proponents of this position, such as Charles Leadbeater, argue that as the economy develops, it is producing a weightless society that is more knowledge based and less reliant on natural resources.[13]

Ecological modernization theorists contend that one of the primary forces driving environmental sustainability within the modern economy is rationality. By allowing the market to develop to its full potential, a new, modern "ecological rationality" will emerge and percolate throughout all institutions of "advanced" societies.[14] As a result, a new focus will be placed on the necessity of maintaining the resources and ecosystem functions upon which societies depend. Ecological rationality will replace the pure economic rationality that prevailed in the early stages of modernization. Ecological modernization theorists are proposing a more fine-tuned economic rationality, not an ecological counterforce to economic hegemony. They assume that more explicitly recognizing the inputs of the environment to the economy will lead to a more ecologically and economically rational system. New technologies will be developed to resolve environmental problems and to enhance the environmental sustainability of society through improvements in efficiency. It is assumed that a green rationality will provide the knowledge of how to properly manipulate nature to meet the economic needs of production within the ongoing development of capitalism.

General Electric Company presented an example of the types of transformations that ecological modernization theorists posit when it announced that it was going to invest over a billion dollars in "greener technologies" between 2005 and 2010. GE's objective is to improve its energy efficiency as a company and to expand its environmental products for the market. Its public-relations spin frames the environment as a problem to be solved through "ecomagination." The General Electric Company proposes that green products are a valuable product line along with its other commodities, and it expects to produce around $20 billion in revenues from environmental products alone.[15] Thus the drive to accumulate capital on a larger scale supposedly embraces an "eco-consciousness." Environmental problems, then, become a source of marketing to expand profit.

Within the ecological modernization perspective, nature remains undertheorized. It is a realm of material input for the economy and society in general. Although environmental degradation is recognized, it is merely a socio-technical challenge, given that further development of the economy and social institutions will resolve the situation. The ecological modernization approach involves the "ecologization of economy" and the "economization of ecology" within the current economic system.[16] The former refers to organizational changes in both the production and consumption processes of society, making them account for prevailing environmental interests. The latter entails the extension of economic valuation to the environment and any natural services that it produces. Natural cycles and processes, as well as ecosystems themselves, remain outside the discussion.

Ecological modernization theory is, at base, a functionalist theory in that it does not see the emergence of ecological rationality as coming primarily from social conflict but rather from ecological enlightenment within the key institutions in societies.[17] Ecological modernization theorists contend, then, that radical ecological reform does not require radical social reform—that is, the institutions of capitalist modernity can avert a global environmental crisis without a fundamental restructuring of the social order. Instead, they are focused on the continuity of the social order, with gradual change in its operations. Social production is simply a machine that interacts with the environment. Humans, in their productive apparatus and in their interactions with the environment, simply need to tinker with operations to tweak any dysfunction back into order. For them, nature will continue to exist for our rational exploits, once we overcome its barriers through our ingenuity.

Embracing the optimism of ecological modernization and the workings of the capitalist economic system, proponents of green capitalism, such as Paul Hawken, propose that the capitalist economy can and should be restructured along environmentally sustainable lines. Hawken argues that if the value of nature were properly accounted for, capitalism would develop in an ecologically benign direction.[18] Thus ecological goods and services are not currently properly accounted for by the market, so nature needs to have a rational price structure applied to it. The environment is then broken down into various commodities, and through an analysis of their contribution to market value a price is assigned to them. Once nature is fully commodified, the operation of the market can take care of the environment. For instance, green capitalism proponents aim to establish whether any particular stand of trees has more value for society in terms of its recreational potential, habitat, and ecosystem services or as a source of timber and the profit that the sale of this commodity generates on the market.

Green capitalists, such as Hawken, argue that achieving sustainability is simply a matter of balancing the accounting books and changing the ethics held by the people directing corporations.[19] He asserts that business exists to service people, not simply to make money. Thus if a change in ethics takes place, capitalism can be directed down the path of sustainable development. Advocates of green capitalism stress that through innovative technological development and appropriate reformist government policy, the economy can be dematerialized, reducing the throughput of raw materials and energy that the system requires.[20] When this is done, they contend, the continued growth of the economy, on whatever scale, poses no threat to the natural world.

Like other variations of economism, nature remains a realm of inputs for the continued operation of an expanding economy. Its degradation and natural cycles only matter to the extent that they serve or interrupt the functioning of the economy. Nature presents obstacles that must be overcome, problems to be solved. And it is assumed that the solution to the "nature problem" will be produced by the ongoing development of the market and an advance of "green ethics." Any real attempt to fundamentally transform the social system to address the ecological crisis is not necessary.

The central problem with this perspective is that the reproduction of the environment does not act in accord with "the rules of the market."[21] A forest cannot be reproduced at the same pace that it can be cut and transformed into commodities. Furthermore, we cannot assume that once an ecosystem has been drastically altered, such as when a forest is cut down, it will simply return to the previous state. And the unity of social production and nature becomes mystified in the operation of capital by "the increasing domination of exchange value over use value."[22] The contribution of nature to the production of use value and maintenance of labor disappears within the capitalist framework. Labor time becomes the measure of value under capitalism, as nature becomes a mere object of labor. The alienation of workers and nature, in a competitive, profit-driven system, increases the exploitation of nature as the natural world becomes ever more organized for the capitalist economic system that requires escalating throughputs for production, given that it is inherently expansionary and continually reproduces itself on a larger scale.[23]

Advocates of green capitalism have grafted an "eco-veneer" on an economic system that is driven by the accumulation of capital. It would be wise therefore to reflect upon how embedded the exploitation of nature is in the operations of the capitalist system. Capitalism freely appropriates nature, as it organizes the environment and labor for the production of commodities for sale on the market. Given the global operations of capital and its short-term focus on profit, which excludes any serious consideration of the environment, there is no means within capital's operations to stop the ruin of ecosystems, short of global collapse.[24]

It is worth noting that the expectations of ecological modernization theorists and green capitalism proponents about substantial ecological reform in modern societies have not to date been confirmed. In addition to the various logical and methodological flaws that have often been associated with this tradition, empirical evidence supports the conclusion that the capitalist modernization project leads to environmental degradation, particularly at the global scale.[25] Although some indicators of local environmental quality (for example, air and water pollution) show improvements in some developed nations, the impact of societies on the global environment—in areas such as greenhouse gas emissions and resource consumption—appears consistently to escalate as the modernization project advances.[26] Thus the economistic conceptualization of nature does not appear to be conducive to environmental sustainability.

Economistic approaches to the environment perpetuate a reductionistic understanding of nature. The natural world simply exists as an input, in the background of their considerations, and as a realm to be managed to meet the needs of business in the pursuit of profit. Any environmental problems created by society can simply be fixed through technological ingenuity, as the economy surmounts any external obstacles to its functioning. Although materialist to a degree, economistic approaches remain mechanistic in their orientation to nature, disregarding the dynamic processes of the natural world. Their stated goal is simply to bring the economy and ecology into accord, where capitalism continues to operate. The earth continues to be converted into a variety of commodities. The market is the measure of all things.

The Balance of Nature: Idealized Harmony

Idealistic conceptions of the world—its meaning, its organization, and its purpose—have long been part of social thought. The specific character of these conceptions is often in reaction to prevailing material conditions in the physical world, sometimes including a longing for a return to some previous idealized state. Within ecological thought, deep ecology and the Gaia hypothesis are representatives of this outlook. Like any other general perspective, deep ecology includes a diversity of opinions, ranging from humans being seen as a virus attacking the earth to humans having the potential to live in a natural harmony with the environment. Nonetheless, a unifying theme for this perspective is the conception that if industrial civilization were removed, the world could return to its natural state, where a balance of nature exists. This perspective tends to idealize traditional societies and indigenous people as living in a harmonious state with nature prior to the intrusion of the "modern" world. The notion of an ordered world is an old theme, found both in natural theology and in mechanistic depictions of the world. Deep ecologists reject mechanistic accounts of nature. They also scuttle the hierarchy that was central to natural theology, by displacing humans from the position just below God. Instead, they insist upon an ecocentric conception, where humans are only one of the many species inhabiting the earth and deserving of no special privilege. Ideal Nature is assumed to be a place of harmony. The real world is measured against this ideal state.

In 1973, Arne Naess highlighted that there were two currents within environmental thought. One was a "shallow ecology" primarily concerned with fighting pollution and resource depletion. The other, deep ecology, included the objectives of shallow ecology but also entailed a shift in thought, where nature is seen and defined not as it relates to human interests but from its own position. A new point of view was required: ecocentrism, as opposed to anthropocentrism.[27] The deep ecology position attempts to shift social perception away from the economistic understanding of the world. Nature is seen as having intrinsic worth rather than as simply being a resource for humans. Deep ecologists insist that the social forces that harm the environment must cease in order to preserve life in all of its forms and to seek a world of harmony. Industrial civilization is deemed to be the primary enemy.[28] To transcend this imbalance, a revolution in values and thoughts is needed. Thus much of deep ecology focuses on establishing its philosophical moorings via Buddhism, strains of Christianity, Rachel Carson, Aldo Leopold, Henry David Thoreau, John Muir, Charles Darwin, and Gary Snyder. Drawing upon this smorgasbord of social thought, an ecocentric paradigm is counterposed to the dominant worldview of nature and is seen as a means of overcoming the current ecologically destructive social order. Rather than a society that stokes the fire of ever-increasing material needs and views the world as simply an object, a world of simplicity and equality among all species is offered.

James Lovelock's Gaia hypothesis, which posits that the earth is an organism in its own right, is based on intellectual foundations similar to those of deep ecology.[29] First and foremost, the Gaia hypothesis is a highly functionalist view, strikingly paralleling Talcott Parsons's conception of society as a superorganism, and has all the attendant problems of such a view.[30] In particular, the assumption that all components of the global ecosystem are interconnected in an ordained functional harmony requires the invocation of teleological forces. As Lovelock asserts, the Gaia hypothesis is "an alternative to that equally depressing picture of our planet as a demented spaceship, forever traveling, driverless and purposeless, around an inner circle of the sun."[31] Thus there is a driver and purpose to the universe. After all, unless some supernatural force mandates that it must be so, why should material forces lead to a natural state of harmony? In this, advocates of deep ecology and the Gaia hypothesis often slip into a spiritualist morass, denying the potential for rational inquiry.

Our concern here is not with deep ecology's emphasis on the subtle interconnections and complexity of nature, its distaste for human arrogance, or its argument for the ethical importance of recognizing that humans are but one among millions of species on earth and not the divinely (or self-) appointed masters of Creation. Indeed, we are fully sympathetic with deep ecology's views on these matters. Rather, it is its philosophical idealism and its conception of nature as an ideal functional system, of the earth as a literal (rather than metaphorical) superorganism, existing in a grand state of balance if unmolested by humanity, that is our focus. Why would there be a grand balance in nature? Natural history is a record of drastic changes and discontinuities in the biophysical world. The assumption of a natural harmony is not consistent with a critical historical understanding of nature. Furthermore, deep ecology is based on an anti-materialist theory of causality—one that posits that our value system, particularly the one emerging with the birth of modernity and a scientific worldview, is, at base, the cause of the environmental crisis. Rather than a discussion of the social forces that drive social production, a critique of the dominant worldview—divorced from its social-material influences—

becomes paramount. Change becomes a matter of adjusting values and developing the proper eco-ethics, and from there, it is assumed, changes in the social structure will follow.

Although values remain an important part of the social world, limiting discussion to this realm prevents a systematic understanding of the material forces that largely contribute to the organization of society and its interactions with nature, not to mention the forces that continue to contribute to the reproduction of the capitalist system. Since the late fifteenth century, an economic system propelled by the accumulation of capital has been the dominant force shaping human society. Deep ecologists do not disagree that an economic system premised on growth leads to conflicts with natural processes and environmental degradation. But little of their analysis is situated to critique the workings of the economic system, as far as what forces drive it. Furthermore, a discussion of material processes is not at the forefront of their analyses. Thus deep ecology's conceptualization of the interaction of society and nature is quite limited. If a sustainable society is only a matter of changing values and ethics, an analysis of environmental problems gets shortchanged. Measuring nature against an idealized notion of balance will hinder our ability to understand both natural processes and the ongoing interactions between society and the environment.

Dialectical Nature: Structural Constraints and Emergent Potential (The Ongoing Dance of Life)

We contend that a full understanding of nature is best realized through a materialistic, dialectical, and historical lens. Both the economistic and the deep ecology views outlined above tend to be ahistorical. The economistic view, though materialist, neglects the complexity of processes in the natural world, whereas the deep ecology view, though concerned with the subtleties of nature, rejects materialism. A dialectical approach to understanding nature is needed—one that overcomes the limitations of economism and deep ecology. Rather than evaluating nature in terms of an idealized state, such as the abstract balance of nature assumed in deep ecology, the world is better understood and explained in terms of its history. A materialist and dialectical approach can account for the interactions that take place at all levels, the structural constraints on change and the forces that facilitate it, the emergence of new properties, and the periods of stasis and discontinuity in history. The dialectical materialist tradition, particularly the strain that developed in the natural sciences, provides a conception and measure of nature different from that proposed by neoclassical economists and deep ecologists.[32] This tradition recognizes that nature includes processes that operate on their own terms and that have no inherent "purpose." At the same time, this tradition recognizes that the production of human society involves a constant interaction with the natural world, which involves a continual transformation of nature and society. Such a recognition of this interaction and continual transformation does not serve as a justification for human efforts to subdue and control nature; rather, it entails the acknowledgment of the inevitability of change and the interaction between elements of the material world—that is, long before humans evolved, nature was in a continual process of transformation due to the interaction of material forces and conditions, such as the origins of the biosphere and the emergence of life.[33]

For Marx, human history remains part of natural history but is not subsumed by it—that is, society is embedded in nature and dependent on it, although there are distinct social and natural processes.[34] A dialectical relationship exists between society and nature, as they continually transform each other in their coevolutionary development.[35] The direction of this relationship is not predetermined; the future remains open.

Natural scientists in the Marxist tradition have been at the forefront of developing a dialectical materialist position for understanding nature via an understanding of the development of life and natural history. The work of dialectical natural scientists, particularly that of Richard Levins, Richard Lewontin, and Stephen Jay Gould—who follow in the tradition of Darwin and Marx, as well as that established by Lancelot Hogben, Hyman Levy, J. D. Bernal, J. B. S. Haldane, and

Joseph Needham—provides a valuable foundation for understanding the natural world.[36] The work of these dialectical scientists dismantles the reification of essentialist and idealist conceptions of nature and avoids mechanical materialist presumptions that the world can be reduced to the workings of a machine and neatly molded to suit the demands of the market. The focus of these dialectical scientists is on interactions at various levels in the natural world—between genes and whole organisms, organisms and the environment—and the dynamic and contingent historical process of evolution. In opposition to the hyper-reductionism of Richard Dawkins and Daniel Dennett, which tries to push the level of causation in evolutionary history and in society to the level of the gene, Levins, Lewontin, and Gould argue that causal forces operate at different levels of aggregation and that a comprehensive causal explanation cannot be reduced to a single level.[37]

In regard to the development of an organism, Levins and Lewontin challenge the notion that life is simply the unfolding of a genetic blueprint that provides the design for our lives.[38] They contend that life cannot be reduced to the mechanistic operations of genetic forces, where change is predetermined, following an ascribed path until death. Instead, organisms remain in a state of making, so long as they live, given that they are the consequence of the relationships and interactions between genes and the environment. To gain a more comprehensive understanding of life, the relationship between the internal and external processes of life must be conceptualized as a whole. Failing to do so neglects the complexity of biological processes and the dynamic character of life. The organism is a site of interaction between the environment and genes.[39] Its development is the unique consequence of the genes it carries, the conditions of the environments through which it passes, the historical context in which it resides, and random events (in the larger world, as well as at the molecular level). Simply stated, an organism does not compute itself from its genes; interactions and the environment must be considered.

Lewontin notes that Darwin took an important step in evolutionary science "by alienating the inside from the outside: by making an absolute separation between the internal processes that generate the organism and the external processes, the environment, in which the organism must operate."[40] Darwin made this distinction to free science from existing tendencies to collapse the entire world into a unified, indistinguishable whole that made life itself unanalyzable.[41] His materialist approach opposed the idealist explanations that life and the organization of the world were a reflection of an ordered plan at the hand of God.[42]

Within evolutionary science, Darwin rejected the Lamarckian notion that variation itself was directed by the environment. He posited that the direction of variation was independent of the environment, effectively random. Changes through evolutionary history, then, were not seen as the product of trends in variation but rather in the nonrandom retention of traits produced through the independent process of variation—the key point of natural selection.

However, like Jean-Baptiste Lamarck, Darwin constructed a functionalist theory—that is, he posited that the process of natural selection fitted organisms to their environments, and that the environment, as the determinant of the selective regime, ultimately largely determined the organism. The Darwinian perspective sees diversity of species as a consequence of diverse environments "to which different species have become fitted by natural selection. The process of that fitting is the process of *adaptation*."[43] The interaction of the organism and the environment involved a selective process, where an organism fit into an ecological niche. The notion of a niche implies a predetermination, a hole in nature, which is filled by an organism, rather than a transformation on the part of either the environment or the organism.[44]

Lewontin argues that though it is true that the internal process of heritable variation is not causally dependent on the environment in which organisms live, "the claim that the environment of an organism is causally independent of the organism, and that changes in the environment are autonomous and independent of changes in the species itself, is clearly wrong."[45] Rather than adaptation, the process of evolution is best described as a process of construction. Organisms actively transform the environment through living (such as collecting food and constructing shelter), although the conditions of the environment are not wholly of their own choosing, given that previously living agents and inorganic forces historically shaped nature. Niches come into being in part as a result "of the nature of the organisms themselves."[46]

The dialectical interchange between the environment and the organism becomes a central tenet of the coevolutionary perspective proposed by Lewontin and like-minded scholars. Levins and Lewontin explain that organisms are dependent on nature for their survival.[47] Although a larger physical world exists, from which organisms receive benefit, such as the atmosphere, organisms make use of only a small part of nature in the creation of their immediate environment.

Independent forces and processes operate in nature. Volcanic eruptions can occur independently, but these are physical conditions beyond any individual organism. They shape the physical world that life confronts. At the same time, the life activities of organisms—for example, gathering food—determine what parts of the world become an immediate part of their environment. In the process of obtaining sustenance, organisms transform the world for themselves and other species. This dynamic holds for all life. Thus, organisms confront a physical world that has been shaped by natural processes and past life, while it is also being transformed by coexisting species.[48]

The characteristics of an organism, such as its metabolism, sense organs, shape, and nervous system, influence how it responds to signals in nature and how it processes materials. For example, ultraviolet light helps to lead bees to food, whereas for humans, it can cause skin cancer. Thus the biology of a species influences interactions with nature. In the process of consumption, interacting with the larger physical world, life is engaged in a process of production as the physical conditions are changed to meet the needs of organisms. New environments are created for life, influencing the conditions that other organisms will confront.[49] Organisms are both a subject and an object in the physical world, creating in part their own environment, given the existing conditions, as well as facilitating their own construction. A dialectical materialist approach provides the means to understand the complex interactions between organisms and the environment.[50]

Although organisms do not perceive all the autonomous processes of the larger world, in their interactions with and transformations of nature, they respond to these conditions. Lewontin explains that in their responsive abilities, such as the rates and forms of reproduction, which vary in invertebrate animals according to changes in space and time (including temperature, weather, and so forth) of the world surrounding them, organisms are influenced by external nature.[51] Life remains immersed in external conditions that are the consequence of the biological activities of contemporary life and all life that has preceded it.

Life, by necessity, involves interaction, which leads to change that is not entirely predictable. Organic processes are historically contingent and, thus, defy deterministic universal explanations of their particulars. Lewontin rejects teleological conceptions of evolution:

> All species that exist are the result of a unique historical process from the origins of life, a process that might have taken many paths other than the one it actually took. Evolution is not an unfolding but an historically contingent wandering pathway through the space of possibilities. Part of the historical contingency arises because the physical conditions in which life has evolved also have a contingent history, but much of the uncertainty of evolution arises from the existence of multiple possible pathways even when external conditions are fixed.[52]

Organisms are emergent, involving both internal and external dynamics. So long as genes, organisms, and environments are studied separately, the advance of our knowledge of the living world will be hindered. Given that life is both a subject and an object in its own historical development, the reductionistic notion that DNA is the sole secret to life is misleading. As Barry Commoner points out, "DNA did not create life; life created DNA."[53]

A key feature of the Marxist view of history is that change is not typically smooth and continuous but rather often occurs very rapidly following periods of stasis (temporary periods, of indeterminate length, of counterbalancing opposing forces leading to relative stability). Throughout history, the worldview of the ruling class has typically been quite different from this, either identifying eternal stasis as the natural condition or change as inevitably smooth and gradual. This is a view, obviously, comforting to those in power because it undermines the idea that revolutions are likely. The discovery of "deep time" by geologists and of organic evolution by naturalists undermined the eternal-stasis perspective, but the notion of slow, continuous change was a key facet of the thinking of Victorian scholars, reflected in Charles Lyell's uniformitarianism and Darwin's gradualism. Of course, neither view, rapid change or gradual change, is absolutely correct; the complexity of human and natural history has ensured that both types of change occur (it goes without saying that the rate of change is not binary, either necessarily rapid or gradual, but this dichotomy is heuristically useful). Furthermore, the rate of change of any particular phenomenon is a factual question and cannot be determined by ideology. However, a key point of scientists in the Marxist tradition is that the ideology of the ruling class often distorts one's perceptions of the world. The Marxist tradition therefore emphasizes the necessity of being particularly skeptical of assertions about the natural world when they conform to ruling-class ideology.[54]

The Marxist view of historical change in the natural world is perhaps best expressed in Niles Eldredge and Stephen Jay Gould's argument that the evolutionary history of organisms is best characterized as "punctuated equilibria," long periods of stasis, punctuated with (geologically) brief periods of rapid change.[55] This is based in part on a literal interpretation of the fossil record, which generally shows fossils of a species remaining quite similar over extended stretches of time and then suddenly (in the geological sense) being replaced by a substantially different, although apparently related, type. Their argument is in no way a rejection of Darwinism in general, only a challenge to Darwin's strong preference for gradualism. They invoke no special mechanisms for change. Rather, they argue that speciation typically happens when a subset of a species becomes isolated. In a small isolated population, mutations can spread rapidly throughout all members of the species, and the rate of change can be further accelerated if the population faces different selection pressures than the parent species. In large populations that are geographically widespread, although connected through breeding, mutations spread slowly, and any mutations that are favorable to organisms in one part of the range are not necessarily retained, because they become watered down by genes from the larger population. For these reasons, Eldredge and Gould proposed that widespread species will generally change little over most stretches of time but may change rapidly around the point of speciation, when a subpopulation becomes isolated.

Gould has also argued that organisms are not mere putty to be sculpted over the course of their phylogeny (evolutionary history) by external environmental forces, but rather, their structural integrity constrains and channels the variation on which natural selection operates.[56] In this, Gould is challenging the notion that variation is isotropic, effectively random in all directions. He notes that the structural nature of the development of an organism throughout its life course (ontogeny) limits the types of phenotypic variation that are possible because changes at one stage of the developmental process have consequences for later stages. Therefore, many characteristics of an organism cannot simply be modified without having substantial ripple effects throughout the whole organism. The inherited patterns of development do not readily allow for all types of modification. Therefore the evolutionary process is a dialectical interaction between the internal (inherited structural constraints) and the external (environmental selection pressure), just as the ontogeny (individual development) of individual organisms is a dialectical interaction between their genes and the environment.

The structural nature of development has consequences for patterns of change. To illustrate this point, Gould makes use of a metaphor, Galton's polyhedron.[57] In true fashion, Gould draws upon the arguments of various historic figures involved in the evolutionary debate to build his own. Francis Galton, who was Darwin's cousin (Erasmus Darwin was grandfather to both), who helped lay the foundations for much of modern statistics, and who is regarded as the father of eugenics, was deeply impressed by his cousin's work on evolution, but he disagreed with Darwin's assumptions about the nature of variation.[58] He developed an analogy to challenge Darwin. While Galton did not appreciate the dialectical position within his own analogy, Gould was never one to miss a conceptual gem even in the most unlikely of places and was always able to bring out the potential of a concept.

Gould explains that in Darwin's idealized formulation, species are metaphorical spheres, such as marbles, that roll freely on any phylogenic course the external world pushes them along—that is, their structure offers no resistance to pressure from the external environment, and thus they move readily wherever environmental forces direct them. Alternatively, in Galton's metaphor, species are polyhedrons, multi-sided solid objects that have flat faces (such as dice), whose structure prevents them from rolling freely when only slightly perturbed and limits the paths they can follow after receiving a sufficient push from the external world. They can switch the facet on which they rest, but they cannot simply rest in any given position. In contrast with a sphere, which may roll smoothly with a light tap, the polyhedron will resist minor perturbations but, given sufficient force, will switch facets abruptly. Thus species cannot perfectly track changing environments, because of the structural interconnections they develop over the course of their phylogeny, which limit and, potentially, direct the type of change that is possible. Note that this metaphor also points to another concept common in the historical materialist tradition: Change does not necessarily happen smoothly but rather can happen rapidly, preceded and followed by periods of relative stability, shaped by opposing forces.[59] The polyhedron contains both structural constraints and the potentiality for new states. Hence it has an affinity with the theory of punctuated equilibria.

This metaphor can also serve as an illustration of the global environment. Ecosystems have resiliency within certain bounds. Their natural cycles and processes continue to operate within certain states. Complex systems, such as the global climate, can maintain a stable state for extended periods, but if sufficiently disturbed, such as by the anthropogenic emission of greenhouse gases, they can change abruptly. The recognition of thresholds and the potential for sudden change in the natural world is central to a dialectical and historical understanding of nature. Natural thresholds can be surpassed, creating a sudden change in the global ecosystem.[60] The polyhedron, in this case the global environment, could be pushed so hard that the changing of the facets results in conditions that cannot sustain societies. A proper understanding of this point undermines economistic approaches to quantifying the value of nature's services to society, because there is no directly linear correspondence between human-generated pressure on the environment and changes in the environment. For example, the "cost" to society and the other creatures that inhabit the earth may be modest for the first several billion metric tons of carbon emitted by societies, but when a natural threshold is approached, the cost may escalate dramatically and in an effectively unpredictable manner.[61] The program to assign economic value to natural processes fails to appreciate both the inherent complexity and unpredictability of natural systems and the lack of a direct correspondence between ecological dynamics and economic dictates.

The Enduring Struggle and the Threat of Extinction

A dialectical materialist approach to nature provides the means for understanding the complex interactions throughout the natural world, the ability to explain the world in terms of itself. It involves both the capacity to recognize that contingency and emergence are inherent aspects of a living world, and the capability to study the structural constraints and the inherent potential for change. In this, a materialist dialectic avoids the mechanistic reductionism of economistic approaches, where nature exists in the background, as simply an input to the economic system. It also avoids the idealized notion that nature exists in a state of balance and that a return to such a state is simply a matter of developing the appropriate moral-ethical system.

The dialectical materialist perspective recognizes that the world is one of constant change but not one where anything goes.

Constraints and possibilities remain in the structural conditions of the world. Abrupt, punctuated change can radically shift life to new pathways or the environment to conditions that present serious challenges to existing life. It is of utmost importance that nature is understood in terms of itself. Human society is dependent upon the environment and must interact with it if it is to continue to reproduce itself. This interaction involves the transformation of the world. The dialectical materialist approach highlights how history involves change. But all change and any change is not good. The interaction between humans and the environment is an enduring struggle to live within a finite world, under emerging conditions. There are social interactions that threaten to push the polyhedron of the global environment toward states of radical change that threaten the world we know with global mass extinction.

The causes of the previous five mass extinctions are not fully understood, but the mass extinction taking place today is clearly being driven by *Homo sapiens* via an economic system that operates at the global level.[62] The constant expansion of the capitalist system has pushed environmental degradation to the planetary level, as habitat destruction decimates the living conditions of species and as ecosystems are radically transformed.[63] Human civilization, under capitalism, is engaged in a process of destroying the future, as "we suck our sustenance from the rest of nature in a way never before seen in the world, reducing its bounty as ours grows."[64]

Eldredge points out that as humans moved beyond isolated ecosystems, to operate at the planetary level, our alienation from nature increased.[65] We developed the illusion that we were not dependent upon the environment. Eldredge warns that the current global mass extinction is quite different from previous ones, in that the source of the extinction remains on the scene: humans destroying habitat for the sake of profit.[66] Thus recovery of ecosystems is not possible so long as the same forces continue to act and change the world as has been the practice. For example, forests continue to be cleared to make room for urban growth and crops. As the environment is simplified and biodiversity declines, the operation of ecosystems is hindered as resource capture through energy, water, and nutrients is diminished. The resiliency of an ecosystem is also hampered, reducing its ability to purify water and its integrity to mitigate floods.[67] Furthermore, given the interdependency of species and the complexity of interactions among species, the remaking of the environment through habitat destruction poses the threat of cascading extinctions. The loss of a specific larval host plant in Singapore led to the loss of tropical butterfly species. Hummingbird flower mites are dependent upon both the hummingbirds that provide transportation to other flowers and the flowers from which they "depend for nectar and pollen." If either the flowers or the hummingbirds are threatened with extinction, so are the flower mites. The potential loss of "irreplaceable evolutionary and coevolutionary history" is very grave, as "species coextinction is a manifestation of the interconnectedness of organisms in complex ecosystems."[68]

The rate of speciation is caught in a time conflict, as the current rate of extinction is faster than the rate of evolution.[69] The mass extinction being orchestrated today is a unique historic event, given that it is being driven by anthropogenic forces that continue to operate. Since 1600, the extinction rate has been 50 to 100 times the average estimated rate of extinction during previous epochs, but the rate "is expected to rise to between 1,000 and 10,000 times the natural rate."[70] Thus a radical change in the operations of human society and its interactions with nature is necessary to stop the ecological crisis that is taking place.[71]

The interaction between human society and nature is a never-ending dance, which always presents a challenge. Because change is the law of life, this does not mean we are helpless or that we should not try positively to influence the conditions of the world. In fact, given the current state of the environment and the ecological crisis in which we live, monitoring and directing how humans interact with nature is a priority. Humans must establish a form of social production that does not alienate people from nature and that interacts with nature in a manner that does not undermine the environment's ability to regenerate. This requires constant vigilance and flexibility to respond to contingency, as the world continues to change. So long as society is driven by short-term goals, such as the drive to accumulate capital, the longevity of both the current global environmental epoch and humanity are threatened.

As Lewontin explains, there is no evidence that organisms are becoming more adapted to the environment.[72] Evolution does not entail a drive toward perfection. All elements of life are changing. Around 99.99 percent of all species that ever existed are extinct. Likewise, there is no evidence for claims of harmony and balance with the external world. Environmental change will continue. Natural and social history are in constant motion. Chance is always present. "What we can do," Lewontin emphasizes, "is to try to affect the rate of extinction and direction of environmental change in such a way as to make a decent life for human beings possible. What we cannot do is to keep things as they are." A dialectical materialist approach provides the means to grapple with an emerging world and helps to further our understanding of the human-environment interaction, pointing the way to a more accurate measure and understanding of nature.

passed the alienation of nature), his understanding of the labor and production process as the metabolic relation between humanity and nature, and his coevolutionary approach to society-nature relations.

Nevertheless, because Marx's overall critique of political economy remained unfinished, these and other aspects of his larger materialist conception of nature and history were incompletely developed—even in those works, such as *Capital*, volume 1, published in his lifetime. Moreover, the relation of his developed political-economic critique in *Capital* to the wider corpus of his work was left unclear. The *Grundrisse* has therefore become an indispensable means of unifying Marx's overall analysis. It not only stands chronologically between his early writings and *Capital*, but also constitutes a conceptual bridge between the two. At the same time it provides a theoretical-philosophical viewpoint that is in some ways wider in scope than any of his other works.

The form of the *Grundrisse*—Marx composed it as a set of notebooks primarily for his own self-edification in preparation for his critique of political economy—has made it a difficult work to interpret. One way to understand his general theoretical approach is in terms of the relation between "production in general"—a conceptual category introduced in the opening pages of the *Grundrisse*, originally conceived as the basis of its "first section"—and specific historical modes of production.[3] The latter included pre-capitalist economic formations and capitalism's immediate historical presupposition, that is, primitive accumulation—together with capitalism proper.

Marx used the concept of production in general as a basis from which to develop his general theory of needs, which encompassed both natural prerequisites and historic developments—the production of new needs manifested in new use values. It was the conflict between production in general (as represented by use value) and specifically capitalist production (as represented by exchange value) that pointed to capitalism's historical limits and necessary transcendence.

The nature-society or ecological dialectic embodied in the *Grundrisse* can be seen in terms of five interrelated realms: (1) the attempt to construct a materialist critique encompassing both production in general and its specific historical forms; (2) the articulation of a theory of human needs in relation to both society and nature—pointing beyond the capital relation; (3) the analysis of pre-capitalist economic formations and the dissolution of these forms through primitive accumulation, representing changing forms of the appropriation of nature through production; (4) the question of external barriers/boundaries to capital; and (5) the activation of capital's absolute limits.

Production in General and Natural-Historical Materialism

The starting point for Marx's critical ontology in the *Grundrisse* was that of production in general. Production in the most concrete sense was always historically specific—production at a definite stage of social development. Nevertheless, an understanding of these specific forms gave rise to a more general, abstract conception, that of the "production process in general, such as is common to all social conditions, that is, without historic character."[4] "All epochs of production," Marx wrote, "have certain common traits, common characteristics. *Production in general* is an abstraction, but a rational abstraction in so far as it really brings out and fixes the common element and thus saves us repetition. . . . For example. No production possible without an instrument of production, even if this instrument is only the hand. No production without stored-up, past labour, even if it is only the facility gathered together and concentrated in the hand of the savage by repeated practice."[5]

Production in general in Marx's analysis was tied to the production of use values. Use value "presupposed matter," and constituted the "natural particularity" associated with a given human product. It existed "even in simple exchange or barter." It constituted the "natural limit of the commodity" within capitalist production—the manifestation of production in general as opposed to specifically capitalist production.[6]

Closely related to production in general, was labor in general. "Labour," Marx wrote in *Capital*, "is, first of all, a process...by which man through his own actions, mediates, regulates and controls the metabolism between himself and nature. . . . It [the labor process] is the universal condition for the metabolic interaction [*Stoffwechsel*] between man and nature, the everlasting nature-imposed condition of human existence."[7]

This approach to nature and production first appeared in the *Grundrisse*, where Marx discussed the metabolic "change in matter [*Stoffwechsel*]" associated with "newly created use value."[8] Just as this metabolic relation constituted the universal condition defining production, so the alienation of this metabolism was the most general expression of both human alienation and alienation from nature, which had its highest form in bourgeois society. As Marx explained: "It is not the *unity* of living and active humanity with the natural, inorganic conditions of their metabolic exchange with nature, and hence their appropriation of nature, which requires explanation or is the result of a historic process, but rather the *separation* between these inorganic conditions of human existence and this active existence, a separation which is completely posited only in the relation of wage labor and capital."[9] It was the historical alienation of human beings from nature under capitalist production rather than their unity in production in general that therefore required critical analysis.

Here Marx was building on an earlier materialist-dialectical conception presented in his 1844 *Economic and Philosophical Manuscripts*, where he had written that "Nature is man's *inorganic body*—that is, insofar as it is not itself human body. Man *lives* on nature—means that nature is his *body*, with which he must remain in continuous interchange if he is not to die. That man's physical and spiritual life is linked to nature means simply that nature is linked to itself, for man is a part of nature."[10] This dialectic of organic-inorganic relations was derived from Hegel's *Philosophy of Nature* and was rooted ultimately in ancient Greek philosophy. In this context organic meant pertaining to organs; inorganic referred to nature beyond human (or animal) organs; the "inorganic body of man" was the extension of the human body by means of tools. (The Greek *organon* encompassed both organs and tools; seeing the former as "grown-on" forms of the latter, whereas tools were the artificial organs of human beings). "In its outwardly oriented articulation," Hegel wrote, "it [the animal] is a production mediated by its inorganic nature."[11]

In the *Economic and Philosophical Manuscripts*, Marx gave this a more materialist reading, arguing that "the life of the species, both in man and in animals, consists physically in the fact that man (like the animal) lives on inorganic nature; and the more universal man (or the animal) is, the more universal is the sphere of inorganic nature on which he lives."[12] This was carried forward into the *Grundrisse* where he referred to "the *natural* conditions of labour and of reproduction" as "the objective, nature-given inorganic body" of human subjectivity. "The earth," he stipulated, is "the inorganic nature of the living individual. . . . Just as the working subject appears naturally as an individual, as natural being—so does the first objective condition of his labour appear as nature, earth, as his inorganic body."[13]

The *Grundrisse* is full of acknowledgments of nature's limits, natural necessity, and the coevolution of nature and society. The planet itself had evolved, taking on new emergent forms, so that the "processes by means of which the earth made the transition from a liquid sea of fire and vapour to its present form now lie beyond its life as finished earth."[14] With the development of industrialized agriculture, Marx argued—foreshadowing his analysis of the metabolic rift in *Capital*—"agriculture no longer finds the natural conditions of its own production within itself, naturally, arisen, spontaneous, and ready to hand, but these exist as an independent industry separate from it." It now requires external inputs, such as "chemical fertilizers acquired through exchange," the importation of Peruvian guano, "seeds from different countries, etc." In this sense a rift had been created in the natural metabolism.[15]

The Theory of Needs and the Transcendence of Capital

There was in Marx's view no exclusively natural character to human needs and identity. But there were nevertheless natural prerequisites to human existence, and a natural substratum to production in general. "Use value," he wrote, is the "object of . . . satisfaction of any system whatever of human needs. This is its [wealth's] material side, which the most disparate epochs of production may have in common."[16] Hence all commodity production necessarily consisted of use value as well as exchange value. The natural prerequisites of production, embodied in use values, could be transformed but not entirely transcended through human production. Human needs, "scant in the beginning," were, in their specifically human character, historically changing needs, developing "only with the forces of production," erected on top of this natural substratum.[17] New needs were produced through the continual transformation of both the human relation to nature and of human beings to each other—and hence of human species being. The development of production was therefore nothing but the historical development of human needs and powers in interaction with nature. "Not only do the objective conditions change in the act of reproduction, e.g. the village becomes a town, the wilderness a cleared field etc., but the producers change, too, in that they bring out new qualities in themselves, develop themselves in production, transform themselves, develop new powers and ideas, new modes of intercourse, new needs and new language."[18]

Neither natural history nor social history could be conceived as static; each was complex and forever changing, embodying contingent, emergent, and irreversible aspects, and above all interconnectedness.[19] The metabolic relation between human beings and nature was thus necessarily a coevolving one, in which the dependence of human beings on nature was an insurmountable material fact. Moreover, the future depended on the *dynamic sustainability* of this historically changing relation, in forms that provided for "the chain of successive generations."[20]

This outlook was integral to Marx's materialist conception of nature and history as developed in his work as a whole. In *The German Ideology*, Marx and Engels observed that "the first premise of all human history is, of course, the existence of living human individuals. Thus the first fact to be established is the physical organisation of these individuals and their consequent relation to the rest of nature. . . . All historical writing must set out from these natural bases and their modification in the course of history through the action of men." From such natural prerequisites of history, Marx and Engels proceeded to human history proper: production, as the specifically human relation to nature, was not only the mere satisfaction of needs but the creation at the same time of new needs.[21] These might be far removed from their original natural bases. "Hunger is hunger," Marx observed in the *Grundrisse*, "but the hunger gratified by cooked meat eaten with a knife and fork is a different hunger from that which bolts down raw meat with the aid of hand, nail and tooth."[22]

Under the regime of capital this dialectic of needs production became inverted, so that the production of use values, reflecting the fulfillment of old needs and the positing of new ones on natural foundations, existed only as a means not an end; and the pursuit of exchange value became the sole object of production. Capitalism created open, endless dissatisfaction, since the pursuit of exchange value as opposed to use value had no natural or social point of satisfaction, but led only to a drive/craving for more. Thus a treadmill of production was generated in which production appeared "as the aim of mankind and wealth as the aim of production." This contrasted with the "loftier" if still "childish world" of the ancients, in which human satisfaction was still the object of production, albeit from "a limited standpoint."[23]

In the alienated, upside-down world of capital, the dominant necessity driving all others was the unquenchable desire for abstract commodity wealth, which was nothing but the limitless desire for more commodity production. This meant that the original conditions of production—land and even human beings—became mere accessories to production. Generalized commodity production disrupted all original human-natural relations, all relations of sustainability and community, in the ceaseless drive for production for production's sake, wealth for wealth's sake. But "when the limited bourgeois form is stripped away," Marx asked, "what is wealth other than the universality of individual needs, capacities, pleasures, productive forces, etc., created through universal exchange? The full development of human mastery over the forces of nature, those of so-called [external] nature as well as of humanity's own nature?"[24] Such "human mastery" was of course not about the robbing of nature but the realization of a wealth of human needs and powers through human production, and not for a single generation, but for successive generations.

Pre-Capitalist Economic-Ecological Formations and Primitive Accumulation

Marx's very detailed (to the extent then possible) treatment of pre-capitalist economic formations in the *Grundrisse* was meant to lead into the analysis of capitalist development itself, as part of a general historical understanding. That section of the *Grundrisse* had the heading: *Forms which precede capitalist production. (Concerning the process which precedes the formation of the capital relation or of original accumulation).*[25] It was preceded by a section headed *Original Accumulation of Capital.*[26] Moreover, the section on pre-capitalist forms ended with the reconsideration of the original, primitive accumulation of capital arising out of these historical precursors, making it clear that the original basis for accumulation and capitalism's simultaneous dissolution of all earlier economic formations was the central issue.[27]

The discussion of pre-capitalist economic formations focused on the communal nature of these formations (already substantially broken down in the class societies of the ancient and feudal worlds). Marx's analysis of "original" or "primitive" accumulation was thus concerned with the *dissolution* of these remaining communal and collective forms and the complete alienation of the land—providing the ground for the emergence of the modern proletariat and the self-propelling process of capital accumulation. As he wrote in *Capital*, "private landownership, and thereby expropriation of the direct producers from the land—private landownership by the one, which implies lack of ownership by others—is *the basis of the capitalist mode of production.*"[28] The main presupposition of capitalism was the dissolution of all previous connections to the land on the part of the direct producers. It was "the historic dissolution of . . . naturally arisen communism" as well as "a whole series of economic systems" separated from "the modern world, in which exchange value dominates."[29]

The *Grundrisse* provided a trenchant analysis of these processes of dissolution. What was primarily at issue was the "*dissolution* of the relation to the earth—land and soil—as natural conditions of production—to which he [the human being] relates as to his own inorganic being."[30] Living labor, which was originally connected to and in community with the land was now defined by the fact that the earth was the worker's "not-property," that is, his (and her) "not-landownership . . . the negation of the situation in which the working individual relates to land and soil, to the earth as his own." This prior communal relation to the earth was now "historically dissolved" in its entirety by capitalist relations of production.[31] The forcible expropriation of the earth

> "clears," as Steuart says, the land of its excess mouths, tears the children of the earth from the breast on which they were raised, and thus transforms labour on the soil itself, which appears by its nature as the direct wellspring of subsistence, into a mediated source of subsistence, a source purely dependent on social relations. . . . There can therefore be no doubt that *wage labour* in its *classic form*, as something permeating the entire expanse of society, which has replaced the very earth as the ground on which society stands, is initially created only by modern landed property, i.e. by landed property as a value created by capital itself.[32]

The result was "a dialectical inversion" in which property was entirely on the side of capital, establishing the right of property over alienated labour, which existed only for (and through) its exploitation.[33] In this dissolution of the traditional relation to the land the labor force was "released" as formally free labor power, without any recourse for survival except to offer itself up for exploitation by capital. "In bourgeois economics," Marx wrote, "this appears as a complete emptying-out . . . universal objectification *as total alienation*, and the tearing-down of all limited, one-sided aims as sacrifice of the human end-in-itself to an entirely external end."[34]

External Barriers and Boundaries

For Marx capital was self-expanding value, inseparable from accumulation. As he explained in the *Grundrisse*, "If capital increases from 100 to 1,000, then 1,000 is now the point of departure, from which the increase has to begin; the tenfold multiplication, by 1,000% counts for nothing."[35] The increase, from whatever starting point, is all, since it is from this increase that profits are obtained.

This meant that capital had constantly to revolutionize its appropriation of both nature and human labor power. "Capital," the *Grundrisse* stated,

> is the endless and limitless drive to go beyond its limiting barriers. Every boundary is and has to be a barrier for it. Else it would cease to be capital—money as self-reproductive. If ever it perceived a certain boundary not as a barrier, but became comfortable with it as a boundary, it would itself have declined from exchange value to use value, from the general [abstract] form of wealth to a specific, substantial mode of the same. . . . The quantitative boundary of the surplus value appears to it as a mere natural barrier, as a necessity which it constantly tries to violate and beyond which it constantly seeks to go.[36]

Here Marx was relying on the dialectical treatment in Hegel's *Logic* of the nature of limits (barriers) to growth or expansion.[37] A seeming absolute boundary that can be completely overcome is in reality a mere barrier. Nevertheless, capital's ability to overcome all spatial and temporal, and all natural, limits, through the "annihilation of space by time"—treating these as mere barriers (rather than boundaries) to its own self-expansion—was more ideal than real, generating constantly expanding contradictions.[38] In perhaps the most penetrating passage ever written on the dialectic of natural limits under capital, Marx stated in the *Grundrisse*:

> Just as production founded on capital creates universal industriousness on one side . . . so does it create on the other side a system of general exploitation of the natural and human qualities, a system of general utility, utilising science itself just as much as all the physical and mental qualities, while there appears nothing *higher in itself*, nothing legitimate for itself, outside this circle of social production and exchange. Thus capital creates the bourgeois society, and the universal appropriation of nature as well as of the social bond itself by the members of society. Hence the great civilizing influence of capital; its production of a stage of society in comparison to which all earlier ones appear as mere *local developments* of humanity and as *nature-idolatry*. For the first time, nature becomes purely an object for humankind, purely a matter of utility; ceases to be recognized as a power for itself; and the theoretical discovery of its autonomous laws appears merely as a ruse so as to subjugate it under human needs, whether as an object of consumption or as a means of production. In accord with this tendency, capital drives beyond national barriers and prejudices as much as beyond nature worship, as well all traditional, confined, complacent, encrusted satisfactions of present needs, and reproductions of old ways of life. It is destructive towards all of this, and constantly revolutionizes it, tearing down all the barriers which hem in the development of the forces of production, the expansion of needs, the all-sided development of production, and the exploitation and exchange of natural and mental forces. But from the fact that capital posits every such limit as a barrier and hence gets *ideally* beyond it, it does not by any means follow that it has *really* overcome it, and since every such barrier contradicts its character, its production moves in contradictions which are constantly overcome but just as constantly posited.[39]

The juggernaut of capital therefore sees all of nature as a mere object, an external barrier to be beaten down, surmounted, or circumvented. Commenting on Francis Bacon's maxim that "nature is only overcome by obeying her"—on the basis of which Bacon proposed to "subjugate" nature—Marx observed that for capitalism the discovery of nature's autonomous laws "appears merely as a ruse so as to subjugate it under human needs."[40] He decried the one-sided, instrumental, exploitative relation to nature associated with contemporary social relations. Despite its clever "ruse," capital is never able fully to transcend nature's limits, which continually reassert themselves with the result that "production moves in contradictions which are constantly overcome but just as constantly posited." No thinker in Marx's time, and perhaps no thinker up to our present day, has so brilliantly captured the dialectical complexity of the relationship between capitalism and nature.[41]

The Activation of Capital's Absolute Limits

This argument takes on added significance for us today, at a time when, as István Mészáros claims, we are witnessing "the activation of capital's absolute limits."[42] This takes various forms but is most apparent in the ecological realm. The problem, as Mészáros explains, is that "neither the degradation of nature nor the pain of social devastation carries any meaning for its [capital's] system of social metabolic control when set against the absolute imperative of self-reproduction on an ever-extended scale."[43] All of this is inherent in the alienating character of capital, which is rooted in the alienation of the human metabolic relation to nature. "Under the capitalist modality of metabolic exchange with nature," Mészáros writes, "the *objectification* of human powers necessarily assumes the form of *alienation*—subsuming productive activity itself under the power of a *reified objectivity*, capital."[44] In the present age of planetary environmental crisis, capital is increasingly giving evidence of its ultimate "destructive uncontrollability," imperiling civilization—or worse, life itself.[45]

Sustainability in relation to the earth was a requirement of production in general for Marx, but one that capitalism was compelled to violate. As he explained in *Capital*, what was required from the standpoint of production in general was "a conscious and rational treatment of land as permanent communal property, as the inalienable condition of the existence and reproduction of the chain of human generations." Instead capitalism brought "the exploitation and the squandering of the powers of the earth."[46] The "total alienation" to which capitalist society tended pulled the rug out from under it, creating ever greater conflicts between production in general and specifically capitalist production.

Such a theory of *total alienation* (*Après moi le déluge!*) required as its negation a theory of total liberation: a revolutionary struggle to unleash human potential in ways that did not contradict the wealth of capacities that resided within all human beings and all generations, and that safeguarded the earth.[47] The goal of production, Marx believed, should be "the cultivation of all the qualities of the social human being," generating a social formation "as rich as possible in needs, because rich in qualities and relations."[48] Yet this was a future that could only be fully materialized in a society in which the associated producers rationally controlled their metabolic relation to nature, transcending total alienation and creating a genuine community with the earth.[49]

14. The Sociology of Ecology

A key dividing line within environmental sociology—even more perhaps than in sociology in general—is the question of "realism" versus "constructionism." To what extent is nature independent of human action and even conceptions, and to what extent is it constructed by society and human thought processes? Realists within environmental sociology tend to materialism and think in terms of nature's ontological independence of human action and conceptions. They emphasize natural limits to human action. In this view, nature can be successfully altered to meet human needs up to a point—but only if nature's laws and limits are first recognized and followed. This view is compatible with a dynamic notion of nature that incorporates evolutionary postulates. Constructionists, in contrast, tend to idealism and skepticism, and they stress the epistemological limits of our knowledge of nature. They underscore the extent to which nature as we know it is constructed by human actions and cognition, and they are suspicious of what they regard as "essentialist" or "positivistic" postulates about nature. In this view, social development is frequently conceived (if only for methodological purposes) as unconstrained by natural forces, which can therefore be set aside in purely social analysis.

A common complaint of realist environmental sociologists is that sociology in the twentieth and twenty-first century has leaned toward a broad, overarching constructionism and human exemptionalism (the notion that human beings are mostly exempt from nature's laws), ignoring or downplaying realist concerns with natural limits, and cordoning social science off from natural science. This has only been heightened by the so-called cultural turn in sociology. Constructionist environmental sociologists, for their part, complain of the naïve view of science as a mirror of nature and the technological Prometheanism that they attribute to realism (here reduced to crude positivism).

There are of course sophisticated versions of each of these views. Critical realists recognize the epistemological obstacles to our knowing the Kantian "thing in itself"—or "intransitive objects of scientific knowledge"—and stress the historical basis of human actions and cognition.[1] Cautious constructionists explicitly acknowledge that there are natural limits (the existence of "a reality out there") within which human beings are constrained and that our cognition is a coproduct of nature and society. Some advocates of the "strong programme" in the sociology of science—a view generally seen as adamantly constructionist—argue that a kind of idealism at the level of categories and truth claims "is compatible with an underlying materialism."[2]

Nevertheless, the division between realism and constructionism continues to bedevil environmental sociology in particular (a field that necessarily transgresses the divide between society and nature at every point), creating quite distinct theoretical emphases. Realists within environmental sociology are concerned primarily with the ontology of environmental crisis and see this as a reason to alter existing social relations. Constructionists focus much more on the epistemological aspects and "reflexivity" of our construction of nature and science, seeing environmental crises first and foremost as discursive constructions and therefore open to diverse interpretations.[3]

In what follows, we attempt to throw further light on the divide between realism and constructionism and to show how this division can be transcended through a "realist-constructionist" or praxis-oriented account rooted in a particular, situated context—the historic formation of ecological science in the early twentieth century. We ask questions such as these: (1) To what extent was the growth of ecology as a science an attempt not simply to construct a new scientific understanding of nature, but a manifestation of developing social relations of production and emerging conflicts within human society that were transferred to the realm of nature/ecology (and then frequently reimported into society as naturalized facts)? (2) How are current conceptions of ecological crisis within ecological science affected by the human-historical conflicts that entered at the outset into the very constitution of ecological science? (3) What form of the sociology of ecological science best complements an environmental sociology concerned centrally with ecological crisis?

What we hope to demonstrate is the importance of both ontological realism and historical constructionism, synthesized within a critical-realist perspective. We argue that within both natural science and social science (and especially within ecology, which increasingly embraces both), it is essential to retain a realist/materialist view that also embraces an understanding of the human-historical construction of the world within limits. Ultimately, there is no contradiction between the Galileo principle ("It still moves") and the Vico principle ("We can understand it because we made it") if each of these is properly understood and delimited. In a slight revision of Marx's principle of historical materialism, we can say human beings make their own history, not entirely under conditions of their choosing but rather on the basis of natural-environmental and social conditions inherited from the past. This is the essence of what has been called "the philosophy of praxis."[4]

For environmental sociologists to raise such matters in the context of the development of ecological science is of course to trespass on the realm of the sociology of science, which specializes in precisely these kinds of difficult inquiries into science—but in ways that are only rarely directly pertinent to environmental sociology, and that seldom address the materialist/realist tradition. In the following analysis, we will make use of insights drawn especially from the early "externalist" (or social-institutional) approach to the sociology of science associated with such thinkers as Boris Hessen, Edgar Zilsel, and Robert K. Merton. In addition, we make use of Marxist and feminist "standpoint theory" and draw on Imre Lakatos's distinction between "progressive" and "degenerative" research programs.[5] Our overall analysis here is influenced by Roy Bhaskar's and Andrew Sayer's critical realism.[6]

Following this excursion into the sociology and philosophy of science, we use some of the analytical tools derived there to elucidate the concrete core of our argument, focusing on the struggle over the foundations of ecology as an emerging scientific field of research from 1926 to 1935. This saw the development of two competing ecological holisms. One of these approaches arose out of a tradition of idealism and organicism and is represented most fully by the ecological holism/racism of Jan Christian Smuts and his followers within ecological science in South Africa. The other flowed out of a tradition of materialism and took its most definitive form in the ecosystem ecology of Arthur Tansley in Britain.

Not only did these two traditions openly war with each other, but they also crossed swords in surprising ways that demonstrate that the development of science can never be cordoned off from the larger social struggles or our conceptions of nature from those of society. The importance of realist-constructionist accounts of science (particularly if conceived in historical and materialist ways) is highlighted. For example, only in this way is one able to deal with the issue, raised by Marx and Engels, of the "double transference" of ideas of nature and society and the erroneous naturalization of social relations that has sometimes resulted.

Our argument calls into question the traditional social science/humanities story, as often depicted by environmental historians and environmental sociologists, that—in opposition to the main line of ecological science—sides with Smuts's idealist, organicist "holism" against the materialist ecosystem ecology of Tansley. As David Pepper says, "The ecocentric interpretation of twentieth-century science tends towards . . . idealism."[7] Indeed, we are frequently led to believe that ecology in following Tansley rather than Smuts simply took the wrong path, choosing reductionism over holism.[8]

Dissenting from this view, we attempt to illuminate why Tansley's ecosystem ecology reflected a "progressive" research program and Smuts's "holism" a "degenerating" research program (following Lakatos's distinction) in terms of the natural-scientific and also the social-scientific implications of these arguments. The materialist/realist view, we argue, was superior in both its ontological realism and its constructionist tendencies. It was more attuned to the difficulties of the uncritical transference of social ideas to the natural realm and then their transference back (in objectivist, naturalist clothing) to the social—the problem of the double transference. Moreover, the situated social context in which these theories were developed ultimately, we believe, favored the evolution of ecological materialism. This suggests that not all ontologies and not all constructionisms are created equal. Our examination of the Smuts-Tansley debate, however, is more than an attempt simply to validate one view over another. Instead, it seeks to illustrate, through a situated case, that the real concern of a sociology of modern ecology should *not* be the strict opposition of realism versus constructionism but rather the proper demarcation of each in the analysis of the coevolution of nature and society.

Navigating the Great Divide: Realism versus Constructionism

From its first appearance in the mid-1970s until the 1990s, environmental sociology was characterized by "an almost universal commitment . . . to a realist epistemology and a materialist ontology."[9] Nevertheless, the social constructionist perspective, which soon gained prominence within the sociology of science, social problems, and the sociology of gender, began strongly to impress itself on environmental sociology by the 1990s. The result was a debate between realists and constructionists that, while resembling similar controversies in other areas of sociology, took on an extremely virulent form.

From the beginning, environmental sociologists have charged that sociology as a discipline has been far more reluctant than other social science disciplines to incorporate natural-environmental postulates into its analysis—a failing, they claimed, that was less evident in the work of sociology's classical founders.[10] In their original formulation, which helped to launch the field, William R. Catton Jr. and Riley E. Dunlap presented environmental sociology as a subdiscipline, embodying a "new environmental paradigm" that opposed the "human exemptionalist" (from nature) assumptions prevalent in much of social science and sociology in particular.[11]

Environmental sociology arose in conjunction with the environmental movement in the 1970s, spurred on like the latter by the warnings of scientists with regard to ecological crisis.[12] Environmental sociologists thus saw themselves as addressing this developing ecological crisis from the standpoint of social systems, institutions, processes, and agents. Because of this focus on the reality of ecological crisis that had defined the field from the start, numerous environmental sociologists saw the sudden intrusion of strong social constructionist views into environmental sociology, roughly a decade and a half after its inception, as a threat to the very constitution of the subdiscipline. Realist environmental sociologists responded sharply to the antirealism, for example, of Keith Tester, who provocatively declared that "a fish is only a fish if it is socially classified as one, and that classification is only concerned with fish to the extent that scaly things living in the sea help society define itself. . . . Animals are indeed a blank paper which can be inscribed with any message, and symbolic meaning, that the social wishes."[13]

For realists within environmental sociology, this kind of strong social-constructionist "antirealism," as it was sometimes called even by those sympathetic to it, only seemed to reinforce, at an even more extreme level, an anthropocentrism with regard to nature that environmental sociology from the beginning had sought to combat.[14]

Realist environmental sociologists were further alarmed by the persistent questioning of not only science in general but also scientific depictions of ecological crises, as the methods and conclusions of sociologists of science influenced by the Edinburgh "strong programme" and the work of Bruno Latour and Steve Woolgar began to filter into environmental sociology.[15] Thus in extending arguments from the sociology of science into the terrain of environmental sociology, Steven Yearly stressed the "*uncertain* basis" of the global warming hypothesis, which he contended rested on questionable scientific authority and scientific framing, concluding that the fact that "we cannot know such things for certain" was the "Achilles' heel" of environmental science, as with science in general.[16] Realist environmental sociologists responded by insisting that in the face of the overwhelming scientific evidence and scientific consensus, with respect to global ecological crisis, to support skepticism in this area was to undermine the moral responsibility of society to nature and to future generations.[17]

Strong social constructionism is concerned with challenging the materialist/realist emphasis of science and gives credence in varying degrees to epistemologically based skepticism, nominalism, solipsism, antirealism, subjectivism, cultural relativism, and idealism in accounts of science and nature. In this, the strongest criticisms have been epistemological in character. Like philosophers who have traditionally seen epistemology as "polishing the mirror" of knowledge, social constructionists in the sociology of science frequently see themselves as polishing the mirror of scientific knowledge.[18] Science—contrary to widespread belief, even within science itself—is, we are told, more a mirror of the mind and of culture than of nature or reality. Such strong social constructionists are thus drawn to what Bhaskar calls the "epistemic fallacy," reducing all being or existence to human knowledge.[19] This leads to such startling claims as "the natural world has a small or non-existent role in the construction of scientific knowledge."[20] Steve Woolgar explicitly sides with nominalism against essentialism (which he associates with realism), arguing for a "reversal" of the realist arrow from nature to cognition, in favor of the nominalist-constructionist arrow from cognition to nature. "Objects," he says, "are constituted in virtue of representation . . . representation gives rise to the object."[21]

Klaus Eder sees the "cultural sociology of nature . . . as a way to expunge the latent naturalism from social theory."[22] Social constructionists in environmental sociology have aggressively questioned the realist tradition in terms that reduce it to an equally one-sided straw argument, devoid of any relation to even a mild constructionism. Thus Phil Macnaghten and John Urry complain that environmental sciences and much of environmental sociology "rest upon what we have termed the doctrine of environmental realism: that the realm of nature is separate and distinct from that of culture."[23] They claim that most surveys of environmental attitudes on issues like global warming and acid rain reflect "tacit assumptions of 'environmental realism': that environmental risks simply exist 'out there' independently of social practices and beliefs and can thus act as the unambiguous object of individual perceptions, attitudes and values."[24]

In this view, environmental realism and the discourse on sustainable development are characterized as "part of a modernist tradition in which the limits of 'natural' processes can be defined unproblematically by science."[25] Realist environmental sociology is therefore seen as immersed in an ontological fallacy—the contrary to Bhaskar's epistemic fallacy.

Nevertheless, realist conceptions of science and of environmental sociology are at their best far more dialectical than this description would suggest. As Stephen Jay Gould—a paleontologist, evolutionary biologist, and historian of science—eloquently expressed it, from the standpoint of materialist natural science:

> Science, since people must do it, is a socially embedded activity. It progresses by hunch, vision, and intuition. Much of its change through time does not record a closer approach to absolute truth, but the alteration of cultural contexts that influence it so strongly. Facts are not pure and unsullied bits of information; culture also influences what we see and how we see it. Theories, moreover, are not inexorable inductions from facts. The most creative theories are often imaginative visions imposed upon facts; the source of imagination is also strongly cultural.
>
> This argument, although still anathema to many practicing scientists, would, I think, be accepted by nearly every historian of science. In advancing it, however, I do not ally myself with an overextension now popular in some historical circles: the purely relativistic claim that scientific change only reflects the modification of social contexts, that truth is a meaningless

notion outside cultural assumptions, and that science can therefore provide no enduring answers. As a practicing scientist, I share the credo of my colleagues: I believe that a factual reality exists and that science, though often in an obtuse and erratic manner, can learn about it.[26]

Gould's view here, which eludes both the Scylla of the epistemic fallacy and the Charybdis of the ontological fallacy, is perfectly compatible with the sociology of science standpoint associated with Robert Merton in particular.[27] In fact, Merton is frequently referred to in Gould's work.[28] This kind of "social institutional constructivism" emphasizes the social contexts and institutions that condition science, and it has long been the bedrock of the history as well as the sociology of science.[29]

Social-institutionalist, even materialist, understandings, as David Bloor explains from the standpoint of the "strong programme," are logically compatible with and integrated into the more sophisticated constructionisms.[30] In addition to "social institutional constructivism," David Demeritt, in his useful fourfold typology of social constructionism, points to two other forms of constructionism that are compatible with realism: "social object constructivism," which emphasizes that social constructs such as gender are just as "real" in their causal effects as reality itself, and "artefactual constructivism," which promotes the non-dualist view that scientific knowledge is the result of a difficult negotiation between diverse human and nonhuman actors.[31]

Where social constructionism becomes antirealist or irrealist—the fourth form of social constructionism in Demeritt's typology—is when it claims (though sometimes as a purely methodological principle) that science and its objects are the product of human action and cognition alone. Demeritt labels this "Neo-Kantian constructivism."[32]

The debate in environmental sociology thus frequently consists of the difficult task of avoiding both overly naturalistic and overly sociologistic arguments.[33] Realists in recent years have placed increasing emphasis on the notion of the coevolution of nature and society, while constructionists have been turning to notions of "co-construction."[34]

Environmental feminists, especially those influenced by the historical-materialist tradition, have been fighting wars on two fronts, drawing dialectically on both constructionism and realism, while rejecting one-sided versions of both. These thinkers have insisted on the social construction of gender in the face of biological determinism and essentialism, yet have refused nevertheless to give up links to materialism, realism, and science.[35] Although insisting that "there is an important sense in which it is correct to speak of 'nature' as itself a cultural product or construction," Kate Soper observes that "it is not language that has a hole in its ozone layer; and the 'real' thing continues to be polluted and degraded even as we refine our deconstructive insights at the level of the signifier."[36] Furthermore, she insists that a dialectical understanding of the real, material opposition as well as unity of nature and society is necessary to address ecological problems: "I have consistently argued that there can be no ecological prescription that does not presuppose a demarcation between humanity and nature."[37]

Recently, Latour, one of the founders of the social constructionist approach to the sociology of science, has also shifted in what some have seen as a more realist direction with his actor network theory, which focuses on the dialectical relation between nature-culture.[38] Indeed, the concepts of "nature" and "culture," he insists, should be replaced by "nature-culture," in which both human actors and nonhuman actors (both now referred to as "actants") are seen as interacting with and mutually constituting one other. This perspective has led Latour at times in a more classically materialist direction. He argues that "by seeking to reorient man's exploitation of man toward an exploitation of nature by man, capitalism magnified both beyond measure."[39] As Latour has moved in this direction, his work has increasingly influenced ecological Marxists.[40]

Marxism in its classical conception was compatible with a sophisticated, critical materialism. Marx was a dialectical thinker, who absorbed much from Hegel and idealist philosophy in general. It is not surprising therefore that his work was highly critical of crude empiricism, mechanism, naturalism, and essentialism (positivism), while remaining materialist in orientation. Contemporary critical realism has developed on these foundations and has helped to inspire much of this environmental-sociological analysis.[41]

What we are calling here the "realist-constructionist" approach to the "sociology of ecology" evolves out of this broad critically informed realist tradition and is an attempt to understand the social construction of ecological science—within the context of a philosophy of praxis emphasizing human attempts to transform (not merely mentally construct) the world. Unlike crude naturalism, it takes into account the human construction of knowledge (albeit embedded in actual historical processes), but unlike the absolutist constructionism of strong idealism, its constructionism always takes account of and is tempered by the "materialist principle," "which derives from the fact that people are themselves material, animal and part of nature such that they are subject to certain of its causal laws and conditions."[42]

This approach can be contrasted to Macnaghten and Urry's conception of the "sociology of nature," which attempts to develop a "sociology of environmental knowledges" from the opposing idealist-constructionist standpoint.[43] Their work emphasizes how culture influences nature by discursively "reading" (or thinking or speaking) it and that this is not necessarily based in material conditions.[44]

In the following argument, as already indicated, we take the Tansley-Smuts debate of the 1930s on ecological holism, which led to the constitution of modern ecosystem science, as our main historical case study. We use a realist-constructionist outlook to explain why a materialist-realist approach to ecology (and environmental sociology) became dominant over its idealist-organicist rival—without in any way denying the historical, contingent nature of this process or that it was inevitably a particular social construction and a product of social struggle. The ecological science that emerged in this period, we suggest, was deeply affected by an "externalist" class-racial conflict and by the competing worldviews of materialism and idealism, as much as it was a direct ("internalist") outcome of the scientific process. The implications of this complex, many-faceted struggle over nature, science, and society remain with us to the present day and have served to shape the contemporary debate on ecology.

From the Sociology of Science to the Sociology of Ecology

The sociology of science is usually seen as having its most important precursor in the work of Boris Hessen. *The Dictionary of the History of Science* depicts Hessen as the originator of the externalist approach to the sociology of science—the broad approach that defined the early history of the field and in which Merton's foundational work was also to fall.[45] As J. G. Crowther, an influential science writer, wrote in his *Social Relations of Science*, Hessen's Marxist sociological analysis suddenly "transformed the history of science from a minor into a major subject."[46]

Hessen was a high-ranking Soviet physicist (director of the Moscow Institute of Physics) and defender of quantum theory and relativity theory at a time in which the Stalinist assertion over science started to call these scientific discoveries into question. In 1931, a Soviet delegation led by Nikolai Bukharin, one of the leading figures of the Bolshevik Revolution and of Marxist thought in the Soviet Union, made a surprise visit to the Second International Congress of the History of Science and Technology. Bukharin was accompanied by Hessen and other major representatives of Soviet Science—most notably Nikolai Vavilov, the foremost agricultural researcher in the Soviet Union and the discoverer of the original centers of world agriculture, constituting the main areas in which the genetic banks for contemporary agriculture are to be found (now known as the Vavilov Areas).

But it was Hessen's paper on "The Social and Economic Roots of Newton's 'Principia'" that was to have the greatest impact on the conference participants, representing a formative influence in the history and sociology of science and becoming one of the central texts for an important group of British "Baconian Marxist" scientists, including J. D. Bernal, J. B. S. Haldane, Hyman Levy, Lancelot

Hogben, Joseph Needham, and Benjamin Farrington.[47] Hessen presented a sophisticated look at how the necessities of economic production and military development under merchant capitalism led to the concentration on specific material-technological problems, such as crucial elements of navigation. These then dominated the thinking and general ethos of scientists at the time. These materialist concerns, he argued, contributed to the mechanistic outlook that played such a large part in Newton's *Principia* and the seventeenth-century scientific revolution in general.

Hessen's analysis took as its critical point of departure the idealist conception of science, which in the view of Karl Marx and Friedrich Engels customarily treated "the history of the sciences as if they had fallen from the skies."[48] For Hessen, a more meaningful understanding of the sources of scientific discovery had to be grounded in an understanding of the material conditions—social and economic, and also natural—in which such ideas evolved. He took as his case Newton's *Principia*—not only the most prestigious work in pre-twentieth-century physics but the symbol of a pure scientific viewpoint. Demonstrating how practical considerations regarding technology—especially with respect to mining, navigation, and optics—contributed to Newton's mechanistic worldview and how Newton's class perspective affected his thinking, Hessen was able to make a powerful case with regard to external materialist influences—both sociological and natural (thermodynamic) on the sciences. Having demonstrated this with respect to Newton, it was easy to establish the same thing with regard to other leading members of the Royal Society, such as Robert Boyle, Robert Hooke, Edmund Halley, and Sir William Petty, whose practical concerns with technology were much more obvious.

Another important contribution to the sociology of science was Edgar Zilsel's classic essay, which emphasized that it was "the rise of the methods of the manual workers to the ranks of academically trained scholars at the end of the sixteenth century [that] is the decisive event in the genesis of science."[49] Just as the origins of these mechanical arts, such as metallurgy, could not be rationally discussed without treating their class basis, they could not realistically be dealt with without some recognition that this was where production and nature met. Zilsel's emphasis on the relation between production, class, and the production of science represented one of the most important openings into the sociology and history of science. Zilsel was associated with the social-democratic politics in Vienna prior to the Second World War, emigrating to the United States after Hitler's rise to power. Steven Shapin sees Zilsel's externalist materialist-class account of the origins of science as overlapping with the insights of such historical-materialist thinkers as Hessen, Bernal, Farrington, and Needham.[50] Recently, the "Zilsel thesis" has been revived.[51]

By far the most important foundational work in the sociology of science originated with Robert K. Merton.[52] Merton frequently indicated his intellectual debt to Hessen, although sharply differentiating his own approach to the sociology of science from Hessen's Marxian analysis and adopting a more eclectic approach.[53] Beginning with his foundational work, *Science, Technology and Society in Seventeenth Century England* (originally published in 1938 based on his doctoral dissertation and reprinted in 1970), Merton carefully distinguished between the cognitive content of science and its sociological-institutional context.[54] His sociology of science deliberately steered away from epistemological questions and from those factors that could be viewed as intrinsic to the scientific method and rational thought generally, which he saw as governed by "universalistic criteria" (in his original formulation, he had said "universalistic facts of nature").[55] He adhered to the broad conception of the sociology of knowledge that "the social relations in which a man is involved will somehow be reflected in his ideas" and carried this over into the sociology of science.[56] But as distinct from the sociology of knowledge (his original area of interest), which was caught up from the beginning in epistemological concerns, Merton attempted to fashion the sociology of science as a largely empirical field concerned with how scientists were influenced by sociological factors, while leaving the actual content of science to be judged by its own universalistic criteria.

In concentrating in this way mainly on the external social-institutional influences, Merton examined both what has been called the "macro-environments of science" (such as political and economic systems and class) and its "micro-environments" (the intellectual milieu, schools of thought, invisible colleges, universities, and colleges, each with their own traditions).[57]

Merton was adamantly opposed to the "ivory tower," or extreme internalist, view of science in which scientists were depicted as "autonomous god-like creatures."[58] But in emphasizing the importance of external influences in his work, he did not thereby create a rigid demarcation between the internal and external but saw them as mediated in complex ways. He made it clear that external influences on science were important but without contending that these were primary with respect to particular scientific discoveries.[59] Merton's views overlapped with Hessen in arguing that historical developments in technology and social organization provided broad concepts and a social ethos out of which scientific developments, such as Newtonian mechanics, evolved.[60] Science, he recognized, was often a response to material-technological challenges. But he also emphasized, particularly in his famous treatment of the effects of seventeenth-century English Protestantism on science, that certain socially embedded values, as in the case of religiously derived views that became embedded in the Royal Society, could affect the progress and direction of science even more directly, creating the institutional basis of a scientific community.

In all of this, Merton defended the cognitive core of science as relatively immune from external influences. These influences were seen as constituting no more than the sociological environment of a science that nonetheless adhered to its own rational, universalistic criteria and hence autonomy. As Shapin explains, "Merton took care to position his thesis between what he saw as the extremes of pure Weberian idealism and the strong materialism that first surfaced in Anglo-American consciousness with Boris Hessen's 1931 [Marxian] account of Newton's *Principia*."[61]

From the formative work of Hessen, Zilsel, and Merton, certain conceptual frames or tools of analysis can be derived, which are used in the following inquiry into the sociology of ecology in the opening decades of the twentieth century. Science is not the product of "autonomous god-like creatures," does not "fall from the sky," and is not simply the product of an ivory tower but takes place in a socially embedded context. It is useful to make a distinction between the cognitive content of science and its social context, recognizing, however, that neither can be separated from each other but are mediated in complex, dialectical ways. The sociology of science has often drawn attention to the external, social-institutional context of science at both the macro-environmental (state, economy, ideology, and religion) and micro-environmental (the "invisible college") levels. For materialist-realist sociologists of science, it is reasonable to assume that science attempts to address "universalistic facts of nature." But our understanding of such "universalistic facts of nature" is nonetheless filtered in all sorts of problematic ways through human cognition and praxis. As Piotr Sztompka points out in his exposition of Merton's views, the objectivity of science is not easily gotten at but is best understood in terms of a "dialectical notion of objectivity"—the product of conflict and passion.[62]

Such a "dialectical notion of objectivity" is perhaps best exemplified today by feminist standpoint theory, growing out of the work of such thinkers as Nancy C. M. Hartsock, Dorothy E. Smith, Sandra Harding, Donna J. Haraway, and Kate Soper.[63] As expounded by Harding, the particular "justificatory approach" identified with feminist standpoint theory "originates in Hegel's insights into the relationship between the master and the slave and the development of Hegel's perceptions into the 'proletarian standpoint' by Marx, Engels, and Georg Lukács. The assertion is that human activity, or 'material life,' not only structures but sets limits on human understanding: what we do shapes and constrains what we know."[64]

In the Marxian view, knowledge (including scientific knowledge) is conditioned by material-historical development and class position. For feminist standpoint theorists—who, according to Fredric Jameson, represent today's "most authentic" heirs to Lukács's insight—this relates to differing material, lived conditions, and hence knowledge claims of women (as an oppressed-exploited group under patriarchy-capitalism) vis-à-vis those of dominant men.[65] Similar claims can be made with respect to those oppressed in racial terms.[66]

According to standpoint theory, knowledge is "socially situated."[67] Such knowledge, as Marxist theory taught, is dependent on the development of social relations and social conflict (and on material-natural conditions). "Standpoint epistemologies," according to Harding, were evident in the materialist approaches to the social construction of science, between Hessen in 1931 and Zilsel in 1942.[68] Such standpoint theory requires what Harding calls "strong objectivity" combined with a recognition of the historical character of knowledge. Harding describes "weak objectivity" as that form of objectivity that attempts to separate the positive from the normative, science from values. Yet because science is a socially embedded and often an elitist activity, such exclusion of values is impossible. "Strong objectivity," in contrast, attends to the social-natural environment and construction of knowledge-science and hence incorporates an interpretation of the historical-material background condition into scientific assessments. Its greater reflexivity gives it a stronger objectivity. This is especially true when the understanding of socially embedded conditions that affect dominant knowledge claims arises from the bottom of society, which has *less* interest in supporting prevailing ideologies, reifications.

As Harding writes:

> The history of science shows that research directed by maximally liberatory social interests and values tends to be better equipped to identify partial claims and distorting assumptions, even though the credibility of the scientists who do it may not be enhanced during the short run. After all, antiliberatory interests and values are invested in the natural inferiority of just the groups of humans who, if given real equal access (not just the formally equal access that is liberalism's goal) to public voice, would most strongly contest claims about their purported natural inferiority.[69]

Strong objectivity for Harding is related to the development of a theoretical perspective that embodies "strong reflexivity," recognizing that what we regard as "nature" is often an embodiment of social relations. Such strong reflexivity "requires the development of oppositional theory from the perspective of the lives of those Others ('nature' as already socially constructed, as well as other peoples), since intuitive experience . . . is frequently not a reliable guide to the regularities of nature and social life and their underlying causal tendencies."[70]

In this view, our ontological concepts of nature are often bound to systems of oppression. Thus Harding, building on the insights of William Leiss, argues that "science's claim to seek to dominate nature in order to control 'man's fate' has actually hidden its real function, and, often, intention: now and in the past, whether scientists intended it or not, science has provided resources for some people's domination of others."[71] The control of nature (and indeed our very concepts of nature) are therefore open to question, as they are connected to the control of society and its Others.

Feminist standpoint theory is to be contrasted to current trends in the sociology of science, Harding insists, partly because of standpoint theory's emphasis on "powerful background beliefs." Such a perspective is not so much concerned with the "microprocesses in the laboratory" as with the "macrotendencies in the social order"—as attendance to the latter is more crucial for the creation of a strong, critical framework.[72]

To address science as a contested realm, as required by notions of strong, dialectical objectivism, requires a view of the development of truth in science (and the demarcation of "good" and "bad" science) that sees this in terms of a dialectical struggle. The sociology of science has generally adhered to a methodological agnosticism with respect to the truth claims of science (though in recent years this has increasingly mutated into an epistemological skepticism and extreme relativism that seemingly undermines scientific knowledge claims). This has served to separate it off from the philosophy of science, which has been concerned in various ways with what distinguishes valid science.

A fully developed critical-historical perspective, however, requires attendance to both sociological background conditions and the truth claims of science. In the following analysis, we rely for heuristic purposes on certain aspects of the philosophy of science view offered by Imre Lakatos in his "methodology of scientific research programmes," and to a lesser extent on the view of scientific revolutions provided by Thomas Kuhn.[73] For Lakatos, the "demarcation problem" raised by Karl Popper with respect to the separation of science from pseudoscience cannot be addressed by falsification, as all scientific theories exist within an "ocean of anomalies."[74] Nor is it sufficient to take the more relativistic view of Kuhn. Rather, the answer, he contends, lies in the examination of "problemshifts" in the methodology of whole scientific research programs that allow one to determine whether a program is progressive or degenerative.[75]

"A research programme," according to Lakatos,

> is said to be *progressing* as long as its theoretical growth anticipates its empirical growth, that is, as long as it keeps predicting novel facts with some success ("*progressive problemshift*"); it is *stagnating* if its theoretical growth lags behind its empirical growth, that is, as long as it gives only *post hoc* explanations either of chance discoveries or of facts anticipated by, and discovered in, a rival programme ("*degenerating problemshift*"). If a research programme progressively explains more than a rival, it "supersedes" it, and the rival can be eliminated (or, if you wish, "shelved").[76]

Of course, the rivalry between research programs can be a very "protracted process," and it is even rational, according to Lakatos, to adhere to a degenerative scientific research program, if it seems possible to turn again into a progressive one. Moreover, the actual fate of scientific research programs cannot be explained simply in terms of their internal logic and development; thus considerations of "scientific rationality must be supplemented by empirical-external history."[77]

Lakatos's approach to science overlapped to some extent with Kuhn's but was clearer in its demarcation of what constituted progressive science. Lakatos, as Michael Burawoy pointed out, attempted to supply a theory of the "dynamics of paradigms" that was lacking in Kuhn.[78] Although Kuhn characterized as "normal science" a situation in which a single paradigm had a kind of monopoly and in which a crisis could develop as anomalies accumulatd—resulting in the growth of a rival paradigm and a full-fledged scientific revolution—Lakatos saw "normal science" in this sense as quite rare.[79] He wrote: "The history of science has been and should be a history of competing research programmes (or, if you wish, 'paradigms'), but it has not been and must not become a succession of periods of normal science: the sooner competition begins, the better for progress. 'Theoretical pluralism' is better than 'theoretical monism.'"[80]

Nevertheless, Kuhn himself insisted that one of the areas in which he agreed with Lakatos was in "our common use of explanatory principles that are ultimately sociological or ideological in structure."[81] In the following analysis, we will try to understand how principles that were "ultimately sociological or ideological in structure" affected, without actually determining, the social construction of ecological science.

The Double Transference

For environmental sociologists who do not err either on the side of the overnaturalizing of society or the oversocializing (overanthropomorphizing) of nature, the relation between nature and society is dialectical and complex. As Raymond Williams famously observed, "The idea of nature contains, though often unnoticed, an extraordinary amount of human history."[82] Conversely, the idea of society is often erected on conceptions of nature.

In any attempt to explore this complex nature-society dialectic and its relation to the rise of ecological science in the 1920s and '30s, it is useful to draw on Marx and Engels's concept of the "double transference" of ideas of nature and society, most evident in their day in social Darwinist thinking. Marx and Engels were strong defenders of Charles Darwin's evolutionary theory, which they viewed as the "death of teleology" in the natural sciences. Yet they were acutely aware that Darwin, as he readily admitted, had drawn some of his inspiration from the bourgeois political economy of Smith and

Malthus, which they saw as reflecting an alienated society. As Marx wrote to Engels in 1862, "It is remarkable how Darwin recognises among beasts and plants his English society with its division of labour, competition, opening up of new markets, 'inventions,' and the Malthusian 'struggle for existence.' It is Hobbes' *bellum omnium contra omnes*."[83]

Neither Marx nor Engels objected strongly in principle to the notion of the "struggle for existence" in nature—though Engels at one time stressed its one-sided character, which excluded cooperation.[84] Still, there were some problems, as they indicated, associated with the reading of the conditions of bourgeois society into nature—thereby producing one-sided conceptions drawn from alienated society and anthropomorphizing nature in terms of these.[85] Much more serious, however, from their standpoint was the re-extrapolation of these ideas—originally derived from bourgeois society and then imputed to nature—back again to society in naturalized, objectified form, and as eternal natural laws, in a kind of *double transference*. As Engels was to write in 1875:

> The whole Darwinian theory of the struggle for existence is simply the transference from society to organic nature of Hobbes' theory of *bellum omnium contra omnes* and of the economic theory of competition, as well as the Malthusian theory of population. When once this feat has been accomplished (the unconditional justification for which, especially as regards the Malthusian theory, is still very questionable), it is very easy to transfer these theories back again from natural history to the history of society, and altogether too naïve to maintain that thereby these assertions have been proved as eternal natural laws of society.[86]

Engels was not criticizing Darwin's entire evolutionary theory, much less Darwin himself, who never carried out such a double transfer. Rather, he was questioning what was subsequently to be labeled *social Darwinism*, putatively based on Darwin's work and rooted in such naturalistic notions as the "survival of the fittest" (a phrase adopted by Darwin in later editions of the *Origin of Species* from Herbert Spencer).

Engels stressed that the main problem of the social application of arguments derived from Darwin's theory of natural selection was due to the fact that much of Darwin's own theory had originally grown out of attempts to extrapolate social concepts to nature—itself a reasonable enterprise if carried out carefully (for example, Darwin's well-known use of the concept of "artificial selection" in introducing "natural selection"). The frequent re-importation back into society of the concepts of an alienated order, now dressed up in naturalized form, only confounded the problem of social analysis. Indeed, the phenomenon of double transference generated distorted, even sometimes deranged, social interpretations. Thus social Darwinists such as Spencer and William Graham Sumner sought to reduce society to eternal biological laws, sometimes of a racist character.[87]

Another instance of such a double transfer, Engels was to point out, was to be found in the development of social energetics. The concept of work had been usefully exported to thermodynamics as exemplified in Sadi Carnot's analysis of the steam engine. Yet attempts "to re-import the thermodynamic concept of work back into political economy" were fraught with problems, as were the attempts already being made at this time "to convert *skilled labour* into kilogram metres and then to determine wages on this basis." These efforts failed to recognize that economic-social conditions were dominant in the organization of human labor and could not be reduced to mere energetics.[88]

In the present day, similar double transferences have been introduced as a result of certain reductionist trends in biology, say in the notion of "the selfish gene." This meant once again taking ideas from society to explain nature and then re-extrapolating these concepts back again from nature to society in naturalized garb.[89]

As Michael Bell has noted: "Metaphors and general patterns of understanding easily flit back and forth between our theories of the realms we label 'society' and 'nature.' . . . Nevertheless, the flitting back and forth of concepts between science and social life deserves special scrutiny because of the way it sometimes allows science to be used as a source of political legitimization."[90]

If the transference of social concepts from the realm of society to that of nature always posed questions (for example, ecologists in the nineteenth century using the notion of "community" to describe the plant world), such a direct transfer of concepts, it is worth repeating, was certainly not to be excluded outright. Less acceptable, in the view of Marx and Engels, was the phenomenon of double transference, which often had the character of sleight of hand.

It is clear that double transference is a major critical issue both for environmental sociologists and sociologists in general. Such transference, especially if metaphorical in nature, can be a source of intellectual inspiration, but it can also if misused lead to the much more questionable anthropomorphization of nature and the naturalization of society. Anthropomorphism is relatively easy to detect, as it usually involves only a single transference rather than going full circle. Here nature is often seen in terms of the attributes of human beings or human communities—or in a religious context in terms of an anthropomorphic God. In contrast, extreme naturalization is often more difficult to detect, as it frequently involves a double transference and thus has come back full circle. Here human society comes to be seen as rigidly determined by irrevocable natural laws, which were conceptually modeled after human society.

Marx and Engels's dialectical critique of double transference in the realm of biology, it should be noted, did not extend simply to those who transferred bourgeois competition (the struggle for existence), to nature, and then in the form of eternal natural law back to human society. They also criticized those such as the Russian social theorist Pyotr Lavrov (a precursor to Kropotkin), who saw nature as one-sidedly cooperative—extrapolating concepts of cooperative human community to nature and then re-extrapolating these in naturalized form to society.[91] Likewise, they objected to the early-nineteenth-century "true socialist" Rudolph Matthäi's anthropomorphic claim that plants "demand soil, warmth and sun, air and rain for its growth" as a naturalistic model from which to argue for a rational human society. The notion of "demand" here was clearly taken from economics. "The plant," Marx and Engels wrote, "does not 'demand' of nature all the conditions of existence enumerated above; unless it finds them already present it never becomes a plant at all."[92] The methodological objection that they raised here was that of the extrapolation of the economic concept of "demand" to nature and then its re-extrapolation back to society—to create a naturalistic argument in this case in the service of a rational "true socialism." Such re-extrapolations were questionable on both dialectical and materialist grounds—regardless of whether the object was to promote bourgeois society or socialism.

Idealism, especially in its absolutist form, Marx believed, was especially prone to such double transferences in its attempts to incorporate science into its ontology of thought, discovering *Geist* in nature and then re-extrapolating this to society. In Hegel's *Philosophy of Nature*, Marx observed that nature/matter "is shorn of its reality in favor of human will" or spirit.[93] Or as Hegel put it, "The purpose of nature is to extinguish itself, and to break through its rind of immediate and sensuous being, to consume itself like a Phoenix in order to emerge from this externality rejuvenated as spirit."[94] Viewed in these terms, however, nature simply becomes the means to reveal the mind, spirit, human personality, and state—it has no independent existence apart from this anthropocentric, teleological goal. This is, in fact, a double transference: the point at which the Hegelian dialectic was at its weakest. As Auguste Cornu stated, the problem that Hegel encountered was that "the assimilation of the real to the rational can only be carried out by extremely arbitrary procedures."[95] As we shall see, the same problem has bedeviled idealist ecological holism, which is forced into arbitrary constructions in its attempt to assimilate ecology to the mind.

Holism as Superorganism

The notion of the sociology of ecology as we are treating it here is the study of the social construction of ecological science—a polymorphic science that aims uniquely at the dialectical unification of the natural and the social. Our concern, moreover, is to understand the origins of ecological science in ways that directly inform environmental sociology (the subdiscipline in sociology), which, viewed in realist-constructionist terms, is concerned primarily with the coevolutionary and often crisis-laden relations between nature and society.

Germany's most prominent Darwinian scientist, Ernst Haeckel, coined the word *ecology* in 1866. However, there was hardly any mention of the new term for a couple of decades. Not until 1885 did a book appear with it in its title.[96] Hence, ecology as an organized discipline cannot be said to have existed prior to the early twentieth century. In its earliest years, ecological science was dominated by a single scientific research program or paradigm, that of Frederic Clements in the United States. He was a professor of botany and wrote the classic textbook *Research Methods in Ecology* in 1905. Clements provided an idealist, teleological ontology of vegetation that viewed a "biotic community" as a "complex organism" that developed through a process called "succession" to a "climax formation." He therefore presented nature-vegetation as an organism or "superorganism" with its own life history, which followed predetermined, teleological paths aimed at the overall harmony and stability of the superorganism.[97] As he put it in his *Plant Succession*, the "climax formation is an organic entity"—the teleological reality of the superorganism.[98] Indeed, "the ecological ideal," Clements later stated, reflecting Jan Christian Smuts's influence, "is one of wholeness, of organs working in unison within a great organism."[99]

The strong teleological character of Clements's analysis gave it necessarily a neo-Lamarckian character. As Ronald C. Tobey explains, Clements "believed that plants and animals could acquire a wide variety and range of characteristics in their struggles to survive, and that these features were heritable. In the 1920s, he thus engaged in Lamarckian experimentalism, which ended in failure by the 1930s."[100] Altogether characteristic of Clements and his followers was the use of the notions of ecological "community" in ways that sometimes seemed to anthropomorphize nature and to impute a teleology to such "communities"—attributing to them "mysterious organizing properties."[101]

Although Clements's organicist perspective dominated plant ecology for half a century, in the 1920s and 1930s rival paradigms arose—those of individualistic, probabilistic, population theory (represented by Henry Gleason) and ecosystem ecology (represented by Arthur Tansley)—that were largely to supersede it. Each of these approaches represented stark alternatives since rooted in fundamentally different ecological ontologies. Clements's teleological model saw the natural environment as a superorganism. Gleason's more reductionist approach based on individuals focused on random processes and probabilistic events in the environment. Tansley's conception of ecosystem projected a materialist holism of a kind radically opposed to the teleological holism of thinkers such as Clements.

The transformation of Clementsian teleological holism in ecology from a progressive to a degenerative research program (in Lakatos's terms) was evident when this theory encountered a huge anomaly in the great Dust Bowl drought of the 1930s, which resulted in a crisis in the theory and a sharp contraction of its empirical content.[102] At the same time, Clements's holism began to merge with a more hierarchical model aimed at human society as well—in the work of Jan Christian Smuts, John Phillips, and John William Bews in South Africa, with whom Clements came to be aligned. A crucial factor in this shift, as in the work of Smuts and Phillips, was the attempt to construct an ecological view that combined teleological holism with ecological racism, in terms that seemed ultimately aimed at justifying the latter in a kind of double transference—what might be termed *social Clementsianism*. This led to an intellectual war between idealist and materialist approaches to ecology, and their respective versions of holism, that has persisted in various fashions in ecological science and ecological thought ever since.

When the smoke cleared in the 1930s and '40s, it was ecosystem ecology that had come to represent the new progressive scientific research program. Idealist, teleological approaches persisted as degenerative programs, marginalized within science, though always threatening to stage a comeback. In what follows, we explore the sociological construction of the understanding of nature in a historical materialist-ecological context and how issues of ecological racism, materialism versus idealism, and the question of double transference came to play a central role in this understanding. The importance of both realism and constructionism—indeed the significance of a realist-constructionist approach—will be highlighted.

Ecological Holism and Racial Hierarchy: Jan Christian Smuts

Inspired by Clements's ecology and by Walt Whitman, Baruch Spinoza, and Charles Darwin, Jan Christian Smuts coined the term *holism* as a means of describing nature-ecology.[103] Smuts shared his vision of nature with South African botanists/ecologists John William Bews and John Phillips and the Clementsian tradition in ecology. Building on Smuts's holism and the notion of succession in grassland evolution, Bews was to write *Plant Forms and Their Evolution in South Africa*, thanking Smuts for guidance and inspiration.[104] Eager to establish his "holism" concept in the scientific world, Smuts then wrote *Holism and Evolution*, a book that was to lead to modern conceptions of deep ecology.[105]

The ecological holism proposed by Smuts emerged out of his position in South African politics. Referred to as General Smuts because of his military role in the Boer War (he fought on the side of the Afrikaners), he was one of the principal figures in establishing the preconditions for the apartheid system. Smuts himself coined the word *apartheid* (meaning literally apartness) in 1917—almost a decade before he coined the word *holism*. Ironically, although Smuts has often been viewed as a "moderate" in the context of white South African racial politics, he has also been referred to as the "architect" of apartheid.[106] He was a strong advocate of the territorial segregation of "the races" and what he called "a grand [white] racial aristocracy."[107] He indicated at one time that he had a simple message: to "defy negrophilists." He is perhaps best remembered worldwide as the South African general who arrested Gandhi. Smuts tried to impede the flow of immigrants from India, imposed martial law against labor strikers, and deported labor leaders from the country.[108]

Smuts was the South African minister of defense from 1910 to 1919, and prime minister and minister of native affairs from 1919 to 1924 and 1939 to 1948. He was sometimes seen as a figure soaked in blood. When the Native Labour Union demanded political power and freedom of speech, Smuts crushed it with violence, killing 68 people in Port Elisabeth alone. When black Jews in Bull Hoek near Queenstown refused to work on Passover, Smuts sent in the police and close to 200 people were killed on his orders. In 1922, when black tribal populations in Bondelswart refused to pay their dog tax and complained about white penetration of their lands, Smuts sent in planes and bombed them into submission.[109] Horrified by these actions, the South African poet Roy Campbell was propelled to write the poem "HOLISM," which included the following lines:

> The love of Nature burning in his heart,
> Our new Saint Francis offers us his book—
> The saint who fed the birds at Bondelswart
> And fattened up the vultures at Bull Hoek.[110]

The "new Saint Francis" was Smuts; "his book" was *Holism and Evolution*. Although Smuts asserted that "we do not want to recreate Nature in our own image," his concept of holism was grounded in the social-political climate of South Africa, and it represented a transfer of social relations to nature and back again to society.[111] It embodied issues of domination and control. He argued that life is a process of change and evolution is a creative process. Rejecting the perceived rigidity of mechanism (or mechanistic materialism), Smuts sought a universal principle to explain the organization of both nature and society. He argued that "all organisms feel the force and moulding effect of their environment as a whole."[112] At the same time, the whole is self-active and operates under its own inherent activities. For him, the world comprised an ongoing, evolving series of wholes, which are constantly interacting. For each whole, the parts are in constant reflexivity, sustaining a dynamic equilibrium. The parts act to fix and repair any damage to the whole, because they are subordinate to the whole.[113]

Holism and Evolution starts with three premises. First, life evolved from matter. Thereafter matter as life (reflecting its emergent evolution) is no longer bound by mechanistic principles of motion and energy. Instead, matter has become a realm of life and the entire world is alive through progressive developments. Second, the natural world is essentially beneficent and moves toward constant improvement, which involves cooperation, service, and order. Third, the universe is concerned and guided by the principle of holism. The production and advancement of wholes is part of the essence of life. For instance, "the creation of wholes, and ever more highly organised wholes, and of wholeness generally as characteristic of existence, is an inherent character of the universe."[114]

In arguing that evolution was a process of creating ever more complex and important wholes and establishing that there was a hierarchy, Smuts was able to organize, order, and divide the world into a hierarchy of wholes, from low to high species. He assumed that evolution was a series of ordered advances toward greater perfection. The organism was the center of control, given that this was the site of the development of personality. As opposed to Darwinian natural selection, Smuts contended that the higher, teleological process of "Holistic Selection is much more subtle in its operation, and *is much more* social and friendly in its activity. . . . Its favours go to those variations which are along the road of its own development, efficiency, and perfection."[115] Nature's hierarchy was then seen as directly "social and friendly"—or cooperative.

Within the hierarchy of wholes there was a hierarchy of personalities (another level of wholes). This was Smuts's famous concept of "personalogy," which he related to his ecology. The notion of superior personalities, such as Johann Wolfgang von Goethe and Walt Whitman (themselves proponents of the organicist vision that appealed to Smuts), as the highest form of life was a view that seemed to present an almost religious striving.[116] Smuts declared "man is in very truth an offspring of the stars"—a quasi-religious view that was meant to stand in opposition to materialism.[117] This outlook was influential with Alfred Adler, Sigmund Freud's great opponent within psychology. Adler argued that "a body shows a struggle for complete wholeness" and saw this as connected to Smuts's emphasis on personality and holism.[118]

The most advanced, complex wholes (personalities), in this view, had greater independence (freedom) from the immediate environment. The less advanced did not have the same degree of freedom and control over their environment, which they could not socially construct to meet their needs and ends. Such people remained at the mercy of nature—they were seen as "children of nature." The hierarchy in the natural and social worlds was the result of natural development. Inequalities between races were the result of natural inequalities rather than social structures and social history. Life tended toward ever-greater perfection and goodness.[119]

Although *Holism and Evolution* was primarily abstract in its discussion, the lectures that Smuts presented at Oxford in 1929 on Africa were much more explicit with respect to his position on natural and racial relationships, and help us to understand the connections between his hierarchical, teleological ecology, in which nature is turned into a hierarchy of wholes—in which a human stratification, based on the notion of personalogy, is erected—and his role in laying the foundations for apartheid in South Africa. There is no doubt that the naturalized hierarchy that constituted Smuts's theory of ecological holism gave seeming philosophical-scientific support to his racial views. Indeed, as one critic of Smuts, the South African ecologist Edward Roux, indicated, in Smuts's holism, which followed on his concept of apartheid, "segregation was raised to a philosophy."[120] Smuts prepared his Oxford lectures to counter those who questioned the presence of Europeans in Africa and their right to influence African development. As a politician centrally involved in the organization of the League of Nations, he framed white European interest in Africa in naturalized, "humanitarian" terms, even when advocating outright racism. W. E. B. Du Bois, many years later, when Smuts pleaded for an article on "human rights" to be adopted by the United

Nations, did not miss the "twisted contradiction of thought" being revealed, given that Smuts had "once declared that every white man in South Africa believes in the suppression of the Negro except those who are 'mad, quite mad.'"[121]

In his Oxford lectures, published as *Africa and Some World Problems*, Smuts presented the colonial explorations of David Livingstone and Henry Morton Stanley as those of early Europeans seeking to bring civilization to the people of Africa. He asserted that their historic mission must be continued to save Africa from barbarism. Smuts in fact argued that Great Britain must take a humanitarian and commercial interest in Africa and that this would further civilize this land. Labor would be recruited from various African nations. But this development would also raise new questions regarding what happens "wherever a superior culture came in contact with a lower, more primitive. We cannot mix the two races, for that means debasement of the higher race and culture."[122]

Smuts argued that blacks naturally lacked an internal impetus for creating the world. In his ecological theory, they were seen as lacking the evolutionary development of a complex (climax) personality—a notion that represented a complex, double transfer from society to nature (via Smuts's holism) and then back again. Thus it was the duty and right of Europeans to organize the social and natural structure of Africa. In an account that drew on the concept of "recapitulation" (ontogeny follows phylogeny) as employed within nineteenth-century biological racism, Smuts wrote:

> It is even possible, so some anthropologists hold, that this was the original mother-type of the human race and that Africa holds the cradle of mankind. But whether this is so or not, at any rate here we have the vast result of time, which we should conserve and develop with the same high respect which we feel towards all great natural facts. This type has some wonderful characteristics. It has largely remained a child type, with a child psychology and outlook. . . . There is no inward incentive to improvement, there is no persistent effort in construction, and there is complete absorption in the present, its joys and sorrows. Wine, women, and song in

> their African forms remain the great consolations of life. No indigenous religion has been evolved, no literature, no art since the magnificent promise of the cave-men and the South African petroglyphist, no architecture since Zimbabwe (if that is African). Enough for the Africans the simple joys of village life, the dance, the tom-tom, the continual excitement of forms of fighting which cause little bloodshed. They can stand any amount of physical hardship and suffering, but when deprived of these simple enjoyments, they droop, sicken, and die. . . . These children of nature have not the inner toughness and persistence of the European, nor those social and moral incentives to progress which have built up European civilization in a comparatively short period. . . . It is clear that a race so unique, and so different in its mentality and its cultures from those of Europe, requires a policy very unlike that which suits Europeans.[123]

Smuts's reference to adult Africans as "children" drew on the recapitulation theory in biology, which had been widely adopted in the late nineteenth century but was already falling out of favor at the time Smuts was writing and has long since been rejected by modern biologists.[124] Recapitulation was the notion that each individual of a species in its development passes through (recapitulates) in telescoped fashion the main stages that the entire species over historical time had previously passed through. The recapitulation theory was often used, as in Smuts's case, to propound theories of biological racism. As Stephen Jay Gould has commented on the frequent racist use of the concept:

> For anyone who wishes to affirm the innate inequality of races, few biological arguments can have more appeal than recapitulation, with its insistence that children of higher races (invariably one's own) are passing through and beyond the permanent conditions of adults in lower races. If adults of lower races are like white children, then they may be treated as such—subdued, disciplined, and managed (or, in the paternalistic tradition, educated but equally subdued). The "primitive-as-child" argument stood second to none in the arsenal of racist arguments supplied by science to justify slavery and imperialism.[125]

Building on recapitulation theory and his own racist-ecological holism, Smuts proposed that separate and parallel institutions and segregation were required to save and retain African wholeness. He argued that this policy would help preserve racial purity, by preventing miscegenation, and would maintain a healthy, good society. Any unnatural mixing of the people, contravening the natural, hierarchical principles, would lead to the moral deterioration of the species. He argued:

> The old practice mixed up black with white in the same institutions; and nothing else was possible, after the native institutions and traditions had been carelessly or deliberately destroyed. But in the new plan there will be what is called in South Africa "segregation"—separate institutions for the two elements of the population, living in their own separate areas. Separate institutions involve territorial segregation of the white and black. If they live mixed up together it is not practicable to sort them out under separate institutions of their own. Institutional segregation carries with it territorial segregation. The new policy therefore gives the native his own traditional institutions on land which is set aside for his exclusive occupation. . . . For urbanized natives, on the other hand, who live, not under tribal conditions but as domestic servants or industrial workers in white areas, there are set aside native villages or locations, adjoining to the European towns. . . . This separation is imperative, not only in the interests of a native culture, and to prevent native traditions and institutions from being swamped by the more powerful organization of the whites, but also for other important purposes, such as public health, racial purity, and public good order. The mixing up of two such alien elements as white and black leads to unhappy social results—racial miscegenation, moral deterioration of both, racial antipathy and clashes, and to many other forms of social evil.[126]

In Smuts's intellectual system of apartheid-holism-apartheid, we therefore find signs of a double transfer. Natural hierarchy (modeled on social hierarchy) is used to justify social hierarchy, and social hierarchy is used to give meaning to natural hierarchy in a never-ending whole. His ecology gives rise to a complex or climax personality that was a manifestation of notions of racial hierarchy—and meant to further justify racial hierarchy. Smuts contended that Africans, in contrast to Europeans, were "children of nature," lacking the drive for social "progress."[127] Therefore Europeans must enact special policies to "conserve what is precious" about Africa and Africans.[128] The racial differences of society were attributed to nature and then re-extrapolated back to society to justify extreme segregation (apartheid). The whole conception is mirrored after a view of dominant personalities as the model for natural-social domination, within a teleological perspective. For Smuts, ecology was the science and justification for this new holism that naturalized social control.

Smuts saw his work as countering materialist approaches to science. The reception of *Holism* within British science was so strong as to catapult him to president of the British Association for the Advancement of Science (BAAS). In his 1931 presidential address to the BAAS, he attacked the physicist John Tyndall's famous 1874 presidential address to the Association (which was much admired by Marx and Engels) as an "unrestrained expression" to the "materialistic creed."[129] For Smuts nature in any truly meaningful sense (beyond mere "brute fact") was to be seen as a construction of the mind:

> Great as is the physical universe which confronts us as a given fact, *no less great is our reading and evaluation of it in the world of values. . . . Without this revelation of inner meaning and significance the external physical universe would be but an immense empty shell or crumpled surface*. The brute fact here receives its meaning, and a new world arises which gives to nature whatever significance it has. As against the physical configurations of nature we see here *the ideal patterns or wholes freely created by the human spirit* as a home and *an environment for itself*.[130]

These views put Smuts in an idealist camp that saw the physical universe as "an immense empty shell" apart from mind and the "pattern of wholes created by the human spirit" as its own environment. He was thus seen as one of the great British Empire idealists, along with such figures as Alfred North Whitehead, Francis Herbert

Bradley, Robin George Collingwood, John Alexander Smith, Lloyd Morgan, and John Scott Haldane.[131] Karl Popper viewed Smuts as a Hegelian evolutionary idealist.[132]

Smuts's ecological holism was enormously influential. In particular, John Phillips, a South African grasslands ecologist, incorporated Smuts's as well as Clements's holism into his own ecological studies.[133] In Phillips's construction of the natural world, humans were part of a biotic community that was filled with cooperation and harmony.[134] At the same time, human beings were naturally organized in a racial hierarchy. The fact that these two ideas coexisted within a single construction is no accident, since as historian Peder Anker notes, "Phillips coined the term 'biotic community' to designate this ecocentric ethics and environmental social policy of segregated ecological homelands."[135] Phillips argued in his scientific writings on ecology that natives should not be granted any autonomy or freedom because it would violate the relations of races within the community. The "ruling races" were to regulate the stock of natives to prevent excess grazing and degradation of the environment. In Phillips's racist biocentrism, miscegenation between the lower European stock and the natives was to be avoided to prevent the degeneration of biological diversity. Women's desire for freedom should be constrained and large families among whites should be encouraged.[136]

In his *Human Ecology*, John Williams Bews, another South African ecologist and follower of Smuts, contended that some humans were determined by the conditions of their environment, whereas other humans were more independent of their environment.[137] This argument developed out of Smuts's theory regarding certain organisms and personalities being more independent and strong, versus those that were affected by the environment. Bews spoke of "the ecological division of mankind" as necessitating "the segregation of the races." In his 1931 article on the "Ecological Viewpoint," he transferred the concept of natural hierarchy modeled on human society back to human society, speaking of a "climax type of men" exemplified by "the small white population in South Africa."[138] Those primitive peoples who were still tied to the "Earth-mother," he argued, should be left as much as possible in their own proper biotic communities. Bews also insisted that marriage was the only natural relation between men and women and that homosexuality was ecologically and morally wrong. Smuts's holism thus reinforced naturalized ecological-racist views. At the same time, Clements, and other ecologists in the United States, became strong defenders of Smuts's and Phillips's ecological holism.[139]

Materialist Ecosystem Analysis: Arthur Tansley

General Smuts's teleological, holistic philosophy together with its racial implications engendered the ire of both socialists and consistent materialists-realists. Smuts's legendary opponent in the great "Nature of Life" debate that took place at the BAAS meetings in South Africa in 1929 was the British Marxist biologist Lancelot Hogben, who at the time occupied a position at the University of Cape Town. Not only did he debate Smuts—opposing his own materialism to Smuts's holism and attacking Smuts for his racial eugenics—but Hogben also reportedly hid black rebels fleeing the racist state (in which Smuts was such a dominant figure) in a secret compartment in his basement.[140] Hogben viewed Smuts's holism as a more sophisticated version (incorporating notions of emergence) of the vitalism ("creative evolution") of Henri Bergson and others. In opposition to this, he presented a mechanistic materialism and agnosticism on the nature of life and the world in general. Although there were deep issues of materialism and science involved, Hogben made it clear that his opposition stemmed in large part from a perception of the dialectical perversion associated with Smuts's ecologically racist holism. As he observed in his book *The Nature of Living Matter*, organized around his 1929 debate with Smuts and his followers: "The benign and tolerant humanism which [the ancient materialist] Epicurus grafted on the soil prepared by the atomists was ill suited to flourish in the stern climate of the [Hellenistic] military state. Like [Smuts's] holism, Aristotle's

[Hellenistic] system was a shrewd blending of science and statesmanship. It enabled its author to combine a personal predilection for natural history with a political partiality for slavery."[141]

Another major opponent of Smuts, and one who would directly influence Tansley, was the British Marxist mathematician Hyman Levy, who, in *The Universe of Science*, developed a critique of Smuts's holism along similar lines to those of Hogben, elucidating a materialist systems theory in response.[142] Levy's own ecological interests were evident in *A Philosophy for a Modern Man*.[143]

However, the central figure in opposing the ideas of Clements, Smuts, Phillips, and Bews in ecology was Arthur Tansley, a moderate or Fabian-style socialist, the first president of the British Ecological Society and the originator of the ecosystem concept.[144] Tansley, significantly, had been a student of biologist Ray Lankester at University College, London.

Lankester was Thomas Huxley's protégé and was considered the greatest Darwinian scientist of his generation. He was also the most adamantly materialist biologist of his day in Britain. When he was a boy, Darwin and Huxley, who were friends of his father, both played with him. As a young professor, Lankester was a close friend of Karl Marx and an admirer of Marx's *Capital*. He was a mourner at Marx's funeral. Lankester considered himself a socialist, though of the more Fabian variety.[145] He was also to become one of the most ecologically concerned thinkers of his time and wrote some of the most perceptive and eloquent essays that have ever been written on species extinction because of human causes, discussing the pollution of London and other ecological issues with an urgency that was not found again until the late twentieth century.[146] Lankester was an adamant opponent of vitalism, authoring a preface to a book criticizing Bergson's *élan vital*.[147]

The young Tansley was deeply influenced by Lankester, along with the botanist Francis Wall Oliver, in his years at University College, London. Like Lankester, Tansley was an adamant materialist. And like Lankester, Tansley was to challenge directly attempts to conceive evolutionary ecology in antimaterialist, teleological terms.

The natural environment that ecologists like Tansley encountered in Britain was overwhelmingly "second nature," in the sense that all of it had been transformed by human beings, and may have brought to mind evolutionary materialist issues of ecological crisis and sustainability—in ways that a more untouched, "pristine nature" as encountered in the colonies (or former colonies) did not. For Clements, Smuts, and Phillips, who drew from their contemplation of the relatively "untouched" grasslands of the United States and South Africa a teleological conception of nature (the historic role of indigenous peoples in the management of these environments was still not understood—or rather denied, especially in Smuts's hierarchy where indigenous peoples remained at the mercy of nature), this all seemed perfectly "natural." But for Tansley, the leading ecologist in the British Isles, the environment with which he had to deal was only "seminatural" at best, affected at every point by human intrusions and transformations.[148]

Although a professional botanist and plant ecologist, Tansley was far from being a narrow specialist and was engaged as well with psychology and philosophy. In 1920, he published his book *The New Psychology* on Freud's psychoanalysis and how the human mind was affected by the laws of biology; it became a bestseller and went through eleven editions.[149] In 1922, with the help of Ernest Jones, Tansley went to Vienna to study with Freud and to undergo psychoanalysis by Freud himself. Freud referred two clinical patients to Tansley whom he treated for years. For many years, Tansley was considered one of the leading British experts on Freud's psychoanalysis. His work on ecology frequently drew analogies from psychology. As late as 1952, in his *Mind and Life*, Tansley continued to attempt to synthesize the basic elements of existence within a framework that encompassed both psychology and ecology.[150]

Tansley considered himself a scientific realist but also one who recognized that our understandings of nature were constructed—what would today be called a "mitigated scientific realism" or "critical realism."[151] Adhering strongly to the materialist principle but recognizing social constructionism, he argued, to a largely idealist audience at the

Magdalen Club at Oxford, that even if one were to suppose "that a large part of the Universe *is* arranged [presumably by God] to fit the scientist's ambition," it was nonetheless true that such natural systems could be considered "real phenomena—that they are *there* and are not mere figments of our fantasy."[152]

In 1935, Tansley found himself increasingly at odds with anti-materialist constructions of ecology that were then gaining influence, and he entered the fray against ecological idealism. It was at this time that he wrote his historic article for the journal *Ecology* titled "The Use and Abuse of Vegetational Concepts and Terms," which declared war on Clements, Smuts, and Phillips and introduced the new concept of "ecosystem."[153] The term *abuse* in the title was meant to convey Tansley's objection to the direction ecological "holism" was taking. The immediate target of Tansley's critique was a series of three essays (and particularly the third of these on "The Complex Organism") by Phillips, in which the latter had attempted to advance the case of Clements and Smuts against materialists like Hogben.[154]

Phillips's organicist constructions raised the ire of Tansley, and in one fell swoop in his *Ecology* article, he attacked a whole set of teleological notions propagated by Clements, Smuts, and Phillips: (1) that ecological succession was inherently progressive and developmental, leading to a climax; (2) that vegetation could be seen as constituting a superorganism; (3) that there was such a thing as a "biotic community" (with members), encompassing both plants and animals; (4) that "organismic philosophy," which saw the whole universe as an organism, was a useful way to understand ecological relations; and (5) that holism could be seen as both cause and effect of everything in nature—and extended to society.

Smuts's holistic teleological view, Tansley pointedly asserted, was "at least partly motived by an imagined future 'whole' to be realised in an ideal human society whose reflected glamour falls on less exalted wholes, illuminating with a false light the image of the 'complex organism.'"[155] This was a polite way of referring to the system of racial stratification, which was built into Smutsian holistic ecology. For Tansley, Clements, Smuts, and Phillips had to differing degrees carried out a questionable extrapolation of anthropomorphic social concepts (in the case of "plant communities" and climax personalities) to nature and then had re-extrapolated these concepts to society. In Tansley's case, the principal objection was how this promoted racist notions. Indeed, as historian Anker contends, Tansley, in the passage quoted above, was "referring to Phillips's racist biocentrism and the politics of holism . . . with its treatment of 'less exalted wholes' at, for example, Rand, Bondlewaart, and Bull Hoek"—the sites of three massacres of native Africans ordered by Smuts.[156]

In combating this type of idealist holism and superorganicism and introducing the concept of ecosystem in response, Tansley turned to the dialectical systems theory used in Levy's *The Universe in Science*, tied to new developments in physics.[157] "The more fundamental conception," represented by his new "ecosystem" concept, Tansley argued, was that of

> the whole *system* (in the sense of physics), including not only the organism-complex, but also the whole complex of physical factors forming what we call the environment of the biome—the habitat factors in the widest sense. Though the organisms may claim our primary interest, when we are trying to think fundamentally we cannot separate them from their special environment, with which they form one physical system. . . . These *ecosystems*, as we may call them, are of the most various kinds and sizes. They form one category of the multitudinous physical systems of the universe, which range from the universe as a whole down to the atom.[158]

Following Levy, Tansley emphasized a dialectical conception of abstraction: "The systems we isolate mentally are not only included as parts of larger ones, but they also overlap, interlock and interact with one another. The isolation is partly artificial, but is the only possible way in which we can proceed."[159]

Moreover, Tansley argued that it was somewhat "arbitrary and misleading" to remove climatic factors from any consideration of the ecosystem and that the relation between organisms and the environment was "reciprocal."[160] Nature, in Levy's and Tansley's conception, was not to be viewed as seamless but, on the contrary, had certain natural seams in its fabric, delineating interactive subsystems of the whole (isolates) that were open to analysis.[161] Tansley thus wrote that "whole webs of life adjusted to particular complexes of environmental factors, are real 'wholes,' often highly integrated wholes, which are the living nuclei of *systems* in the sense of the physicist."[162]

Rather than seeing such ecological "wholes" in terms of a natural, teleological order, he emphasized contingency and constant disruptions to any kind of natural stasis, referring to "the destructive human activities of the modern world" and presenting human beings especially as an "exceptionally powerful biotic factor which increasingly upsets the equilibrium of preexisting ecosystems and eventually destroys them, at the same time forming new ones of very different nature."[163] Thus human beings were capable of "catastrophic destruction" in relation to the environment.[164]

Tansley's view of ecosystem disruption by human beings thus introduced a notion of widespread crisis of ecosystems emanating from anthropogenic causes. "Ecology," he argued, "must be applied to conditions brought about by human activity," and for this purpose, the ecosystem concept, which situated life within its larger material environment, and penetrated "beneath the forms of the 'natural' entities," was the most practical form for analysis.[165] In his comprehensive study, *The British Islands and Their Vegetation*, Tansley put forward a dynamic point of view, in contrast to Clements's model of succession and climax. He explained:

> The position of relative equilibrium, corresponding with what I have called the mature "ecosystem," is the fundamental ecological concept. . . . "Positions of equilibrium" are seldom if ever really "stable." On the contrary, they contain many elements of instability and are very vulnerable to apparently small changes in the factor-complex. Recognition of "positions of stability" is a necessary first step in the understanding of vegetation. The more important sequel is study of the factors which maintain or disturb and often upset them.[166]

Tansley's ecosystem concept was, arguably, more genuinely holistic and more dialectical than the relatively rigid superorganicism and "holism" that preceded it because it brought both the organic and inorganic world within a more complex materialist, Darwinian-style synthesis.

Tansley recognized that the conditions of nature were a product of both natural and human history.[167] Through analyzing pollen deposits in layers of peat, he studied how the advance and retreat of glaciers influenced the distribution of plants in a given geographic area.[168] Disturbance was recognized as an important factor in plant composition—as were historical changes in climate, soil conditions, and animal populations. Tansley highlighted how all of these relationships interacted and influenced the historical succession (or regeneration) of plants in a particular environment. Thus nature, through systematic study, influenced Tansley's conception of ecosystems.

Natural processes and cycles operate, influencing the growth of plants within a particular historical context, and at the same time human encroachment increasingly transformed nature. Tansley explained:

> With his increasing control over "nature" the human animal became a unique agent of destruction of the original ecosystems, as he cleared and burned natural vegetation and replaced it with his pastures, crops and buildings. Limited at first to the regions where civilization originally developed, this destructive activity has spread during recent centuries, and at an increasing rate, all over the face of the globe except where human life has not yet succeeded in supporting itself. It seems likely that in less than another century none but the most inhospitable regions—some of the extreme deserts, the high mountains and the arctic tundra—will have escaped. Even these may eventually come, partially if not completely, under the human yoke.[169]

Draining the fens and deforestation—to create pastures—radically modified the natural conditions, such as soil fertility, and changed the distribution of plants. Tansley indicated that such alterations led to the "establishment of a new ecosystem, [which was] the result of the original factors of climate and soil together with the modifying factors which [humans] introduced."[170] Through studying the disturbance, transformation, regeneration, and destruction of nature, Tansley developed a dynamic conception of ecosystems.

Tansley's dialectical rejection of a system of natural-holistic harmony was rooted, as it had been for other thinkers before him, such as Darwin and Marx, in his materialism. "The Use and Abuse of Vegetational Concepts and Terms" referred to Lucretius (Epicurus), citing Lucretius's *On the Nature of the Universe.* Tansley, like many other thinkers in the materialist and scientific traditions, was inspired by the ancient Epicurean materialist critique of teleology and religion and transformed this into a critique of Smuts's holism.[171]

For Tansley, Phillips's articles, which sought to develop the teleological ecology of Clements and Smuts, "remind one irresistibly of the exposition of a creed—of a closed system of religious or philosophical dogma. Clements appears as the major prophet and Phillips as the chief apostle, with the true apostolic fervour in abundant measure." Phillips, according to Tansley, had "recourse to scientific *arguments*" only "here and there," relying for the most part on "the pure milk of the Clementsian word."[172]

Nowhere was the existence of a closed, idealist-teleological model of nature more evident than the insistence on the inherently progressive nature of succession, always pointing toward the climax state. In contrast, Tansley argued, in contradistinction to Clements, that there was also such a thing as "retrogressive succession"—a succession (in time) that led away from the climax system. In this way, Tansley's ideas paralleled those of his mentor, Lankester, who had rejected all teleological interpretations of evolution, famously arguing that degeneration was possible in the evolutionary process. Nonetheless, Tansley continued to argue from a systems perspective, for a "dynamic ecology" that was organized around the general tendency toward dynamic equilibrium in ecosystem development. This, he claimed, was "the ecology of the future."[173]

As in the case of Hogben, Tansley's rejection of the ecological ideas of Smuts and Phillips seems to have been motivated as much by the dislike of ecological racism as by his opposition to ecological idealism. When Smuts delivered his lectures on Africa and race at Oxford in 1929, Tansley was in the audience and given what we know of the latter's views was almost certainly not impressed. Tansley belonged to a group of thinkers, including H. G. Wells and Julian Huxley, who saw ecology as standing for a more rational approach to human society and nature. Wells and Huxley had coauthored (with G. P. Wells) the important work *The Science of Life*, first appearing in 1929, which had provided a materialist ecological vision.[174] Both Wells and Huxley were close friends with Hogben, while Tansley was closely connected to Huxley.[175] In this context, Wells's judgment on Smuts's ecological racism doubtless reflects the view of all of these thinkers. As Wells wrote in *The Fate of Man*:

> It is one of the good marks in the checkered record of British Imperialism that in Nigeria it has stood out against the development of the plantation system and protected the autonomy of the native cultivator. . . . But against that one has to set the ideas of white-man-mastery associated with Cecil Rhodes and sustained today by General Smuts, which look to an entire and permanent economic, social and political discrimination between the lordly white and his natural serf, the native African. And this in the face of the Zulu and Basuto, the most intelligent and successful of native African peoples. The ethnological fantasies of Nazi Germany find a substantial echo in the resolve of the two and a half million Afrikanders to sustain, from the Cape to Kenya, an axis of white masters . . . with a special philosophy of great totalitarian possibility called holism, lording it over a subjugated but more prolific, black population. The racial antagonism makes the outlook of South Africa quite different from that of most of the other pseudo-British "democracies." Obviously it is not a democracy at all, and plainly it is heading towards a regime of race terrorism on lines parallel and sympathetic with the Nazi ideal.[176]

Smuts, as a leading figure within the British Empire, opposed the Nazis and led South Africa in declaring war on Germany. But Wells was not off base in seeing in the philosophy of Smuts's holism, with its ecological justification of developing apartheid, the foundation of what might be called a philosophy of ecological apartheid—akin in some ways to the "ethnological fantasies" and extreme racial oppression (and exterminism) of Hitler's Germany. Without question, such concerns about ecological hierarchy and racism brought out the very sharp differences between the two scientific research programs and their respective constructions of ecology.

Tansley stuck adamantly to materialist-realist constructionism. He insisted that ecology must be seen in terms of dialectically interrelated, dynamic systems (ecosystems) that were free as much as possible from the prior imposition of teleological concepts and the smuggling in of social concepts that were meant to reinforce rigid social hierarchies through the re-extrapolation of these concepts in naturalized form back into society. All forms of anthropomorphization of nature (the direct, as opposed to metaphorical, transferences of human-social characteristics to nature) were suspect. Ecology should not be seen as a reflection of the glamour of grand human personalities "on less exalted wholes." In his materialist-realist construction of an ecological worldview, Tansley thus rejected both anthropomorphism in the mental "construction" of nature and naturalistic justifications for racial oppression in human society. At the same time, he insisted that human beings could be destructive forces in nature—as revealed in his studies of changes in plant distribution and transformations in ecosystems—undermining ecological systems. Rejecting an idealist-teleological approach to ecology, he directly challenged these theories in terms of their empirical realism: "What researches," Tansley rhetorically asked, "have been stimulated or assisted by the concept of 'the complex organism' *as such*?"[177]

Both sides of this debate, it should be noted, were concerned with promoting conservation in the face of human ecological destruction. Tansley became the first chairman of the British Nature Conservancy and the most consistent advocate for the creation of nature reserves in Britain.[178] Smuts proposed a system of nature reserves in South Africa. But in line with the ecological racism that was an intrinsic part of Smuts's ecological holism, Smuts naturally saw this as carrying over into a conception of reserves for native Africans themselves. Smuts's holism was, as H. G. Wells intimated, "a special philosophy with great totalitarian responsibility," rooted in a double transference.[179] Smuts's clear ambition was to be a grand legislator over both society and nature within what was to become the apartheid system.

Materialism versus Idealism in Ecology's Formative Period

In many respects, the conflict between teleological holism in ecology and ecosystem ecology may be regarded as inevitable, quite apart from questions of race. The conflict over the meaning of ecological holism in one form or another has been one of the crucial tasks of ecological thought. Ironically, one well-known treatment of ecological paradigms has argued that "the materialistic revolution in ecology" associated with Tansley's ecosystem concept carried forward "the first ecological ideal, Clements's superorganism, [which] is not dead, but rather transmogrified into a belief that holistic study of ecosystems is the proper course for ecology."[180] Although more reductionist approaches to ecology existed, ecological science gravitated toward holistic answers of one kind or another.

Tansley's own approach has been described as a "nonteleological mechanistic holism."[181] Ecology demanded either an idealist organicism and holism, such as what was provided by Clements and Smuts, or a materialist holism, which grew out of Tansley's concept of ecosystem. The evolution of Clementsian ecology into a form propounded by Smuts and Phillips—a development supported by Clements himself—however, marked the distorted bias of teleological holism when it sought to expand into a truly holistic vision connected to human society.[182]

Clements and Smuts developed rigidly hierarchical constructions of the ecological world, drawn from society and then reimposed, in the case of Smuts and his followers, back on society in the form of a system of ecological apartheid. The resulting intellectual system was incoherent to an extreme, hamstrung by its own teleology. Phillips sought to link Smuts's teleological holism directly with that of social Darwinists such as Spencer and Sumner, with whom there were many political-social similarities, but was thwarted by the possessive individualism of the social Darwinists, which did not fit easily with a more Aristotelian holistic perspective.[183]

The holism of Smuts and Phillips (and Clements) ran up against problems in incorporating empirical observations, limited as this perspective was by its essentially linear, teleological thrust. As Phillips put it, "succession is progressive only."[184] But the explicit insistence on the teleological "progressive only" in relation to ecological succession marked the degeneration of the organicist research program, as it went against Darwinian conceptions of evolution, which explicitly reject such teleological notions. It also declared by mere fiat what needed to be determined empirically. Not surprisingly, then, Phillips wrote: "General Smuts' need for refinement and for extension of his theme is more and ever more facts, interrelated facts, suggestions and soundly based ideas regarding organisms, communities and the changing stage—the habitat—on which they play their part."[185]

But the incorporation of empirical considerations and even more the generation of a research program that would anticipate novel facts within a dynamic, natural-historical set of relationships was certainly not a strength of teleological holism within ecology, as it assumed that all empirical data would have to fit within its abstractly constructed Procrustean bed. After Clements's early analyses, the intellectual progress of the teleological paradigm stalled. The failed attempt to merge ecological holism of this kind with sociology, particularly of a racist variety, and to implement this as policy in Smuts's South Africa contributed to the degeneration of the entire research program associated with holism, if only by encouraging the development of a considerable materialist ecosystem ecology in opposition. Smuts's attempt to carry out a double transfer had the effect of destroying whatever genuine insights the notion of ecological "holism" contained.

In the Popperian philosophy of science, the downfall of a paradigm is attributed to a crucial experiment or crucial anomaly. In

Clements's case, the anomaly is supposed to be observations on the development of the prairies in the United States during the drought of the 1930s, which followed a pattern other than what Clementsian succession had supposed. The severe drought associated with the Dust Bowl of the 1930s, dramatized in John Steinbeck's *The Grapes of Wrath*, was experienced as a refutation of Clementsian teleological ecology and resulted—in the process of trying to account for the anomaly—in a severe contraction of the paradigm's empirical content, the sign of a degenerative research program in Lakatos's terms.[186]

Popperian disproofs, based on anomalies, are, however, rarely conclusive. Clementsian ecology might have absorbed that anomaly, as it at least attempted to do, as just one of what Lakatos called "an ocean of anomalies" affecting all sciences. The real reasons for the organicist-holist paradigm's downfall are more directly related to its ultimate research objectives. The paradigm projected a holism that pointed to the existence of a "superorganism." The very teleological orientation of this perspective created insoluble problems that became ever more apparent as the hierarchical-teleological analysis was extended into the social realm. The organicist-holist paradigm became "theoretically exhausted" and unable to grapple with the complex, contradictory changes taking place in the real world.[187] The teleological, equilibrium-oriented, neo-Lamarckian approach to ecology represented by the hierarchical model of superorganicism and holism was ill equipped, as we have seen, to deal with the advent of ecological crises, or the destructive aspects of human intervention in the environment (and society). Its linking to philosophic apartheid in the work of Smuts and Phillips likely only added to its problems. Hence, this research program increasingly took on the form of a degenerative research program, in Lakatos's sense—one that was hindered by its philosophical idealism, its tendency to extreme constructionism, its theoretical anthropocentrism, and its hierarchical social content. As Leiss explained, "the domination of nature" has always been about the domination of society (and domination within society).[188] Nowhere was this more evident than in Smutsian holism.

In contrast, the progressive research program associated with Tansley's ecosystem ecology gained ground as part of what Simberloff called a "materialistic revolution."[189] Beginning with the ecosystem concept, it evolved into a general systems theory in ecology. Tansley's ecosystem analysis, arising within plant ecology, was rooted in studying the disturbance, transfer, destruction, and growth of plants in relation to dynamic, historical-natural conditions, such as changes in climate and soil fertility. His science and ecosystem concept was fundamentally informed by his studies of plant history and emergent nature. Tansley's ecosystem ecology easily connected with work on animal ecology developed by Tansley's friend Charles Elton. This approach was given an enormous boost by the publication of Raymond Lindeman's important article "The Trophic-Dynamic Aspect of Ecology," which incorporated energy flows into the ecosystem model.[190]

Increasingly, ecosystem analysis merged with thermodynamic perspectives coming out of classical physics. In the work of Eugene and Howard T. Odum (sons of sociologist Howard W. Odum), ecosystem analysis was integrated with a more general systems ecology evolving out of the notion of metabolic interactions between organisms and their environments.[191] Indeed, the emphasis on metabolism lined up (though not explicitly) with Liebig's proto-ecological reflections on capitalist industrialism and the robbing of the soil, which inspired Marx's ecological-materialist concept of the "metabolic rift."[192] Ecosystem analysis was also broadened to take into account the analysis of the biosphere that had emerged in the work of the Russian biogeochemical-ecological thinker Vladimir Vernadsky. When Rachel Carson and Barry Commoner stormed on to the public stage in the 1960s and '70s, their analysis was rooted in a materialist understanding of ecosystems, which had become the new holism, reflecting its greater heuristic power and its greater attention at the same time to fundamental disjunctures and crisis.[193]

Yet the victory for materialist ecosystem analysis was never really complete and the defeat for teleological holism never irreversible. The two general paradigms continued to struggle in different ways. In the case of the ecosystem program, there was always the danger that it would degenerate as all mechanical materialisms are wont to do into a form of mechanical reductionism.[194] Its very technical facility was seen as leading it in a reductionist and hence ultimately unecological direction. In the case of the organismic approach of teleological holism, it derived new strengths within the ecology movement and on the margins of science through its influence on deep ecology, which also adopted aspects of ecosystem ecology. Here the influence of Smuts persists.[195]

Smuts's holism, as we have seen, was embraced in the psychology of Alfred Adler, who sided with Smuts over Hogben in relation to the "Nature of Life" controversy in South Africa. He arranged for *Holism and Evolution* to be translated into German. Adler seized on Smuts's notion of "wholes as self-acting, self-moving organisms," embodying a particular purpose and direction to become more complete wholes, as his own basis for formulating an "internal principle of action" (a "law of movement") as a natural basis for "human goal striving."[196] He actively promoted Smuts's perspective to colleagues and friends.[197] Smuts's ideas were thus incorporated into the discussions around Gestalt psychology.

It was no doubt in relation to Adler that Arne Naess, the cofounder of deep ecology, was introduced to Smuts's ideas, having studied in Vienna in 1934, when Smuts's ecological holism and its psychological connections were being promoted by Adler.[198] Deep ecology carried forward many of the essentialist, vitalistic, and organismic traditions of the idealist side of the ecological debate. It ended up being more influential in environmental ethics than ecological science, though frequently crossing over into the latter.

The continuing tensions around the social construction of ecological science are revealed by the fact that the influential environmental historian Donald Worster presents Tansley in his magnum opus *Nature's Economy* as the principal source of ecological error and the founder of not only mechanism but also reductionism in ecology. Worster comes out strongly in favor of the "organismic trends in science" represented by Bergson, Morgan, Whitehead, and Smuts: the whole "resurgence of philosophical idealism" in this area in response to mechanistic materialism.[199] For Worster, the "New Ecology" stretches from Tansley and Elton to the Odums and represents a massive extension of the technical means for controlling nature. In this respect, he argues, the organicist tradition is needed as a kind of ethical counterbalance.[200]

A similar position is taken by leading ecofeminist historian and theorist Carolyn Merchant, who writes: "Holism was proposed as a philosophical alternative to mechanism by J. C. Smuts (1926) in his book *Holism and Evolution*, in which he attempted to define the essential characteristics of holism and to differentiate it from nineteenth-century mechanism. . . . Smuts saw a continuum of relationships among parts from simple physical mixtures and chemical compounds to organisms and minds in which the unity among parts was affected and changed by the synthesis."[201]

Uncritically embracing Smuts's holism, Merchant identifies it with the development of ecology itself: "The most important example of holism today is provided by the science of ecology."[202] In contrast, Tansley's work, though occupying a far more important place in the history of ecological science, is dismissed by Merchant as giving rise to purely mechanistic, computer-driven models: "The reductionist ecology of Arthur George Tansley, developed in the 1950s [*sic*], has matured into the 'Club of Rome's' computer model, which predicts the 'limits to growth' for the entire world system."[203]

The contradiction between teleological organicism and mechanistic materialism was in fact insurmountable and narrowly constraining on both sides. Although the materialist tradition provided the more powerful scientific research program, it was often weakened by mechanism and reductionism. And although the teleological holist tradition often seemed more dialectical (though its teleology ultimately undercut that), it had little in the way of a solid material grounding. The answer, we believe, lies in a non-teleological, dialectical materialist ecology. During the 1930s, '40s, and '50s, figures such as Haldane, Needham, Bernal, Hogben, Levy, Farrington, and Zilsel in the historical-materialist tradition struggled over these issues, which can be shown as foreshadowing the later work of American Marxist contributors to biology and ecology, such as Stephen Jay Gould, Richard Lewontin, and Richard Levins.[204] Probably the best example of this is Levins and Lewontin's *The Dialectical Biologist*.[205]

Toward a Realist Constructionism in Ecology and Environmental Sociology

Our argument is that environmental sociology cannot afford to embrace a strong social constructionism/idealism and to expel realist views. Indeed, a sociology of ecology that can serve as a counterpart to environmental sociology must be rooted in realism/materialism. At the same time, a crude positivism that ignores epistemology and social construction in favor of naturalistic determinism is worse than useless. Neither the epistemic fallacy, which reduces ontology to epistemology, nor the ontological fallacy, which reduces epistemology to ontology, is acceptable. Ecological analysis in general depends on the development of a strong dialectical objectivism or dialectical critical realism—what we have termed here *realist constructionism*.[206] To address earthly questions regarding the social-nature relation requires such an approach—essential for engaging with an emergent, ever-changing world of material existence.

A sociology of ecology, geared to the needs of environmental sociology in particular, will be most effective, we have argued, if it takes the form of a realist constructionism. This necessitates what Harding calls a "strong objectivism," which does not simply include the adherence to certain objectivist criteria within science but also recognizes that knowledge including science is socially situated—and hence can only fully be understood and evaluated in its broad development (through historically specific analysis).[207] Today's attempts to evaluate the relative significance of Smuts's and Tansley's contributions to environmental social thought have frequently suffered from a lack of knowledge of the historical construction of these ideas—how they arose in a process of conflict and contradiction and how this affected their respective worldviews.[208]

In treating the ecological world, Marx insisted that human beings had transformed the original nature that preceded human history, which was now almost nowhere to be found on the planet. Nevertheless, he also argued on material-realist grounds that there were fundamental ecological constraints (for example, soil metabolism) on which human society depended. The materialist principle remains crucial for all ecological and ecological-social analysis. Hence both historical constructionism and realism were essential in an ecological materialist analysis. Marx and Engels considered Darwin's evolutionary theory to be a breakthrough in the materialist-realist interpretation of natural history. But they warned of the effects that certain bourgeois ideas, such as the "struggle for existence," competition, and overpopulation, might have if transferred to nature and then transferred back to society as eternal natural laws.

In the end, there can be little doubt that the presence of a teleological conception of nature and a double transference (particularly with respect to race) constituted Tansley's main realist-constructionist objections to Smuts's ecological holism. Smuts, the coiner of the words *holism* and *apartheid*, used his concept of ecological holism to provide a philosophical-scientific justification for the apartheid system for which he helped lay the foundations. For Tansley, Smuts not only transgressed against a materialist conception of nature, but he also transgressed against a materialist-humanist conception of society. Smuts's idealism saw nature/ecology not only as a reflection of the human mind, but also as a reflection of dominant personalities (and races), which represented the apex in a new hierarchical scale of nature. As an ecological research program, Smuts's idealistic holism was unable to compete with Tansley's materialism, as the latter sought to construct/explain nature in terms of its own complex systems and processes, linked to close empirical analysis, rather than as a prefigured teleological philosophy of succession (Clements) or a philosophy of segregation (Smuts). Tansley was not philosophically naïve, recognizing that ecosystems were "isolates" on the model of physics but nonetheless ones that were not entirely arbitrary, as following nature's seams.

The Tansley-Smuts conflict in the construction of ecological science points not only to the importance of realism but also to the value of "externalist" or social-institutional approaches to the sociology of science, concentrating on relations of class, production, power, ideology, and the general social ethos, as exemplified in different ways by the approaches of Hessen and Merton. For all the advances made in recent years in the sociology of science through the examination of the microcontext of the laboratory, a broad social-institutional approach that deals with the larger social background conditions of science remains crucial. Rather than focusing on assorted reifications that distort any conception of science, we must focus on the big issues of alienation, exploitation, and oppression—the reflexive issue, as Harding says, of "Others." For example, where there is a struggle over race in society, this struggle is likely to be replicated in science.[209] And although this in itself does not give us the means of judging science, it aids us in developing more dialectical conceptions of scientific knowledge production that allow us to understand the ways in which the confrontation of reality with reason can be distorted.

In the light of Smuts's idealist dialectic of holism and apartheid, one is reminded of Andrew Sayer's realist-constructionist methodology, which argues that "it is not just the *ideas* (of racial differences etc.) behind *apartheid* in the abstract that are wrong but the actual practices (enforcement of pass laws, etc.) and material structures (segregated and materially deprived townships, etc.) which reciprocally confirm, legitimate and are legitimated by those ideas."[210]

Likewise, "criticism," as in the case of apartheid, "cannot reasonably be limited to false ideas, abstracted from the practical context in which they are constitutive, but must extend to critical evaluation of their associated practices and the material structures they produce and which in turn help to sustain those practices."[211] A sociology of ecology thus has to be forever attuned to the ways in which nature is used in struggles over human society and the consequences of this, as well as to the human exploitation of nature in the service of such social exploitation.

The critical-realist constructionism we have been defending here can be contrasted to the strong (idealist) constructionism or irrealism, sometimes presenting itself as skepticism, that has lately come to influence environmental sociology. One manifestation of this strong constructivism/skepticism has been a tendency to argue that the global ecological crisis, including global climate change, is culturally/epistemologically constructed and is therefore subject to varying interpretations based on different conceptual schemes or discourses. The truth claims of scientific reports in this area are thus declared to be discursive and "uncertain."[212]

Nevertheless, from the standpoint of critical-realist constructionism, it might be said, in Vician terms, that we can understand the reality of global climate change because we have to some extent made it. The physicist Tyndall first discovered as a result of laboratory experiments in 1862 that carbon dioxide created a kind of greenhouse effect, heightening the temperature near the earth's surface.[213] Throughout the century and a half since, we have learned that the phenomenon of global climate change is occurring because of anthropogenic causes. The reality of global climate change as well as our reflexive historical awareness of it is the outcome of the dialectical process of the coevolution of human society and nature, of which science is a part.[214] Such a perspective requires that we avoid giving too much power to our mere conceptions while neglecting extradiscursive reality. As Soper has said, "It is not the discourse of 'global warming' or 'industrial pollution' that has created the conditions of which it speaks."[215] The agenda of a new sociology of ecology concerned with the historical-sociological roots of our scientific understanding of ecology derives its imperative (as does environmental sociology) from the need to confront the planetary crisis that "surrounds us," one which is a product of our own social juggernaut.

There is an unavoidable tension between those who argue that nature is principally to be viewed as mentally constructed (often in idealist terms) and those who claim that nature is a reality that is in some sense independent of our constructions—one that we come to know only through praxis, that is, the attempt to transform the world. These two views, as Soper claims, must be held in "productive tension," allowing us to engage in the realist-constructionist critique necessary for a sociology of ecology.[216]

15. Imperialism and Ecological Metabolism

The concept of ecological imperialism is seemingly unavoidable in our time. Obvious cases are all around us. The invasion and occupation of Iraq was, at least in part, over oil. Instances of ecological imperialism do not, however, stop with Iraq. Whether it is the renewed scramble for Africa, the flooding of the global commons with carbon dioxide, or biopiracy aimed at third world germplasm, ecological imperialism is endemic within a global economy predicated on accumulation. While the appropriation of resources from distant lands has taken place throughout human history, the origins and ongoing growth of capitalism are dependent upon further ecological exploitation and unequal ecological exchange. It takes different forms, depending upon the historical context and the demands of economic production, but it continues to operate in order to funnel resources—land, raw materials, and/or labor—into the process of capital accumulation.

The concept of ecological imperialism has scarcely been visible, unlike notions of economic, political, and cultural imperialism.[1] Most historic studies of imperialism, although appreciating the importance that the imperial countries placed on control of third world resources, have tended to analyze this primarily in terms of its effects on the flows of economic surplus, rather than in terms of ecological damage wrought by the robbing of third world countries of their resources and the destruction of their environments. Although the latter effects have often been recognized they have been treated as geopolitical problems or as factors affecting economic development and not in terms of ecological imperialism per se, which would require that systematic asymmetries in the exploitation of the environment be acknowledged.

Although Marxist theory cannot be said to have approached the issue of ecological imperialism systematically in the past, Marx's own analysis provided the analytical basis for such a treatment, due to his simultaneous concerns with economic expansion, imperialism, and ecological exploitation. Nonetheless, ecological problems are complex, especially as they emerge under capitalism.[2] Ecological degradation is influenced by the structure and dynamics of the world capitalist system, arising from the fact that a single world economy is divided into numerous nation-states, competing with one another both directly and via their corporations. The global economy is divided hierarchically, with nations occupying fundamentally different positions in the international division of labor and in a world system of dominance and dependency.[3] To further complicate matters, the extraction, processing, and consumption of raw materials—an inevitable part of any mode of production—entails constant interactions with dynamic, integrated natural processes and cycles.[4] In this, earthly conditions are transformed, potentially creating various forms of ecological degradation. The exact ramifications, of course, will be determined by the particulars of any situation.

Transfers in economic values are shadowed in complex ways by real material-ecological flows that transform ecological relations between town and country, and between nations, especially the core and periphery.[5] Control of such economic and material flows is central to the forces of competition and the accumulation of capital, and generates social and environmental inequalities throughout the global economy—both within and between nations. Sociologist Stephen Bunker highlighted how the extraction and export of natural resources from peripheral countries involved the vertical flow of not only economic value, but also value in terms of energy and matter, to more developed countries.[6] These trade arrangements, influenced by the dynamics of the global economy and positions within the world system, negatively affected and undermined the socio-ecological conditions in the extractive countries. Recent scholarship on "ecologically unequal exchange" has drawn on Bunker's seminal work, as well as the theory of unequal exchange, to demonstrate the disproportionate (and undercompensated) transfer of matter and energy from the periphery to the core, and the exploitation of environmental space within the periphery for intensive production and waste disposal.[7] The environmental footprint of economically advanced nations involves appropriation of land, resources, and labor in lesser-developed countries, increasing the environmental degradation in the latter for the benefit of the former.[8]

In this chapter we consider how ecological imperialism, which entails control over natural resources, creates asymmetries in the exploitation of the environment and unequal exchange. In particular, the international guano trade in the nineteenth century highlights the emergence of a global metabolic rift, as guano and nitrates were transferred from Peru and Chile to Britain (and other nations) in order to enrich their diminished soils. This global ecological rift involved the decline of soil fertility in Britain, the transfer of Chinese labor to Peru to work the guano islands, the export of natural fertilizer to core nations, the degradation of the Peruvian/Chilean environment, the creation of debt-laden economies, and the War of the Pacific as Chile (backed covertly by Britain) and Peru fought against each other to control resources desired by Britain. It also allowed Britain and other core powers to carry out an "environmental overdraft" within their own countries by drawing imperialistically on natural resources from abroad.[9]

However, before turning attention to the international guano trade and its ecological relations, it is necessary to address how the rise of the capitalist world economy itself was synonymous with the emergence of a hierarchical division of nations through the appropriation of distant lands, labor, and resources. Ecologically, capitalism operates globally as a particular social metabolic order that generates rifts in underlying metabolic relations between humanity and the earth and within nature itself.

The Social Metabolic Order of Capital: Accumulation and Rifts

Humans depend on functioning ecosystems to sustain themselves. Marx noted that there is a necessary "metabolic interaction" between humans and the earth, and labor serves as "a process between man and nature, a process by which man, through his own actions, mediates, regulates and controls the metabolism between himself and nature."[10] A metabolic relationship involves regulatory processes that govern the interchange of materials. Natural systems, such as the nutrient cycle, have their own metabolism, which operate independently of and in relation to human society, allowing for their regeneration and/or continuance. For Marx, the concept of social metabolism captured the complex interchange of matter and energy between human beings and nature.[11] Each mode of production generates a particular social metabolic order that influences the society-nature relationship, regulating the ongoing reproduction of society and the demands placed on ecosystems.[12]

The transition from feudalism to capitalism ushered in a new social metabolic order that shaped the interpenetration of society and nature. As Marxist philosopher István Mészáros has explained, the "innermost determination [of] the capital system is *expansion-oriented* and *accumulation-driven*," which pushes it to subsume the entire world to its logic of accumulation. In this, it attempts to impose a "'*totalizing*' framework of control" where everything must prove its "productive viability" and its ability to generate profit within a desired time frame in order to be seen as useful.[13] Spurred on by competition and constant growth, capitalism is not capable of "self-sufficiency." It must be constantly renewed, replenished, but on a larger scale. It cannot be stationary; thus it is "fundamentally unrestrainable" and cannot "recognize boundaries," whether social or natural, regardless of "how devastating the consequences." As a result, it creates an "*uncontrollable mode of social metabolic control*" focused on acquisition of profit that runs roughshod over the regulatory processes that govern the complex relationships of interchange within natural systems and cycles.[14] The internal dynamics of this social metabolic order produce various global inequalities and ecological contradictions.

A new division of both labor and nature took shape with the development of capitalism as a world system. The bounty of the land was "pumped out of one ecosystem in the periphery and transferred to another in the core. In essence, the land was progressively mined until its relative exhaustion fettered profitability."[15] The process of primitive accumulation established divisions between core and periphery nations, as the wealth of distant lands was appropriated through various mechanisms. As Marx observed: "The discovery of gold and silver in America, the extirpation, enslavement and entombment in mines of the indigenous population of that continent, the beginnings of the conquest and plunder of India, and the conversion of Africa into a preserve for the commercial hunting of blackskins, are all things which characterize the dawn of the era of capitalist production. These idyllic proceedings are the chief moments of primitive accumulation."[16]

Capital constantly seeks to overcome whatever social and natural limits it confronts, "tearing down all the barriers which hem in the development of the forces of production, the expansion of needs, the all-sided development of production, and the exploitation and exchange of natural and mental forces."[17] Distant lands and ecosystems became mere appendages to the growth requirements of the advanced capitalist center.

Nothing so illustrated this unequal ecological exchange in the nineteenth century as the global guano trade that arose to compensate for the "environmental overdraft" that characterized industrial agriculture in Europe and the United States. In the 1840s, Germany's leading chemist, Justus von Liebig, along with other agricultural chemists and agronomists, sounded the alarm with respect to the loss of soil nutrients—such as nitrogen, phosphorus, and potassium—through the transfer of food and fiber to the cities. Rather than being returned to the soil to replenish it, as in traditional agricultural production, these essential nutrients were shipped hundreds, even thousands, of miles and ended up as waste polluting the cities.

John Chalmers Morton, who studied the application of mechanical power in agriculture, noted that agricultural improvements increased the uniformity of land, making it easier to increase the scale of operations and to employ industrial power to agricultural operations.[18] Marx was a devoted student of Liebig's work and studied Morton when writing *Capital.*[19] He incorporated a metabolic analysis into his critique of political economy and saw capitalism as generating a form of industrialized agriculture that industrially divided nature at the same time that it industrially divided labor. He determined that an economic system premised on the accumulation of capital led to intensive agricultural practices to increase the yield of food and fiber for markets. Marx lamented how capitalism degraded labor and nature under these conditions:

> Large-scale industry and industrially pursued large-scale agriculture have the same effect. If they are originally distinguished by the fact that the former lays waste and ruins labour-power and thus the natural power of man, whereas the latter does the same to the natural power of the soil, they link up in the later course of development, since the industrial system applied to agriculture also enervates the workers there, while industry and trade for their part provide agriculture with the means of exhausting the soil.[20]

The transfer of nutrients was tied to the accumulation process and increasingly took place on national and international levels. As a result, this type of production, along with the division between town and country "disturbs the metabolic interaction between man and the earth, i.e. it prevents the return to the soil of its constituent elements consumed by man in the form of food and clothing; hence it hinders the operation of the eternal natural condition for the lasting fertility of the soil."[21]

In other words, capitalist agriculture created a metabolic rift in the nutrient cycle, squandering the riches of the soil. Horrified by the scale of soil degradation, Liebig exclaimed, "Truly, if this soil could cry out like a cow or a horse which was tormented to give the maximum quantity of milk or work with the smallest expenditure of fodder, the earth would become to these agriculturalists more intolerable than Dante's infernal regions."[22]

According to Liebig, British high farming (early industrialized agriculture) looted the soil of its nutrients and then sought to compensate for this by robbing other countries of the means needed to replenish their own soil. "Great Britain," he wrote, "deprives all countries of the conditions of their fertility. It has raked up the battle-fields of Leipsic, Waterloo, and the Crimea; it has consumed the bones of many generations accumulated in the catacombs of Sicily.... Like a vampire it hangs on the breast of Europe, and even the world, sucking its lifeblood without any real necessity or permanent gain for itself."[23]

Marx, too, referred to the imperialist exploitation of the soil nutrients of whole countries—developing out of the rift in the metabolism between human beings and the earth. "England," he observed, "has indirectly exported the soil of Ireland, without even allowing the cultivators the means for replacing the constituents of the exhausted soil."[24] As capitalism expanded, increasingly importing food and fiber from abroad, so did the metabolic rift. Marx indicated that capitalist growth serves the interests of the "main industrial countries, as it converts one part of the globe into a chiefly agricultural field of production for supplying the other part, which remains a pre-eminently industrial field."[25] In this, the abuse and "misuse" of "certain portions of the globe . . . depends entirely on economic conditions."[26] August Bebel, a close friend, mainly by correspondence, of Marx and Engels, and "the outstanding founder and leader of the German socialist movement,"[27] captured the ecological transfer and contradictions of the global economic system, stating:

> All those countries which principally export produce of the soil, but receive no materials for manuring in return, are being gradually but

> inevitably ruined, Hungary, Russia, the Danubian Principalities, and America. It is true, artificial manure, especially guano, replaces that of men and cattle, but few farmers are able to buy it in sufficient quantities on account of its price, and in any case it is reversing the natural order of things to import manure from a distance of many thousand miles, whilst that which one has close at hand is wasted.[28]

Ecologically, a key fact was the capacity of the core capitalist states to compensate for the degradation of their own environments through the even more rapacious exploitation of the natural resources of periphery economies. As Mark Elvin noted, in *Retreat of the Elephants: An Environmental History of China*, core capital in Europe (as opposed to China) possessed "imperial overseas resources...that could be drawn on like an environmental overdraft without any need for further [ecological] restoration."[29]

The Tale of Guano and Nitrate Imperialism

In the nineteenth century, the guano/nitrates trade united China, Peru, Chile, Britain, and the United States in a global metabolic rift. Guano was deemed a precious commodity that would help replenish lost soil nutrients in advanced countries. Capitalist farming practices, and the division between town and country, confronted the natural limits of the soil, while constantly trying to increase agricultural yields for short-term economic gain. The tale of guano and nitrates, which is rooted in soil depletion, involves the advance of soil science, transformation of landscapes, transfer of human populations, exploitation of nature and peripheral nations, and integration of the global economy. This case helps illustrate the workings of ecological imperialism and the emergence of a global metabolic rift that involved environmental degradation and unequal ecological exchange. It helps us understand the environmental overdraft that contributed to European prosperity while hiding the extent of the ecological degradation of industrial capitalism.

The existence and use of guano as fertilizer had been known for centuries in Europe, but its importance to European and U.S. agriculture was not immediate, given the particular economic conditions and the state of agricultural science. In 1604, an English translation of Father Joseph de Acosta's book, *The Natural and Moral History of the Indies*, was published. De Acosta described how heaps of bird dung covered Peruvian islands as if it were snow, and how indigenous peoples mined this powerful material to fertilize their lands.[30] In the seventeenth century, fascination surrounded the use of guano for agriculture. However, an international trade in guano was not established. Nor is it certain that such a trade was possible at this point in time. Furthermore, the advances in the science of soil chemistry, specifically the nutrient relationship between soil and plants, did not occur until the nineteenth century.

At the beginning of the nineteenth century, the German explorer Baron Alexander Humboldt observed how Peruvian farmers used guano to enrich their dry farm lands.[31] He took samples of guano back to Europe in 1803, but there was no drive then to study this particular substance. However, as soil depletion increased, so did the need for fertilizers, stimulating business interests in the application of guano as a fertilizer. In the 1820s, tests were conducted to assess the chemical composition of guano in comparison to the requirements of plants and the nutrients lost through crop production. Guano contained high concentrations of phosphate and nitrogen. In 1835, a few cases of guano were imported to Great Britain to test the dung on crops. Guano proved to be a powerful fertilizer. The possibility of high returns seemed promising, given that the high yields surpassed what was calculated as the likely costs of guano importation.

Advances in soil science furthered interest in guano. In 1840, Liebig detailed how modern farming practices and the division between town and country contributed to the loss of soil nutrients. In the same year, Alexandre Cochet, a French scientist, discovered that valuable quantities of nitrate of soda could be extracted from guano and nitrates (saltpeter) both of which were abundant in Peru, helping stimulate the rush for guano.[32] Guano was soluble, so it was fast-acting, and provided an immediate influence on the growth of plants.

In the 1850s and '60s, Liebig described the intensive agricultural methods of Britain as a system of robbery, opposed to rational agriculture. Numerous social and ecological problems were created given these methods. The soil required specific nutrients to produce crops; however, food and fiber (which took up nutrients) were shipped from the countryside, over long distances, to cities.[33] Increasingly, the material-ecological transfer took place both on the national and international level: "In the large towns of England the produce both of English and foreign agriculture is largely consumed; elements of the soil indispensable to plants do not return to the fields,—contrivances resulting from the manners and customs of the English people, and peculiar to them, render it difficult, perhaps impossible, to collect the enormous quantity of phosphates which are daily, as solid and liquid excrements, carried into the river."[34]

The riches of the soil were squandered. As a result, the soil was depleted of its necessary nutrients. The degradation of the soil hastened the concentration of agriculture among a smaller number of proprietors who adopted even more intensive methods of production, including the mass importation of manures and eventually the application of artificial fertilizers. Marx indicated that capitalist agriculture, and by extension capitalism in general, generated an antagonism between human beings and nature, creating an "irreparable rift" in their "metabolic interaction."[35] The expansion of capitalist operations had international implications, as the British circumnavigated the globe to furnish "raw materials in bulk to the mother country in the centre."[36]

Soil degradation in Britain and the United States sparked the international guano rush, as agriculturists sought the precious fertilizer to compensate for the soil nutrients that they were losing.[37] Peru had the largest deposits of high-quality guano. The mining of this product involved the importation of Chinese "coolies." Shifts in fertilizer ushered in a war between South American nations, while Britain maintained access to the supply of nitrate fertilizer. As Eduardo Galeano noted in relation to guano and nitrates, the resource curse has long plagued the periphery: "The more a product is desired by the world market, the greater the misery it brings to the . . . peoples whose sacrifice creates it."[38]

The Guano Rush

Peru had the largest deposits of high-quality guano and an abundant supply of nitrates. Its guano contained the highest concentration of nutrients that were useful to crops. It rarely rained on the coast of Peru. As a result, the nitrogen in the guano was not washed away, as it was on other islands and coasts throughout the world. The mountains of guano that de Acosta described were on the Chincha Islands off the coast of Peru. These islands served as habitat to numerous species of sea birds. The ocean currents surrounding these islands created an upflow of decayed matter, sustaining a massive population of anchovies, which the birds ate and deposited as waste on the rocks. The anchovy diet greatly enriched the usefulness of the dung produced by the birds. The guano deposits, hundreds of feet deep, had accumulated over thousands of years.[39]

In the 1840s, Peru was still in debt to Britain for monies borrowed during the fight for independence from Spain. Guano offered an avenue for Peru to meet its debt payments and gain foreign exchange through the sale of guano contracts. Lima was at the time the richest city in South America. Although there were a number of contracts between the Peruvian government, acting on behalf of the Lima oligarchy, and European businesses (primarily British, but also French) during the duration of the guano trade, which thrived for forty years, the dominant trade agreement was between Lima and the British firm Anthony Gibbs & Sons. The company holding the contract with the government had exclusive rights over the sale of guano on the global market. As a result Britain dominated the global guano trade.

The government of Peru claimed ownership of the guano.[40] Peruvian subcontractors, who were granted contracts from the government, were placed in charge of the digging and loading process. Lima repeatedly renegotiated the Peruvian guano contracts, trying to get a better deal. In addition to receiving a specified amount of money per ton of guano shipped, the government borrowed money against the contracts. Much of the money made in the sale of guano was directed toward paying off the existing and accumulating debt taken out by the Lima oligarchy, in a classic case of imperial dependency.

In 1841, the first full cargo of guano arrived in Britain. The manure was quickly sold on the market, stimulating a drive to secure more guano. An extensive advertising campaign was conducted to promote the use of guano. Gibbs & Sons published *Guano: Its Analysis and Effects*, detailing the various techniques of guano application, praising the powers of guano to make plants grow taller, stronger, and more productive.[41] While this book served as a marketing ploy, its conclusion was clear: increased yields using a "cheap" fertilizer. Other publications tested guano against other fertilizers, employing Liebig's work on the loss of soil nutrients.[42] These tests heralded the triumphs of guano as far as its ability to meet the nutrient needs of crops. Guano became an obsession, seeming to offer an escape from the ecological contradiction that had been created.

Marx noted that the "blind desire for profit" had "exhausted the soil" of England, forcing "the manuring of English fields with guano" imported from Peru.[43] Industrialized capitalist agriculture had fundamentally changed the nutrient cycle. Agriculture was no longer "self-sustaining" as it "no longer finds the natural conditions of its own production within itself, naturally arisen, spontaneous, and ready to hand."[44] Britain was not the only country confronting severe losses in soil nutrients. Farms in upstate New York and plantations in the southeastern United States were in desperate need of powerful fertilizers.[45] Thus both merchants and agriculturalists from Britain and the United States sought the fertilizer to compensate for the soil nutrients that they were losing.[46]

Given the British trade monopoly on Peruvian guano supplies, the United States pursued imperial annexation of any islands thought to contain guano deposits. In 1856, Congress passed the Guano Islands Act, allowing capitalists to seize ninety-four islands, rocks, and keys around the globe between 1856 and 1903.[47] "In the last ten years," Liebig observed in 1862, "Britain and American ships have searched through all Seas, and there is no small island, no coast, which has escaped their enquiries after guano." But, in the end, the deposits on the islands of Peru were the best, given the ideal natural conditions to preserve the nutrients.

For forty years, Peru remained the most important country for meeting European and North American fertilizer needs. During this period, millions of tons of guano were dug, loaded, and shipped from Peru. In 1850, Britain imported over 95,000 tons of guano.[48] The following year, almost 200,000 tons were imported; by 1858, over 302,000 tons. From 1863 to 1871, the imports per year ranged from 109,000 tons to 243,000 tons. As noted above, guano was not only exported to Britain; from 1866 to 1877, Peru exported from 310,000 to 575,000 tons a year to the world as a whole.[49]

The Chincha Islands, with deep guano deposits, were a site of constant activity. In the early 1850s a British officer reported witnessing the simultaneous loading of guano on a hundred ships, representing 11 different countries (44 United States, 40 English, five French, two Dutch, one Italian, one Belgian, one Norwegian, one Swedish, one Russian, one Armenian, and three Peruvian), from a single island off the coast of Peru.[50] Additionally, hundreds of other large ships would be waiting at sea for a turn to be loaded.[51]

Despite the millions of tons of guano that were exported from Peru, international demand could not be met. Inferior guano deposits on islands throughout the world were mined and sold on the market. Off the African coast, an island with substantial guano deposits had 460 ships on one day, simply waiting to fill their ships with the cargo. In a short period of time, the "island [was] reduced to nothing but a plateau of bare rock."[52] The guano trade suffered setbacks, as inferior guano was packaged and sold with false labels, claiming that it was Peruvian guano. Farmers became leery of guano on the market, but the necessity for fertilizer remained, given the metabolic rift in the nutrient cycle.

The guano trade transformed Peru in a number of ways. In the early 1800s, silver was the primary export of Peru. After Peru's independence, Britain quickly forged trade relations, importing wool and cotton. While Peru desired trade protection, Britain worked to reduce tariffs and duties, desiring free trade. Once the guano trade was established, this resource became the primary export commodity. Guano supplied 5 percent of state revenue in 1846–47. In 1869 and 1875, 80 percent of state revenues came from the guano trade.[53] The terms of trade continued to decline, as Peru was forced into accepting liberal policies which favored metropolitan capital in the imperial states.[54] The export economy failed to help the domestic economy. The Lima oligarchy spent money on luxury items, rather than social development, and on paying interest on loans. Much of the infrastructure, such as its irrigation systems and roads, fell into disarray.[55] The country was dependent on foreign nations for general commodities.

During this period, Peru remained the most important country for meeting British and North American fertilizer needs. At the same time, the country remained in debt to bondholders. The Peruvian ruling class profited heavily from the guano trade. Some of the money was used to help rich landowners enlarge their sugar and cotton operations. In particular, Domingo Elías, who handled contracts related to the extraction of guano, purchased more land and extended his plantation operations. He helped transform the agricultural sector into a producer of cash crops (such as cotton and cochineal) for export to Europe and the United States, transferring the riches of the soil to more developed nations.[56] Liebig and Marx noted that through incorporation into the global capitalist market and long-distance trade, the earth was robbed of its richness, the soil was depleted of its nutrients, and the separation between town and country increasingly became international. These conditions and consequences were only exacerbated through the exportation of guano and the production of cash crops, increasing the global metabolic rift. In spite of this trade, Peru remained a country in debt and one with vanishing resources.[57]

The guano trade transformed the natural landscape of Peru, especially the islands where guano was mined. In *Peru in the Guano Age*, A. J. Duffield, who took measurements to estimate the remaining guano deposits, describes the changes that had taken place: "On my return from the south [part of Peru] we passed close to the Chincha islands. When I first saw them twenty years ago, they were bold, brown heads, tall, and erect, standing out of the sea like living things, reflecting the light of heaven, or forming soft and tender shadows of the tropical sun on a blue sea. Now these same islands looked like creatures whose heads had been cut off, or like vast sarcophagi, like anything in short that reminds one of death and the grave."[58]

The guano deposits that took thousands of years to accumulate were being depleted. Boussingault, a French soil scientist, noted that since guano had become "a subject of the commercial enterprise of mankind" its reserves were quickly disappearing.[59] The rate of extraction was faster than the rate of natural accumulation. To make matters worse, the prospect for additional excrement was questionable, given that the extraction of guano was executed without regard to the needs of the birds, which were driven away and/or slaughtered in some cases.[60] The natural fertilizer that had been used for hundreds of years in Peru was being exported and diminished, as the social metabolic order of the capitalist world system expanded.

"Worse than Slave Labor": Chinese Coolies and Guano Extraction

The guano trade not only involved the shipping industry and the distribution of manure on fields but necessitated a labor regime to extract the materials from the islands. In the pursuit of profit, both Peru and Britain contributed to the global movement and exploitation of labor. In the 1840s Peru had a labor shortage for its plantations and mines. The government passed "an immigration law subsidising the importation of contract labourers."[61] Anyone who imported "at least fifty workers between the ages of 10 and 40" was paid 30 pesos per head. Exploiting the decades of social disruption, due to the Opium Wars and the Taiping Rebellion in China, European merchants began the systematic importation of Chinese laborers to Cuba and Peru.[62] Through coercion, deceit, and even kidnapping—often by the same individuals and companies who had engaged in the slave trade—tens of thousands of Chinese "coolies" were contracted for through Macao and Hong Kong.[63] The voyage to Peru took approximately five months. During this passage, the

Chinese coolies were provided with a meager ration of rice. The mortality rate during the first fifteen years of the trade was between 25 and 30 percent. To escape the horrible conditions, some Chinese in passage "jumped overboard [if and when allowed on deck] to put an end to their sufferings."[64] Marx and Engels characterized the labor of "Indian and Chinese coolies" as "disguised slavery," and they reveled in stories of "the very coolies" on ships destined for the Americas and elsewhere rising "in mutiny," as happened a number of times during passage.[65]

The first Chinese coolies or indentured manual laborers arrived in Peru in 1849. Between 1849 and 1874, over 90,000 Chinese coolies were shipped to Peru. Around 9,700 died during passage.[66] The majority of coolies were forced to work on the sugar plantations and to build the railroads. However, many were forced to work on the guano islands. Of the three realms of employment, the guano islands had the worst labor conditions. For many years, Domingo Elías, who held the contract for operating the extraction of guano, employed coolies, but also used convicts, army deserters, and slaves to work the guano islands. The workforce on these islands varied through the years, but often involved between 200 and 800 individuals.

The extraction of guano required digging into mounds of excrement that covered rocky islands. The capital outlay for extraction was minimal. The most expensive items were the bags into which guano was loaded. Using picks and shovels, coolies were required to dig through the layers of guano, filling sacks and barrows. Each worker had to load between 80 to 100 barrows, close to five tons, each day. Once the barrows were filled, the workers hauled the guano to a chute to transfer it to the ships. If the workers failed to move five tons during the day, they were physically punished. On occasion over 20,000 tons were said to have been extracted from the islands in a day.[67]

George Peck visited the islands and noted that the Chinese were "over-worked beasts of burden," forced to "live and feed like dogs."[68] Their emaciated bodies struggled to carry sacks of guano and to push the barrows. Acrid dust penetrated the eyes, the nose, the mouth of a worker, and the stench was appalling. A. J. Duffield noted: "No hell has ever been conceived by the Hebrew, the Irish, the Italian, or even the Scotch mind for appeasing the anger and satisfying the vengeance of their awful gods, that can be equalled in the fierceness of its heat, the horror of its stink, and the damnation of those compelled to labour there, to a deposit of Peruvian guano when being shoveled into ships."[69]

Infractions by workers were met by severe punishment, such as flogging, whipping, or being suspended for hours in the sun. In some cases, workers were tied to buoys in the sea. Prison sentences would have meant substantial losses in regard to lost labor, so physical punishment was preferred. Suffering from an inadequate diet, physical cruelty, and the inability to escape from the stench of the guano, many Chinese committed suicide by jumping off the cliffs and into the ocean. Peruvian employers attempted to stymie revolt by working with the British to import opium to pacify Chinese workers.[70]

Although coolies were not legally slaves, they lived in de facto slavery or worse. As prisoners, unable to leave the islands, they received minimal monetary returns. In an account of the Chincha Islands, Alanson Nash noted, "Once on the islands" a coolie "seldom gets off, but remains a slave, to die there."[71] The cruelty imposed upon the Chinese laborers was inseparable from reports regarding the guano trade. The coolies were driven as expendable beasts: "As fast as death thins them out, the number is increased by new importations" of coolies who are thus "sold into absolute slavery—*sold by Englishmen into slavery*—the worst and most cruel perhaps in the world."[72] Working under the whip, the cruelties were "scarcely believable, and very few, if any, of the Chinese survived more than a few months." "Those Chinese who did not commit suicide by some means or other speedily succumbed to overwork, breathing the guano dust, and a want of sufficient food."[73]

The connection between the fertilized fields of Britain and the exploitation of Chinese workers did not escape the British consciousness. Writing in the *Nautical Magazine* in 1856, a correspondent observed that the powers of guano as a fertilizer were well known, "but few probably are aware that the acquisition of this deposit, which enriches our lands and fills the purses of our traders, entails an amount of misery and suffering on a portion of our fellow creatures, the relation of which, if not respectably attested, would be treated as fiction."[74] The *Morning Chronicle* wrote that the conditions of labor on the guano islands "seems to realise a state of torment which we could hardly have conceived it possible for man to enact against his fellow man."[75] The *Christian Review* ran a story about the Chinese coolie trade, noting that "the subtle dust and pungent odor of the new-found fertilizer were not favorable to inordinate longevity," creating a constant demand for more workers, given that guano labor involved "the infernal art of using up human life to the very last inch."[76] For Marx, writing in the *New York Daily Tribune* on April 10, 1857, Chinese coolies were being "sold to worse than slavery on the coast of Peru" as a result of British imperialism. Even some shipmasters, upon delivering their cargo of coolies in 1854, were "horrified at the cruelties they saw inflicted on the Chinese, whose dead bodies they described as floating round the islands."[77]

Despite British outrage regarding the treatment of the Chinese coolies on the guano islands and British attempts to end the coolie trade, British merchants continued to transport "hundreds of thousands of Indian indentured servants to British colonies" around the world.[78] Ironically, in Peru, the success of the guano trade and the cheapness of importing Chinese coolies as workers made it possible for slavery to be abolished in the 1850s. Coolies were simply acquired to replace the slaves. Slaveholders, such as Domingo Elías, were compensated for the loss of the slaves that were now free. At the same time, Elías and other businessmen profited from the importation of coolies.

The labor process on the guano islands was quite simple, depending primarily on human labor to make the guano useful. In order to sustain the large profits and control over the workers, the process was not modernized. Despite the millions of tons of guano that were being exported from Peru, international demand could not be met. The asymmetrical movement of natural resources, the unequal exchange of resources, to meet imperial interests was intimately connected to the exploitation of human labor under inhuman conditions.

The War of the Pacific: Control of Nitrate Fields

In 1821, in the Tarapacá desert province of Peru, Mariano de Rivero discovered immense deposits of nitrate, which could be used as fertilizer. At this point, Peru had massive quantities of two resources (guano and nitrates) that soon became the most important fertilizers in the world. In 1830, Peru exported over 8,000 tons of nitrates. The importance of nitrate only increased. In 1853 a process for efficiently mining the nitrate fields in Tarapacá was discovered, and soon after rich deposits were also found in the adjacent Bolivian province of Atacama. As a result of the great guano rush, the availability of guano had started to decline. Plus, "in 1857, DuPont secured a patent for blasting powder made from nitrate."[79] These nitrate fields began to displace guano as a source of fertilizer by the late 1860s, and became important for the production of TNT and other explosives, crucial to the expanding war industries of the industrial-capitalist states.[80] By 1875, British investments, primarily in the nitrate industry, in Peru totaled £1,000,000.

The Peruvian ruling class became very wealthy as a result of the guano and nitrates trade. This wealth did not, however, lead to economic development to any significant extent, apart from the building of railways. Instead, Peru became heavily indebted primarily to British investors, with its guano exports mortgaged well into the future.

From 1864 to 1866 the Peruvian guano trade was interrupted by what is sometimes known as the Chincha Islands War (also the War of the Quadruple Alliance) between Spain and the quadruple alliance of Peru, Chile, Ecuador, and Bolivia, following Spain's seizure of the guano-rich Chincha Islands, which provided between two-thirds and three-quarters of Peru's annual income. An important factor in obstructing Spain's repeated attempts to come out of the war with control of the Chincha Islands was the position of the United States. U.S. Secretary of State Seward adamantly declared that Washington would not remain neutral if Spain attempted to expropriate the islands permanently. The Chincha Islands War undoubtedly contributed to Peru's growing dependence on Britain.[81]

Following the war with Spain, Peru returned to a guano-induced prosperity that seemed to grow by leaps and bounds. "The country," as Galeano observed, "felt rich . . . and the state carelessly used up its credit, living prodigiously and mortgaging its future to British high finance."[82] As guano exports waned and the emphasis shifted to nitrates, Peru in 1875 attempted to get out of its growing debt trap by imposing a state monopoly in its nitrate zones in Tarapacá, expropriating the holdings of private investors (many of them foreign, particularly British) and offering them government certificates of payment. Subsequently, the Peruvian government sought to regulate the output of guano and nitrates so that they would not compete against one another. These actions angered foreign creditors who relied on a high level of guano exports and owned a substantial portion of the nitrate industry.

To further complicate matters, in 1879 Bolivia attempted to raise taxes on nitrate exports from its Atacama province. Together these changes in nitrate extraction led to the War of the Pacific (sometimes called the Nitrate War), four years after the Peruvian state's expropriation of the nitrate industry. "The Antofagasta Nitrate and Railroad Company, a Chilean concern fully controlled by English capital and counting among its shareholders the Gibbs banking and trading house [the same commercial house that dominated the guano trade in Peru]," operated out of the Atacama Desert, and used the port of Valparaiso in Chile for exports and Chilean merchants as intermediaries in the nitrate trade.[83] Thus Chile, backed by British investors, declared war on Bolivia together with Peru, with which Bolivia was allied. The two main goals of the Chilean army were to gain control over nitrate and guano deposits and to undermine Peru's economic ability to prevent occupation of these areas. With its more modern, British-built navy and French-trained army, Chile was soon able to seize Bolivia's Atacama province and Peru's Tarapacá—never to leave. Armaments dealers from the North sold weapons and used the War of the Pacific as a testing ground for new weapons such as torpedoes. In 1881, José Manuel Balmaceda, then Minister of Foreign Affairs in Chile, expressed that "the real and direct causes of war" were "the nitrate bearing territories of Antofagasta and Tarapacá."[84] Before the war Chile had almost no nitrate fields and no guano deposits. By the end of the war in 1883 it had seized all of the nitrate zones in Bolivia and Peru and much of Peru's guano territory.[85] Capturing these resources served as a means to address the mounting foreign debt and other political and economic problems.[86]

During the war British speculators bought up the government certificates issued by the Peruvian government during its expropriation of the nitrate industry—certificates that were then selling at fire sale prices. As Galeano wrote: "While Chileans, Peruvians, and Bolivians exchanged bullets on the field of battle, the English bought up the bonds, thanks to credits graciously afforded them by the Bank of Valparaiso and other Chilean banks. The soldiers were fighting for them without knowing it."[87]

Immediately after the war the Chilean government at the urging of British investors decided that ownership of the nitrate operations in Tarapacá properly belonged to those owning the government certificates. Before the war the British controlled 13 percent of the Tarapacá nitrate industry; immediately after the war this rose to 34 percent, and by 1890 to 70 percent.[88]

Former U.S. secretary of state James G. Blaine, who had been involved while in office in peace negotiations regarding the War of the Pacific, testified in April 1882 to a congressional committee investigating the U.S. diplomatic role during the war. According to Blaine, the War of the Pacific had been a case of British-instigated and Chilean-executed aggression against Peru and Bolivia with the sole object of seizing the guano and nitrate territories. For some time before the war, he contended, Peru had been prevented from buying armaments from Britain. He declared the war was about

> the guano and the nitrates . . . nothing else. It was to get possession of it. . . . The ironclads that destroyed the Peruvian navy were furnished by England, and the Peruvian agent came to this country to see whether they could find a good ship to go out there in anticipation of this war, when they knew it was coming. They said they didn't dare to apply in England to get

> it, and we were not able to furnish it. I do not speak of the government; I mean the manufacturers of this country. . . . It is an English war on Peru, with Chili as the instrument. . . . Chili would never have gone into this war one inch but for her backing by English capital, and there was never anything played out so boldly in the world as when they came to divide the loot and the spoils.[89]

Blaine noted that the Chilean capitalists were so compliant with the British desire to divide up and loot the guano and nitrate territories of Peru and Bolivia that

> the Chilean Government has put up by advertisement 1,000,000 tons of guano, which I suppose is worth $60,000,000 in Liverpool, and they pledge themselves in the advertisement to pay one-half of it into the Bank of England for the benefit of the English bondholders who put up the job of this war on Peru. . . . It [England] had not as much excuse in this as Hastings and Clive had in what they did in India. The war on Peru has been made in the same interest that Clive and Hastings had in India, and England sweeps it all in.

Blaine's contention that the British government would have refused to provide armaments and naval vessels to Peru was no mere speculative claim, since during the war Chile's own Minister of Public Works referred explicitly to an armaments blockade against Peru organized by foreign creditors.[90] Nor were such actions beyond the pale for British imperialism, which at that time was fighting expansionist wars in Afghanistan and Zululand, and was soon to invade Egypt. Blaine himself was speaking as one of the chief architects of U.S. imperialism in the late nineteenth century. As he made clear in an interview for the *New York Tribune*, his issue with British capitalists was not so much what they did in the War of the Pacific, as that they were intruding on an imperial domain in the Americas that properly belonged to the United States.[91]

Having lost its two principal resources for export, the Peruvian economy collapsed after the war. Peruvian Marxist José Carlos

Mariátegui explained that defeat in the War of the Pacific increased Peruvian dependence on British capital:

> Very soon [after the war] the capitalist group that had formed during the period of guano and nitrates resumed its activity and returned to power. . . . The Grace Contract [which they negotiated] ratified British domination in Peru by delivering the state railways to the English bankers who until then had financed the republic and its extravagances.[92]

The Peruvian government no longer had access to guano and nitrates to exploit, so it had no way to pay off the foreign debts with which it was still encumbered, except by handing its railroads over to British investors who had themselves clandestinely backed Chile in its seizing of much of Peru's territory and its most valuable natural resources. Bruce Farcau explained that the guano and nitrate deposits in Peru turned out, "like the Midas touch," to be "a curse disguised as a blessing," first in the creation of a debt-laden economy, then in a war and a loss of these resources, and finally in the further loss of control over the Peruvian economy.[93]

In this case, the metabolic rift in the nutrient cycle in Britain created a demand for fertilizers that were abundant in Peru. Through ecological imperialism, including various forms of unequal exchange, the bounty of the latter country was usurped, while contributing to the global metabolic rift and environmental degradation. Mariátegui explained that with "the loss of these resources came the tragic realization of the danger of an economic prosperity supported or held together almost solely by the possession of natural wealth at the mercy of the greed or aggression of foreign imperialism or vulnerable to the continual changes in industrial needs arising from scientific invention."[94]

In the end, Peru was left "bleeding and mutilated, the country suffered from a terrible anemia."[95] The export economy encouraged by the expansion of the global economy drained Peru of its resources and exploited its people. At the same time, the country became entangled in the debt trap, which deepened the blood-letting of the country.

Chile and the Curse of Nitrates

The War of the Pacific allowed Chile to seize control of the nitrate territories. In the decades that followed, the curse of the nitrates followed, as it was Chile's turn to bleed due to the ecological imperialism of the core nations. Europe still needed guano and nitrates in vast quantities to maintain its agricultural productivity. Moreover, as nitrates became crucial to the manufacture of explosives, Britain sought to control this trade for the benefit of its own capitalists, exploiting these ecological resources to their zenith while siphoning off the bulk of the economic wealth generated through the extraction of nitrates.

Before the war Chile was one of the poorest countries in South America, carrying lots of foreign debt; however, gaining control of the nitrate fields brought foreign investment, providing needed capital, and created an economic boom for this struggling nation. In 1880, Chile exported 275,000 tons of nitrate; in 1890, one million tons were exported.[96] But British capital had almost complete control over nitrate operations.[97] In 1888, Chilean president José Manuel Balmaceda, who had carried out modernization reforms in that country, including extensive state spending on public works and support of education, announced that the nitrate areas of Chile would have to be nationalized through the formation of Chilean enterprises, and he blocked the sale of state-owned nitrate fields to the British. Attempts by the state to control the extraction and distribution, as well as the wealth, of nitrates angered foreign capitalists. Three years later a civil war broke out within Chile with British capital and other foreign investors supporting the opponents of Balmaceda with money and armaments. John Thomas North, the British "nitrate king," was said (according to an 1891 report from the U.S. ambassador to the U.S. Secretary of State) to have contributed £100,000 to the anti-Balmaceda forces in the Chilean Congress.

The press in London characterized Balmaceda as a "butcher" and a "dictator of the worst stripe." The London *Times* referred to Chile under Balmaceda as "a communist government." British warships blockaded the Chilean coast. When the defeated Balmaceda committed suicide in 1891, the British ambassador wrote to the Foreign Office: "The British community makes no secret of its satisfaction over the fall of Balmaceda, whose victory, it is thought, would have implied serious harm to British commercial interests." State control of industries and economic infrastructure in Chile quickly receded after the war, as the British extended their investments.[98]

Demand for nitrate fertilizer remained of utmost importance to Britain, though at the same time the depletion of these resources was a grave concern. As German socialist August Bebel explained in the early twentieth century: "Chile salt-petre deposits, as also the guano deposits, are being quickly exhausted, while the demand for nitrogenous compounds is constantly growing in Germany, France and England, and in the past ten years in the U.S.A. as well. The English chemist William Crookes brought up this question as far back as 1899, and referred to it as a matter of much greater importance than the possibility of an imminent exhaustion of the British coal mines."[99]

Britain siphoned whatever monetary and material wealth it could from Chile. By the early 1890s, as a result, Chile was delivering three-quarters of its exports to Britain while obtaining half of its imports from that same country, creating a direct trade dependence on Britain greater than that of India at that time. When the First World War broke out in Europe two-thirds of Chile's national income was derived from nitrate exports primarily to Britain and Germany. The British monopoly of the nitrate trade through its control of the Chilean economy had put Germany at a serious disadvantage in its competition with Britain, since nitrates were necessary for explosives as well as fertilizer. Germany in the opening decade of the twentieth century accounted for a third of Chilean nitrate exports. Like Britain, Germany had worked to have Balmaceda ousted. But Chile remained largely under British control, creating a huge geopolitical problem for Germany. Just prior to the First World War the German chemist and nationalist Fritz Haber devised a process for producing nitrates by fixing nitrogen from the air. The result within a few years was to destroy almost completely the value of Chilean nitrates, creating a severe crisis for the Chilean economy.

The Nature of Ecological Imperialism

The economic development of capitalism has always carried with it social and ecological degradation—an ecological curse. Moreover, ecological imperialism has meant that the worst forms of ecological destruction, in terms of pillage of resources and the disruption of sustainable relations to the earth, fall on the periphery rather than the center. Ecological imperialism allows imperial countries to carry out an "environmental overdraft" that draws on the natural resources of periphery countries. As the material conditions of development are destroyed, third world countries are more and more caught in the debt trap that characterizes extractive economies. The principles of conservation that were imposed partly by business in the developed countries, in order to rationalize their resource use up to a point, were never applied to the same extent in the third world, where imperialism nakedly imposed an "after me the deluge" philosophy. The guano and nitrates trade during the mid to late nineteenth century highlights the unequal exchange and degradation associated with the ecological contradictions of Britain and other dominant countries in the global economy.

Indeed, it is rather misleading to dignify with the word *trade* what was clearly robbery of ecological and economic resources on a very high order, rooted in one of the most exploitative labor processes in history and backed up by war and imperialism. The result for Peru and Chile (and also Bolivia, which lost its nitrates in the War of the Pacific) was not development, but rather, as explained by critics from Mariátegui in the 1920s to Frank in the 1960s, constituted the "development of underdevelopment."[100] All of this, following Marx, needs to be understood in terms of the larger theory of global metabolic rift, which captures the underlying nature of the capitalist relation to the environment. In the case of the guano trade, the development of ecological imperialism necessitated not only an enormous net flow of ecological resources from South to North, but also gave added impetus to the importation of foreign labor, particularly coolie labor from China, under conditions that, as Marx said, were "worse than slavery." Within the world system of capital, the robbing of the soil in

Europe necessitated the importation of guano from Peru, and in the process fed into the robbing of human labor on a truly global scale. This might even be referred to as the "triangle trade" of mid-nineteenth-century ecological imperialism.

Ironically, the exploitation of guano in Peru in recent years is raising the question once again of the complete exhaustion of this natural resource, with the price in export markets in the United States, Europe, and Israel between 2007 and 2008 doubling to $500 a ton (as opposed to the $250 a ton it sells for in Peru). Peruvian guano is now highly prized as an organic fertilizer for organic farms around the world. But this new global demand, which has increased the rate of guano extraction, is pointing to "the end of guano" with supplies likely to run out in a decade or two, negating decades of successful sustainable development of this resource. In the Chincha Islands, where 60 million seabirds once deposited guano at the height of the nineteenth-century guano boom, there are now about four million birds. The guano deposits, which were once 150 feet high, reach on some islands, such as Isla de Asia, south of Lima, "less than a foot or so." The once abundant anchoveta (of the anchovy family) that once constituted the main food for the seabirds has been depleted by commercial fishing, since it is sold globally as fishmeal for poultry and other animals. Where Chinese laborers once dug guano, it is now worked by impoverished Quechua-speaking native laborers from the Peruvian highlands. In all respects this shows the absolute devastation that constitutes the natural end state of ecological imperialism.[101]

The nature of ecological imperialism is continually to worsen ecological conditions globally. Capital in the late twentieth and early twenty-first century is running up against ecological barriers at a biospheric level, barriers that cannot be so easily displaced, as was the case previously, through the spatial fix of geographical expansion and global labor and resource exploitation. Ecological imperialism—the growth of the center of the system at unsustainable rates, through the more thoroughgoing ecological degradation of the periphery—is now generating a planetary-scale set of ecological contradictions, imperiling the entire biosphere as we know it. Only a social solution that addresses the rift in ecological relations on a planetary scale and their relation to global structures of imperialism and inequality offers any genuine hope that these contradictions can be transcended. More than ever the world needs what the early socialist thinkers, including Marx, called for: the rational organization of the human metabolism with nature by a society (or societies) of freely associated producers, in order to establish a social metabolic order no longer predicated on capital accumulation, ecological imperialism, and the degradation of the earth.

PART FOUR

Ways Out

16. The Ecology of Consumption

> But think, I beseech you, of the product of England, the workshop of the world, and will you not be bewildered, as I am, at the thought of the mass of things which no sane man could desire, but which our useless toil makes—and sells?
>
> —WILLIAM MORRIS[1]

Environmentalists, especially in wealthy countries, have often approached the question of environmental sustainability by stressing population and technology, while deemphasizing the middle term in the well-known IPAT (environmental Impact = Population x Affluence x Technology) formula. The reasons for this are not difficult to see. Within capitalist society, there has always been a tendency to blame anything but the economic system itself for ecological overshoot. Yet if the developing ecological crisis has taught us anything, it is that even though population growth and inappropriate technologies have played important roles in accelerating environmental degradation, the ecological rift we are now facing has its principal source in the economy.

Although the population of the world has grown dramatically over the past century, the global economy has expanded much faster, and the most affluent nations devour natural resources at a much higher rate than do poorer nations with equal or larger population sizes. In China and India in recent years the effects of population growth (which has slowed in these nations) have been dwarfed by rapidly rising per capita economic output coupled with the *already* immense population sizes. All of this has been effectively demonstrated by ecological footprint analysis, which has allowed us to picture more fully the role of affluence in generating ecological overshoot.[2]

Consequently, there has been a gradual shift in environmentalism from "demographic Malthusianism" (a strict focus on the number of people) to a kind of "economic Malthusianism" (a focus on the number of consumers). In this new economic Malthusianism the emphasis is not so much on population control, but on consumption control.[3] We are led to believe that if consumers—meaning the mass of the population—can be restrained or their appetites rechanneled all will be well. As Worldwatch's *State of the World, 2010* report asserted: "Like a tsunami, consumerism has engulfed human cultures and Earth's ecosystems. Left unaddressed we risk global disaster. But if we channel this wave . . . we not only prevent catastrophe but may usher in a new era of sustainability."[4]

Ironically, the new economic Malthusianism comes closer in some ways than demographic Malthusianism did to the intent of Thomas Robert Malthus in his classic *Essay on Population*. Malthus's argument was principally a class one, designed to rationalize why the poor must remain poor, and why the class relations in nineteenth-century Britain should remain as they were. His greatest fear was that due to excessive population growth combined with egalitarian notions "the middle classes of society would . . . be blended with the poor." Indeed, as Malthus acknowledged in *An Essay on Population*, "The principal argument of this *Essay* only goes to prove the necessity of a class of proprietors, and a class of labourers."[5] The workers and the poor through their excessive consumption, abetted by sheer numbers, would eat away the house and home (and the sumptuous dinner tables) of the middle and upper classes. He made it clear that the real issue was who was to be allowed to join the banquet at the top of society:

> A man who is born into a world already possessed, if he cannot get subsistence from his parents on whom he has a just demand, and if the society do not want his labour, has no claim of *right* to the smallest portion of food, and, in fact, has no business to be where he is. At nature's mighty feast there is no vacant cover for him. She tells him to be gone, and will quickly execute her own orders, if he do not work upon the compassion of some of her guests. If these guests get up and make room for him, other intruders immediately appear demanding the same favour.... The order and harmony of the feast is disturbed, the plenty that before reigned is turned into scarcity.... The guests learn too late their error, in counteracting those strict orders to all intruders, issued by the great mistress of the feast, who, wishing that all her guests should have plenty, and knowing that she could not provide for unlimited numbers, humanely refused to admit fresh comers when her table was already full.[6]

A counterpart to Malthus's argument in this regard was his treatment, in his *Principles of Political Economy*, of the problem of effective demand, the lack of which was in his view the explanation for economic gluts or crises. In Malthus's aristocratic view—he was a Protestant parson aligned primarily with the landed classes of the aristocracy and gentry—this did not provide a justification for increasing the wages and consumption of the poorer classes. Rather the answer to the shortage of effective demand was to be found largely in the excess luxury consumption of the landed classes, who supported the overall economy through their overconsumption.[7] If Malthus's theory of population and his theory of effective demand seemed somewhat opposed to each other in their economic logic, the overall consistency of his analysis lay in its unitary class perspective.

Today's economic Malthusianism within environmental discourse is similar in the unitary class perspective that it offers. It is all about making mass consumption and hence the ordinary consumer (not the wealthy few) the culprit. It insists that the average consumer be encouraged to restrain his/her consumption or else that it be rechanneled toward beneficial ends: green shopping. It is thus the masses of spendthrift consumers in the rich countries and the teeming masses of emerging consumers in China and India that are the source of environmental peril.[8]

A good example of this switch from demographic to economic Malthusianism can be found in the Worldwatch Institute's *State of the World, 2010*, subtitled, *Transforming Cultures: From Consumerism to Sustainability*. Based in Washington, D.C., the Worldwatch Institute was established in 1974 by environmentalist Lester Brown and earned a reputation as one of the premier mainstream environmental think tanks, particularly with the publication of its annual *State of the World* reports beginning in 1984. Funding for its operations has come primarily from foundations, such as (most recently) the Heinrich Böll Foundation, the Bill and Melinda Gates Foundation, and the W. K. Kellogg Foundation. *State of the World* reports over the years have normally been constructed in a similar way: a series of chapters on different aspects of the developing world ecological crisis, followed by a final chapter offering suggestions on ways out of the crisis, usually focusing on sustainable business and sustainable technology.

Yet the *State of the World, 2010* marks a sharp divergence in this respect. In this year's report there are no chapters on the developing environmental problem. Instead the entire report, aside from a six-page chart on the historical chronology of the ecological threat, is devoted to what would previously have been consigned to the conclusion: a strategy of change, focusing on sustainable consumption. Moreover, this reflects a larger transformation in the basic thrust of Worldwatch that has been going on for some time: a shift from its previous demographic Malthusianism, emphasizing the mass population problem, to its current economic Malthusianism, emphasizing the mass consumption problem.

Much of what constitutes the *State of the World, 2010* is of course unobjectionable from a radical environmental standpoint. Yet the overall thrust is to suggest that the principal environmental problem today can be traced to the economy *via consumers*. Indeed, the environmental impact of society is viewed as beginning and ending with "The Rise and Fall of Consumer Cultures"—the title of the main thematic essay of *State of the World, 2010*, by Erik Assadourian, project director for the report. Assadourian sees consumption, even more than population and technology, as the driver of today's planetary environmental crisis.[9]

It is this notion of consumer culture as the beginning-and-end-all of the environmental problem that we will question in the analysis that follows. A genuine ecological critique of the role of consumption in contemporary society, we will suggest, necessitates the transcendence of capitalist commodity production—as the main precondition for the emergence of a new system emphasizing human and ecological needs.

The Enigma of Consumption

Today's dominant economic Malthusianism takes advantage of the semantic confusion generated by two different definitions of consumption. In environmental terms, consumption means the using up of natural and physical resources. It is sometimes referred to in terms of the *throughput* of energy and materials from natural resource tap to environmental sink (the eventual depositing of the waste in air, land, and sea). It thus stands for all economic activity. For example, environmental social scientists Thomas Princen, Michael Maniates, and Ken Conca argue that "all decision makers along the chain from extraction to end use are 'consumers.'" It is therefore possible to "construe [all] economic activity as consuming." Therefore "producers are consumers; production is consumption."[10]

In economics, in contrast, consumption is only one part of aggregate economic demand—that part accounted for by the purchases of consumers. In a given national economy (abstracting from exports/imports) total demand consists of consumption plus investment plus government spending. Government spending can be seen as consisting of public consumption and public investment. In its simplest terms, then, income equals consumption plus investment. From an economic standpoint, therefore, consumption is merely consumer demand, in contradistinction to investor demand.

Looked at from the other side of the national accounts, that is, output or production, consumption equals the output of the consumer goods sector (Department 2 in Marx's reproduction schemes) as opposed to the output of the investment goods sector (Department 1). Consumer goods thus represent only one part of total output or production.[11]

Indeed, from a deeper economic standpoint, there are more fundamental material and temporal distinctions between production and consumption. Production is the transformation of nature through human labor. A good must be produced before it can be consumed. Investment goods, as opposed to consumption goods, are specifically aimed at the expansion of this capacity to produce, and hence at the growth of the economy. A steady-state economy, or a system of "simple reproduction" without growth, as Marx and Kalecki referred to it, is essentially one in which everything is consumed, nothing invested.[12]

By ignoring the difference between these two very different notions of consumption, it is easy to insinuate that the problem of the consumption of environmental resources is to be laid at the door of consumers alone. Yet to neglect in this way the impact of investors on the environment is to exclude the motor force of the capitalist economy. Spending by investors is logically just as much a part of overall environmental throughput as is the spending of consumers. To lose sight of investment in the environmental equation is to deemphasize the role of production, profits, and capital accumulation.

The confusion that the misuse of these two different definitions of consumption generates in the environmental discourse is evident in the common fallacy that by not consuming but rather saving income one can somehow protect the environment.[13] Yet in a properly functioning capitalist economy savings are redirected into investment or new capital formation designed to expand the scale of the entire economy. And it is such expansion that is the chief enemy of the environment.

Another error arising from the blending of these two different concepts of consumption is to be seen in the frequent conflation of total environmental waste in society with waste related to direct household consumption, that is, taking the form of municipal solid waste (garbage). All too often garbage is treated as a problem mainly associated with the direct consumption of consumers. But municipal solid waste in U.S. society is estimated to be only some 2.5 percent of the total waste generated by the society, which also includes: (1) industrial waste, (2) construction and demolition waste, and (3) special waste (waste from mining, fuel production, and metals processing). This other 97.5 percent of solid waste disposal, outside of households, is invisible to most individuals who, in their role as consumers, have no direct part in either its generation or disposal.[14] As Derrick Jensen and Aric McBay observe: "If we divide municipal waste by population, we get an average of 1,660 pounds per person per year. But if we include industrial waste, per capita waste production jumps to 16.4 tons per person, or 52,700 pounds." If an individual were somehow to cut out 100 percent of his/her household waste, that person's per capita share of total waste would be largely untouched.[15]

Orthodox or neoclassical economists generally assume the existence of what is called "consumer sovereignty" in modern society, or the notion that all economic decisions are driven by the demands of consumers, who then become responsible for the entire direction of the economy. All stages of economic activity, in this view, are aimed simply at final consumption, which drives the entire process. Yet there is a long history of powerful critiques by heterodox economists—such as Karl Marx, Thorstein Veblen, John Kenneth Galbraith, Paul Baran, Paul Sweezy—of the consumer sovereignty thesis.[16] "The consumer," Marx wrote in the mid-nineteenth century, "is no freer than the producer [the worker]. His judgment depends on his means and his needs. Both of these are determined by his social position, which itself depends on the whole social organization. . . . Most often needs arise directly from production or from a state of affairs based on production. World trade turns almost entirely round the needs, not of individual consumption, but of production."[17] Galbraith referred to this as the "dependence effect," whereby "wants depend on the process [of production] by which they are satisfied."[18] In opposition to the orthodox axiom of "consumer sovereignty," Galbraith pointed to the reality of "producer sovereignty," exercised by corporations, dominating both production and consumption.[19]

Mainstream environmentalists often acknowledge the force of modern marketing in shaping consumer decisions, and have therefore gone a considerable way in recognizing that the consumer is subject rather than sovereign in our society. Nevertheless, much of the environmental critique of "consumer culture" within today's environmentalism still falls prey to the "innocent fraud" of consumer sovereignty, as Galbraith called it. Mohan Munasinghe began a recent article in the *Journal of Industrial Ecology* with the words "Household consumption drives modern economies." Similarly, Randall Krantz declared in a recent article in the same journal: "Consumers are ultimately [the] drivers of demand and consumption." Solving this problem of the consumer driver "will require," he insists, "driving consumer attitudes toward demanding better value, rather than resource-intensive 'stuff.'"[20]

The whole problem of the conflation of the environmental and economic meanings of consumption, addressed here, can be referred to as the *enigma of consumption*. Its significance lies in the fact that while the environmental problem arises primarily from production, in the transformation of nature by human labor, it is increasingly attributed entirely to consumption, which then becomes its own cause and effect. The fantastic world of consumer society, conceived in the abstract, and divorced from the production of commodities in capitalist society, resembles, as we shall see, nothing so much as the "misty realm of religion," deriving its essential meaning from its thorough rejection of all material explanations.[21]

Economic Malthusianism

In the early 1990s Alan Durning published an important study for Worldwatch titled, *How Much Is Enough? The Consumer Society and the Future of the Earth*. Durning put forward the standard IPAT-formula view that the three factors in environmental impact were population, consumption (in the environmental sense, standing for "affluence," or all economic activity), and technology. He then claimed that consumption was "the neglected god in the trinity."[22] Today, in contrast, consumption is increasingly seen by economic Malthusians as the preeminent god of the trinity.

There is perhaps no better example of this than Erik Assadourian's recent "Rise and Fall of Consumer Cultures" in the *State of the World, 2010*. In advancing his consumer culture thesis, Assadourian adopts a notion of "culture" characteristic of philosophical idealism and various contemporary forms of culturalism. Here culture is an autonomous realm that has no direct relation to material-productive conditions (but may indirectly determine the latter). This contrasts strongly with the "cultural materialism" of a thinker like Raymond Williams, for whom culture was dialectically connected to material conditions, that is, to the human transformation of nature through production.[23]

The cultural systems in which human beings are embedded, in Assadourian's more culturalist view, are to be defined entirely by "cultural norms, symbols, values, and traditions a person grows up with [that] become 'natural.'"[24] History is thus the story of the rise and fall of cultural patterns. From this it is a small step to the definition of consumerism as an independent cultural formation, and the explanation of the entire course of Western economic development as a product of this consumerist culture. "Consumerism" itself is defined as "the cultural orientation that leads people to find meaning, contentment, and acceptance through what they consume." To ask "people who live in consumer cultures to curb consumption," Assadourian tells us, "is akin to asking them to stop breathing."[25] In this view, the economy becomes simply a manifestation of culture so economic questions ultimately reflect symbolic choices, as exemplified by consumer culture.[26]

How did such a consumer culture emerge and how are we to understand the history of economic development in these terms? Assadourian gives us the following thumbnail account of the historical developments of the last five centuries:

> As long ago as the late 1600s, societal shifts in Europe began to lay the groundwork for the emergence of consumerism. Expanding populations and a fixed base of land, combined with a weakening of traditional sources of authority such as the church and community social structures, meant that a young person's customary path of social advancement—inheriting the family plot or apprenticing in a father's trade—could no longer be taken for granted. People sought new avenues for identity and self-fulfillment, and the acquisition and use of goods became popular substitutes.
>
> Meanwhile, entrepreneurs were quick to capitalize on these shifts to stimulate purchase of their new wares, using new types of advertising, endorsements by prominent people, creation of shop displays, "loss-leaders" (selling a popular item at a loss as a way to pull customers into a store), creative financing options, even consumer research and the stoking of new fads. For example, one eighteenth-century manufacturer, Josiah Wedgwood, had salespeople drum up excitement for new pottery designs. . . .
>
> Over time the emerging consumerist orientation was internalized by a growing share of the populace—with the continued help of merchants and traders—redefining what was understood as natural. The universe of "basic necessities" grew, so that by the French Revolution, Parisian workers were demanding candles, coffee, soap, and sugar as "goods of prime necessity" even though all but the candles had been luxury items 100 years earlier.
>
> By the early 1900s, a consumerist orientation had become increasingly embedded in many of the dominant social institutions of many cultures.... And in the latter half of the century, new innovations like television, sophisticated advertising techniques, transnational corporations, franchises, and the Internet helped institutions to spread consumerism across the planet.[27]

As even a brief sketch of the development of modern capitalist production and the world economy, this account is found wanting.[28] The shift from one mode of production to another is completely absent, and instead we are simply told that in the seventeenth century "young people" pursuing "social advancement" left the land and the town guild system, choosing instead the consumerist "acquisition and use of goods" as "new avenues for identity and self-fulfillment." No mention is made of class dynamics, land enclosures, the poor laws, the Industrial Revolution, the steam engine, the new factory system, capitalism, commodity production, colonialism, the world market, and so forth. Wedgwood's very exceptional role as an eighteenth-century pottery manufacturer, who made use of nascent advertising techniques to reach his aristocratic and middle-class customers, is taken as exemplary of the new system, while Watt and his new steam engine are noticeable in their absence.

This ahistorical account of economic development is replicated in the analytical framework that pervades *State of the World, 2010* as a whole. Missing from the index to the volume, despite its pretensions to describing the rise of consumerism over the last few centuries, are the following terms: economy, class, feudalism, capitalism, commodity, production, proletariat (working class), bourgeoisie (capitalist class), capital, corporations, accumulation, investment, savings, surplus, profit, wages, colonialism, slavery, imperialism, credit, debt, and finance. The growth of consumption, we are led to believe, can be viewed in complete abstraction from such historical concepts and economic developments.

Assadourian's account of historical agency focuses not surprisingly on elites, specifically on the role of "cultural pioneers" in shifting cultural paradigms, from which economic/market relations are derived. Hence a cultural pioneer such as Wedgwood (an entrepreneur) was, in this view, able to help initiate our current consumer culture, while new "networks of cultural pioneers" (also presumed to be entrepreneurs) can be expected to shift this consumer culture of the future in the direction of sustainable consumption.[29]

To be sure, room is left in this analysis for individual action in the form of all sorts of acts of voluntary simplicity (the moral equivalent of the earlier Malthusianism's "moral restraint" in propagation).[30] But the real emphasis is on elite business managers as cultural pioneers. Assadourian devotes considerable space to how business marketing and the media have stoked the fires of consumerism in the United

States—although he fails to perceive the relation of this to the historical rise of monopoly capital, that is, the economy of the giant firm.[31] In carrying out the transition from the "consumer paradigm" to a "sustainability paradigm," he contends that private marketing, which pushed consumers toward profligate consumption by taking advantage of their acquisitive natures, can be replaced today with social marketing that could drive consumers to green consumption through the greening of their appetites. At the same time, he myopically proposes that corporations will move away from the view "that profit is the primary or even the sole purpose of business." Indeed, "a sustainable economic system will depend," he imagines, "on convincing corporations . . . that conducting business sustainably is their primary fiduciary responsibility."[32]

The main example provided in *State of the World, 2010* of a corporation that is supposedly moving from an exclusive focus on profits to a sustainable business model as its "primary fiduciary responsibility" is Wal-Mart. A chapter in the volume written by Ray Anderson, the chairman of Interface Corporation, and his associates, focuses on work that Interface (the modular carpet-making firm sometimes seen as a leader in green business) has done in working with Wal-Mart in promoting sustainability. Anderson lauds as a cultural pioneer Wal-Mart CEO (now board chairman) Lee Scott, and the Wal-Mart Corporation in general. Scott is quoted as committing his company in October 2005 to "100 percent renewable energy, to create zero waste" (while admitting, at the same time, that he had no idea how Wal-Mart could achieve such goals). This, Anderson tells us, was followed by a "cocooning" stage in which Wal-Mart confronted all of its networks and suppliers, looking for what it could do to become more sustainable. Wal-Mart is said to have now gone through a "metamorphosis," reemerging as a green company on a "sustainability journey"—bringing its green values into the "personal lives" of all of its 1.8 million employees, who are being taught to be more sustainable consumers: recycling and eating more healthy meals.

Another essay in *State of the World, 2010*, written by Michael Maniates, focuses on "choice editing" by corporations (cutting out the undesirable decisions of consumers by removing these individual choices from those offered by business). In this respect, Maniates lauds Wal-Mart for its decision to market only wild-caught fresh and frozen fish that have been certified by the Marine Stewardship Council as sustainably harvested.[33]

Yet this emphasis on Wal-Mart as a symbol of sustainable business and an exemplary proponent of sustainable consumption could hardly be more absurd. Wal-Mart is the world's biggest retailer, accounting for 10 percent of U.S. retail sales, with over 2,700 supercenters in the United States alone, each taking up some 200,000 square feet and occupying (including parking spaces) some 20 acres of land. Its chief concrete environmental commitment, made in 2005, was to become 20 percent more energy efficient by 2013. This would, it claimed, result in it cutting the carbon dioxide emissions associated with its current stores by 2.5 million metric tons by 2013. Yet all of this turned out on closer examination to be a case of corporate greenwashing, since Wal-Mart was at the same time expanding its total operations in the United States and abroad. Its total U.S. greenhouse gas emissions, by its own accounting, rose by 9 percent in 2006. The new stores being added in the United States in 2007 alone were expected to consume enough electricity to add one million metric tons to its overall greenhouse emissions, far exceeding any efficiency gains. As environmental writer Wes Jackson put it, "When the Wal-Marts of the world say they're going to put in different lightbulbs and get their trucks to get by on half the fuel, what are they going to do with that savings? They're going to open up another box store somewhere. It's just nuts."

The Marine Stewardship Council, the seafood certification program adopted by Wal-Mart, has, according to Food and Water Watch, a history of accrediting fisheries with very dubious environmental records; nor is it likely that fish could be sourced sustainably on the scale that Wal-Mart demands. Wal-Mart is facing fines for violating hazardous waste dumping laws in a number states. Although it has announced that it will eliminate non-biodegradable plastic bags for shoppers in its stores, utilize more efficient lightbulbs, and promote greater fuel efficiency in its trucking fleet, it remains notorious for its extreme exploitation of workers and its virulent anti-union stance. In the end, Wal-Mart is an economic juggernaut—anything but a representative of a new sustainable economic order.[34]

The implications of today's economic Malthusianism are nowhere more apparent than in its truncated treatment of class issues. This can be seen in Assadourian's comparison of consumption in India and the United States. He takes the richest 1 percent of the population in India and compares their carbon dioxide emissions to that of the per capita emissions of all Americans, which are five times higher. This per capita level of emissions/consumption is referred to as "the American way of life."[35] Yet this ignores the existence of sharp class differences in wealth, income, consumption, and greenhouse gas emissions in the United States (and other rich economies), since per capita figures are simply a statistical average of widely divergent class levels, and thus wildly distorting. A similar approach was presented in Durning's earlier Worldwatch study. He divided the entire world into three broad classes (based on per capita income in different countries): the consumer class, the middle-income class, and the poor. For Durning all of the individuals in the rich countries (or at least those above the poverty line) belong equally to the "consumer class," which is eating up the world's resources. Although he briefly notes that the top fifth of income earners in the United States have more income than the bottom four-fifths combined, this is thereafter ignored, since the focus is on a common membership in the "consumer class."[36]

The reality is the higher the class/income level the bigger the ecological footprint. In 2008, Americans in the highest income quintile spent three to four times as much on both housing and clothing, and five times as much on transportation, as those in the poorest quintile.[37] In Canada where consumption data is available in deciles, ecological footprint analysts have found that the top income decile has a transportation footprint nine times that of the bottom decile, and a consumer goods footprint four times that of the bottom decile. (All such statistics are invariably distorted by the underrepresentation of the wealthy in the statistical samples.)

Indeed, the class reality in the United States and the discrepancies in environmental impact that result are far more startling than official consumption figures suggest. A relatively small portion of the population (around 10 percent) owns 90 percent of the financial and real estate assets (and thereby the productive assets) of the country, and the rest of society essentially rents itself out to the owners. The wealthiest 400 individuals (the so-called Forbes 400) in the United States have a combined level of wealth roughly equal to that of the bottom half of the population, or 150 million people.[38] The top 1 percent of U.S. households in 2000 had roughly the same share (20 percent) of U.S. national income as the bottom 60 percent of the population. Such facts led a group of Citigroup researchers and investment counselors to characterize the United States as a "plutonomy," a society driven in all aspects by the rich. In this view, the "average consumer" is a meaningless entity, since consumption is increasingly dominated by the luxury consumption of the rich, who also determine production and investment decisions.[39]

Such a realistic class perspective (even emanating from the financial establishment itself) is, however, anathema to today's elitist environmentalism. In discussing "sustainable consumption," ecological modernization theorists like Gert Spaargaren systematically avoid the notion of "relations of production" (class relations) and choose to speak instead in bland systemic terms of "provisioning" and the relation "between providers [green entrepreneurs] and citizen-consumers." The obvious objective is to reify the fundamental issues—removing all dimensions of historicity and power.[40]

An Immense Collection of Commodities

Karl Marx began his critique of political economy in *Capital* with the words: "The wealth of societies in which the capitalist mode of production prevails appears as an 'immense collection of commodities.'"[41] A commodity is a good produced for sale on the market for a profit. It has both a use value and an exchange value. But it is the exchange value of a commodity, out of which profits are generated, which is of primary interest to the profit-seeking capitalist or corporation. Capitalism can thus be described as *generalized commodity production* with the individual commodity as its cell form. It is a society and economy characterized by "the production of commodities by commodities"—in the sense that labor power is reduced to little more than a commodity, to be bought and sold on the market and manipulated for the sake of profit.

Once the commodity form is analyzed, and its social relations depicted, it becomes clear that production, exchange, distribution, and consumption under the regime of capital is the production, exchange, distribution, and consumption of commodities. Production and consumption, constituting the beginning and the end of this process, are therefore elements of a single process, dialectically connected, and like any organic whole mutually interacting—production, however, is "predominant" since its conditions are "the real point of departure" for the various moments.[42] Moreover, in a capitalist society production is directed to the generation of commodity values (exchange values). Nothing could be more absurd then than the conception of a market society in which commodities were autonomous entities, abstract things or "stuff," simply demanded and then consumed by individual consumers—as if consumption existed independently of production, and use value not exchange value constituted the object of production.[43]

Commodity fetishism, as described by Marx, is the generation within human consciousness of an illusory, inverted realm of commodities, which become mystical entities in their own right, like the gods of mythology, seemingly dominating over human beings—separated from the fact that they are manifestations of *definite material-productive relations*. In today's festishized world cultivated by marketing, godlike material powers and symbolic values (love, wealth, power, immortality) are attributed to commodities. Consumption, the market, exchange, and the consumer are thus systematically fetishized. "Commodity reification," as Fredric Jameson has noted, "has become the central phenomenon in the enlargement and spread of capitalism around the world, taking the social form of what has come to be identified as consumerism."[44] To detach the reality of consumerism from its fundamental basis in capitalist commodification is thus a major analytical error, a case of what Alfred North Whitehead called "the fallacy of misplaced concreteness."[45]

Such reification affects even social scientists: particularly mainstream economists and establishment-oriented environmental sociologists (ecological modernization theorists), who begin and end their analyses with household consumption. "If you proceed from production," Marx and Engels observed in the *German Ideology*, "you necessarily concern yourself with the real conditions of production and with the productive activity of men. But if you proceed from consumption, you can set your mind at rest by merely declaring that consumption is not at present 'human,' and by postulating 'human consumption,' education for true consumption and so on. You can be content with such phrases, without bothering at all about the real living conditions and the activity of men."[46]

A realistic approach to the ecology of consumption would start with Raymond Williams's remark that far from being "too materialistic. . . . Our society is quite evidently not materialistic enough."[47] What people are taught to value and consume in today's acquisitive society are not use values, reflecting genuine needs that have limits, but symbolic values, which are by nature unlimited. Marketing has completely taken over production. Salesmanship, as Thorstein Veblen pointed out nearly a century ago, has so penetrated into the production process that most of the costs of production, associated with designing and producing a commodity, are concealed sales costs, aimed at marketing the product. The goal of the system is what Veblen called "the quantity-production of customers" and not the satisfaction of needs.[48] Products are turned into brands systematically removed from human needs—all for the purposes of economic expansion, profits, and accumulation. Advertising alone (not including other forms of marketing, such as targeting, motivation research, product development, sales promotion, direct marketing, etc.) accounts for between 4 and 12 percent of the sales price of common commodities such as a Dell laptop, Palmolive soap, Levi's jeans, and a GM pickup truck.[49]

The fact that commodities in today's capitalist society are promoted on the basis of abstract qualities other than their materiality or usefulness, constitutes the basis for what Juliet Schor has called "the materiality paradox." In Schor's words, "The materiality paradox says that when consumers are most hotly in pursuit of nonmaterial meanings, their use of material resources is greatest."[50] The recognition of such a materiality paradox can be traced all the way back to antiquity. As Epicurus put it: "Wealth based on nature is delimited and easily provided, whereas that based on empty beliefs plunges out to infinity."[51] Today this "bad infinity"—to use a Hegelian term—is omnipresent.[52] A society dominated by such nonmaterial meanings (abstract value) encourages economic and environmental waste, a throwaway culture, a fashion cycle extending to more and more commodities, and so forth. Under genuine materialism products would be desired, as William Morris insisted, solely for their usefulness and their beauty—taking into account also their relation to the physical environment.[53]

The entire system of marketing, in which trillions of dollars are spent persuading individuals to buy commodities for which they have no need, and no initial desire, would have to be dismantled if the object were to generate a genuine ecology of consumption. Today's gargantuan marketing system (which now includes detailed data on every U.S. household) is the most developed system of propaganda ever seen, a product of the growth in the twentieth century of monopoly capitalism. It is not a system for expanding choice but for controlling it in the interest of promoting ever-greater levels of sales at higher profits. It is therefore aptly referred to by Michael Dawson as "the consumer trap." The production of high-quality goods increases production costs and decreases sales (since the products thereby do not have to be replaced as often) and this goes against the goals of capital. The general thrust is the production of commodities that are inexpensive and low quality and frequently replaced.[54] In recent decades, the consumer trap has merged with the debt trap in which ordinary working people are more and more enmeshed—part of the growth in our time of monopoly-finance capital—in their attempts simply to maintain their "standards of living."

Socialism and Plenitude

Society is an organic whole, from which we abstract for purposes of investigation. Production and consumption, as elements of this unity, belong to a single metabolism. In a non-alienated social and ecological condition, production and consumption represent a mutually reinforcing metabolism (in a relation of reciprocal exchange with nature) constituted at all times by sustainable human needs.

The recognition of the necessary dialectical relation between production and consumption (encompassing as well their natural preconditions) is the key to the creation of a more ecological society. A vision of sustainable development means focusing on human relationships, sensuous experience, and qualitative development. A genuine ecology of consumption—the creation of a new system of sustainable needs-generation and satisfaction—is only possible as part of a new ecology of production, which requires for its emergence the tearing asunder of the capitalist system, and its replacement with a new human whole. The goal would be a society that is mindful of natural limits in which production and consumption would be focused on collective needs and human development. Moreover, this could only be achieved in a context of community, that is, in a relation of reciprocity—and thus in metabolic relation with nature as a whole.[55] The latter aspect is crucial. "We abuse land," Aldo Leopold wrote, "because we regard it as a commodity belonging to us. When we begin to see land as a community to which we belong, we may begin to use it with love and respect."[56]

It follows that real sustainable development is not about sustaining economic accumulation and growth but about sustainable human development and the maintenance of the conditions of life for the millions of other species on Earth. This necessitates the rational, scientific regulation of the human metabolism with nature so as to maintain the fundamental conditions of life.[57] It involves planning to ensure that the basic human needs of nutritious food, adequate housing, clean water, and the conditions of a healthy existence are available for all.[58] But the ultimate goal is the rich development of individual human powers, which is possible for each and every person only through the development of the wealth of human capacities and needs in accord with natural conditions. This requires the creation of free, disposable time, and a distancing of society from the treadmill of production. Time would no longer be regarded as money but as lived experience.

The reality of ecological overshoot in modern society tells us that the wealthiest countries and the world as a whole must enter what classical economists called a "stationary state," that is, become no-growth (even degrowth) economies. This means bringing mere quantitative growth (in aggregate terms as currently measured) to a halt in the rich countries, and then reversing growth, while at the same time qualitatively expanding the range of human capacities and possibilities and the diversity of nature.[59] More specifically, in terms of the present system, the imperative is to bring capital accumulation as such to a halt. But such a cessation of accumulation as a necessary prerequisite to solving the ecological problem is not possible for the regime of capital itself, since it contradicts its internal logic. A drastic slowdown or cessation of net capital formation under the present system means economic crisis, with the heaviest costs being imposed on those at the bottom.[60]

The necessary shift from a quantitative to a qualitative economy—which must start immediately if the world is to move away from the present "business-as-usual" path toward planetary ecocide—requires the progressive dismantling of the regime of capital, and the construction brick by brick of a new organic social and ecological system in its place. This means a radical confrontation with the logic of capital at all points in the society. Capital's private domination over disposable time needs to be turned into a *new social allocation of disposable time*—increasing the numbers of those with gainful employment by reducing average working hours, thereby making possible human development. Such changes will require a revolutionary transformation in the dominant social values associated with current material relations, and indeed a transformation of these material relations themselves.

"Nothing is enough," Epicurus wrote, "to someone for whom enough is little"—thereby questioning the entire spirit of capitalism almost two millennia before its emergence. It is this irrational drive of capital, the unlimited quest for ever-greater private riches, divorced from and even at the expense of real public wealth, that needs to be transcended.[61]

Juliet Schor has employed the concept of "plenitude" in describing a sustainable ecological future. Plenitude is meant to stand for the wealth and diversity of relationships, divorced from our usual way of looking at development in terms of GDP growth and the increase in financial assets. It recognizes that the social goal should be to gain greater distance from the current treadmill of accumulation. For Schor a social economy geared to plenitude is one that focuses on: (1) "diversify[ing] out of the market," (2) "self-provision," (3) "true materialism," and (4) "investments in one another and our communities." More specifically, her analysis emphasizes individuals voluntarily diversifying out of the corporate economy by concentrating on their "time wealth" rather than monetary wealth; choosing to *self-provision* (even homesteading) in basic areas such as food, housing, and clothing; focusing on *true materialism* by seeking useful, durable goods, and avoiding waste; and seeking voluntary, sustainable *investments in one another and our communities*.[62]

A socialist concept of ecological plenitude would embrace many of these same notions of a rich, diverse, and more qualitative existence, focusing on social needs satisfaction and human development, and at the same time abandoning the largely voluntaristic perspective, which is a hindrance to their achievement. Instead it would promote a genuine, mass-based structural transformation in the workings of modern society in the interest of human communities. Here it is not a question of individuals seeking simply to withdraw from the capitalist economy, but rather one of creating *a new ecological hegemony* within civil society aimed at transforming the entire structure of production and consumption in the context of capitalism's accelerating structural crisis. In this context, it makes sense to discuss not just "diversifying out" of the dominant economy (a strategy aimed mainly at the middle class in Schor's analysis, which fails to challenge the dominant relations of production and economic system and lends itself to co-optation) but the creation of new core of non-alienated productive relations within society as a whole. A modern, sustainable, steady-state economy is only possible through the transformation of the social formation itself with the rise of a new hegemonic ecology. Its governing principle must be a new "culture of substantive equality," without which the creation of a genuinely sustainable society, based on the reciprocity between human beings and humanity and nature, is impossible.[63]

But from whence is this new ecological hegemony to arise? The only conceivable answer is through the organization on socialist principles of an ecological and social counter-hegemony, deriving its impetus from various social actors. A new ecological materialism arising in the revolt against the global environmental crisis must merge with the old class-based materialism of socialism—a synthesis made possible by the deep ecological, as well as economic, roots of classical Marxism and socialism. Such a new historic bloc, in the Gramscian sense, would unite the contradictory and discordant ensemble of relations of resistance flowing out of "superstructures" and productive "bases" (revolutionizing at one and the same time culture and material conditions).[64] It would draw on various classes and class fractions (including the critical intelligentsia), but would depend fundamentally on the working class(es)—though not so much today on the *industrial proletariat* as such, but on a wider *environmental proletariat*, giving rise to a much broader, and at the same time more unified, material-ecological revolt. This revolutionary possibility was already implicit in the earliest works of Marxism, such as Engels's *Condition of the Working Class in England*.[65]

The socialist struggles of the twenty-first century in Latin America and elsewhere are taking place based on the emergence of a revolutionary historic bloc that draws on both traditional worker struggles and a wider ecological and communal consciousness. More collective forms of production and more communal forms of consumption are being advanced, aimed at the genuine needs of communities, while increased emphasis is being placed on ecological sustainability. As María Fernanda Espinosa, Minister of Heritage in Ecuador, stated in Cochabamba on April 20, 2010, the world ecological crisis is a "symptom of the development model of the world capitalist system and its logic and destructive relationships." A realistic attempt to address ecological problems thus involves finding new ways of building an economy and interacting with nature, based on socialist and indigenous principles, in which we "accumulate no more," while at the same time improving the human condition. In any true ecological revolution, "the answers to the structural crisis of climate change and the problems of the world must be of the same order—structural, revolutionary, deep."[66] An ecology of consumption is only possible based on an ecology of production aimed at sustainable human development. Herein lies the future of socialism and the earth in our time.

17. The Metabolism of Twenty-First Century Socialism

One of the most important aspects of Marxist scholarship in recent decades has been the recovery and subsequent development of Karl Marx's argument on social and ecological metabolism, which occupied a central role in his critique of political economy. So central was this that Marx defined the labor process itself in metabolic terms. As he wrote in *Capital*: "Labour is . . . a process between man and nature, a process by which man . . . mediates, regulates and controls the metabolism between himself and nature."[1] Such a conception was two-sided. It captured both the social character of labor, associated with such metabolic reproduction, and its ecological character, requiring a continuing, dialectical relation to nature.

Although the key role played by the concept of metabolism in Marx's thought has been recognized for a long time, its full significance has rarely been grasped until recently. For example, in the 1920s, Georg Lukács emphasized the "metabolic interaction with nature" through labor as a key to Marx's dialectic of nature and society. However, he went no further in elaborating the concept.[2] Present-day attention to this theme has developed mainly along two lines: (1) Lukács's younger colleague István Mészáros's analysis of capital as a historically specific system of *social metabolic reproduction*, and (2) the work of the present authors and others who have built on Marx's notion of a *metabolic rift* in the relation between nature and society.[3] These two strands of Marxist analysis of the nature-society metabolism are dialectically linked. Mészáros's work has been primarily concerned with issues of *social* metabolic reproduction, but this has nonetheless generated some of the most penetrating and prescient analyses of the ecological problem. Recent Marxist work on *ecological* metabolism has converged with the dialectic of social metabolic reproduction, as outlined in Mészáros's *Beyond Capital*, in delimiting the conditions of a sustainable future society.[4] Mészáros, in particular, emphasizes that the qualitative changes in the social order demanded by *ecology* are dialectically connected to a wider set of qualitative challenges—such as the necessity of social control and substantive equality—defining the struggle for a socialism for the twenty-first century.

Marx and Metabolism

The concept of metabolism was established within both chemistry and biology in the early nineteenth century for studying the chemical processes within organisms and the biological operations of organisms. It captures the complex biochemical process of exchange, through which an organism (or a given cell) draws upon materials and energy from its environment and converts these by various metabolic reactions into the building blocks of growth. The metabolism concept allowed scientists to document the specific regulatory and relational processes that direct interchange within and between systems—such as organisms digesting organic matter. Marx incorporated this concept, but in a much broader context, into all of his major political-economic works from the 1850s on, using it to analyze the dialectical relationship between society and nature.[5] By necessity there is a "metabolic interaction" between humans and the earth, as the latter supports life. Labor is "an eternal natural necessity which mediates the metabolism between man and nature, and therefore human life itself."

Through the labor process, humans transform the world and themselves, creating history in relation to the conditions of life.[6]

As a metabolic process, labor is thus, in Marx's conception, a life-giving process. This general approach conforms to modern science. As the great physicist Erwin Schrödinger wrote in his *What Is Life?* (1945):

> How does the living organism avoid decay? The obvious answer is: By eating, drinking, breathing and (in the case of plants) assimilating. The technical term is *metabolism*. The Greek word . . . means change or exchange. Exchange of what? Originally the underlying idea is, no doubt, exchange of material. (E.g. the German for metabolism is *Stoffwechsel*). . . . What an organism feeds upon is negative entropy. Or, to put it less paradoxically, the essential thing in metabolism is that the organism succeeds in freeing itself from all the entropy it cannot help producing while alive.[7]

In a manner consistent with such conceptions, Marx's metabolic analysis viewed socio-ecological systems as depending for their regeneration upon specific metabolic processes involving complex historical relationships of interchange and reproduction.[8] Due to the interpenetration of society and nature, humans have the potential to alter the conditions of life in ways that surpass natural limits and undermine the reproduction of natural systems. In assessing actual metabolic interactions, Marx examined the constantly evolving set of needs and demands that arose with the advent and development of the capitalist system, which transformed the social interchange with nature, directing it toward the constant pursuit of profit. He highlighted this change in *A Contribution to the Critique of Political Economy*, noting that "the exchange of commodities is the process in which the social metabolism, in other words the exchange of particular products of private individuals, simultaneously gives rise to definite social relationships of production, into which individuals enter in the course of this metabolism."[9] Use of the concept of metabolism here was meant to draw attention both to the metabolic exchange between nature and humanity—the underlying condition of human existence—and also to the reality of social-metabolic reproduction. The latter expresses the fact that social formations as organic systems have to be seen as continuing and developing processes. They therefore need to be analyzed in terms of the totality of the relations of exchange, (and relations of social production/reproduction) that constitute them.

The constant reproduction of capital on an ever-larger scale intensifies the metabolic demands on nature, necessitating new social relations and forms of socio-ecological exchange. It is here that Marx's analysis throws light on the complex, developing forms of the estrangement and degradation of labor/nature in capitalist society. This is rooted, he tells us, in the alienation of human labor power (itself a natural agent) and, through this, the entire human-nature metabolism.

The Soil Nutrient Cycle and the Metabolic Rift

Marx coupled his metabolic analysis with his critique of political economy, illuminating how industrialized capitalist agriculture created a metabolic rift, which reflected the unsustainable practices of the system as a whole. Drawing upon the work of the great chemist Justus von Liebig and other scientists, Marx noted that the soil nutrient cycle necessitated the constant recycling of nitrogen, phosphorus, and potassium, as plants absorbed these nutrients. Plant and human wastes in pre-capitalist societies were generally returned to the soil as fertilizer, helping replace lost nutrients. But the enclosure movement and the privatization of land that accompanied the advent of capitalism created a division between town and country, displacing much of the population from the land and expanding the urban population. Intensive agricultural practices were used to increase yields. Food and fiber—along with soil nutrients—were shipped hundreds or even thousands of miles to distant urban markets. The essential soil nutrients accumulated as waste, which polluted cities and rivers. These practices undermined the natural conditions that were necessary for reproduction of the soil. Marx pointed out that capitalist agriculture "disturbs the metabolic interaction between man and the earth, i.e. it prevents the return to the soil of its constituent elements consumed by man in the form of food and clothing; hence it hinders the operation of the eternal natural condition for the lasting fertility of the soil."[10] In other words, it was a robbery system exhausting natural wealth for the sake of private profit.

Large-scale, mechanized agriculture and long-distance trade intensify the metabolic rift in the soil nutrient cycle. Marx indicated that capital creates "the universal appropriation of nature," as it attempts to subject natural laws and systems to the whims of accumulation. "It is destructive towards all of this [nature], and constantly revolutionizes it, tearing down all the barriers which hem in the development of the forces of production, the expansion of needs, the all-sided development of production, and the exploitation and exchange of natural and mental forces." Intensive, industrial agricultural practices are employed to sustain and increase production, as well as to overcome the limitations imposed by the nutrient cycle. Marx warned that the incorporation of industry into agriculture supplied the latter "with the means of exhausting the soil," hastening the rate of environmental degradation.[11]

In the mid-nineteenth century, intensive agricultural production in England and other core nations contributed to the global metabolic rift, as millions of tons of guano and nitrates—as well as various agricultural goods—from Peru, Chile, and elsewhere were transferred to the North to enrich depleted soils. Imported labor from China, "coolies," worked under harsh conditions extracting guano from islands off the coast of Peru. These "beasts of burden" choked on guano dust, were physically beaten, and lived short lives to enrich the soils of the global North.[12]

The international fertilizer trade ushered in decades of civil unrest, war, debt, and global asymmetries in the international hierarchy of nations. The Haber-Bosch process, developed in Germany just prior to the First World War to overcome Britain's monopoly of Chilean nitrates, allowed for the fixation of nitrogen to produce ammonia on an industrial scale, facilitating the development of artificial fertilizers by capitalists in the global North. This attempt at a technological fix increased the industrialization of agriculture, without attending to the source of the metabolic rift in agriculture. This shift in the socio-ecological relationship introduced additional ecological problems over the course of the twentieth century, such as the accumulation of nitrogen in waterways, contributing to the formation of dead zones.

Capital's insatiable appetite for ever-higher levels of profit and accumulation is reinforced by the domination of exchange value over use value, competition, and the concentration and centralization of capital. The impulse of incessant accumulation amplifies the social metabolism of society, increasing the demands placed on nature. New technologies are used above all to expand production and to lower labor costs. Capital's social metabolism is increasingly in contradiction with the natural metabolism, producing various metabolic rifts and forms of ecological degradation that threaten to undermine ecosystems.

Part of revealing the inherent destructiveness of capital is to lay bare the social relations of the system, emphasizing the possibility and necessity of social transformation in the mode of production. Marx argued that socialism offered the opportunity to pursue genuine human needs. At the same time, he emphasized that the transformation of property relations must also entail a systematic reorganization of the interchange with nature. He argued that a society of associated producers was necessary to "govern the human metabolism with nature in a rational way, bringing it under their collective control instead of being dominated by it as a blind power; accomplishing it with the least expenditure of energy and in conditions most worthy and appropriate for their human nature."[13] An ecology that would maintain the earth for "succeeding generations," as Marx put it, would thus require transition to a new social order—presenting human civilization with its greatest and most urgent challenge.[14] Although capitalism served to promote science, its rational application, he suggested, was only possible in a society of associated producers.

The Necessity of Social Control

The centrality of the human-social relation to nature, and the fact that it is mediated by the alienated labor that characterizes social existence under the regime of capital, is graphically illustrated in Mészáros's *Marx's Theory of Alienation*, winner of the 1971 Isaac Deutscher Prize. In a remarkable series of diagrams he provided a description of not only Marx's conception of the complex relation between humanity, nature, and labor, but also how humanity was doubly alienated, in terms of both alienated labor and alienated nature.[15] In his Deutscher Prize lecture of that same year, Mészáros presented his emergent understanding of "the structural crisis of capital" as well as a powerful ecological critique that anticipated (but on far more radical foundations) the *Limits to Growth* argument unveiled by the Club of Rome in 1972.[16] He criticized the advocates of capitalist development for their shortsighted promotion of the U.S. model of "high mass-consumption," pointing out that this approach was oblivious to natural limits—not to mention completely absurd given the inner dynamics of an economic system that generated wealth through the immiseration of most of humanity. He stressed a full four decades ago that this pattern could not be replicated throughout the world without causing immense environmental degradation and exhausting "the ecological resources of our planet."[17]

The ecological and social challenges that confront us are often minimized as the logic of capital goes unquestioned and various reforms are put forward (such as improving energy efficiency via market incentives) under the assumption that the system can be tamed to accommodate human needs and environmental concerns. Such positions fail to acknowledge that the structural determinations of capital will inevitably grind onwards, threatening to undermine the conditions of life, unless systematic change is pursued to eradicate the capital relation entirely. It is here that Mészáros presents a scathing critique of capital and its persistently destructive proclivities—all the while focusing on the necessity of a new social order.

Venezuelan President Hugo Chávez has referred to Mészáros as the "Pathfinder of Socialism," emphasizing the importance of *Beyond Capital* for proposing a theory of transition.[18] While Mészáros's work is firmly rooted in Marx's *critical method*, it stands apart as a foundational contribution in its own right. In *Beyond Capital* (as well as his other books), he establishes the basis for envisioning a future beyond the system of capital. He does this by pointing dialectically beyond Marx's *Capital* to the necessity of a new theory of socialist transition for the twenty-first century. The "capital system" is conceived as a "social metabolic order" that permeates all aspects of society and that activates "absolute limits," making this potentially the most dangerous period of human history. In focusing, like Marx himself, on the "capital relation," rather than simply capitalism itself, Mészáros is able to account for the collapse of post-capitalist societies in terms of their failure to eradicate capital in its totality. In relation to the present structural crisis of capital, he illuminates both the anarchic forces that are undermining the social metabolic reproduction of the system and the necessity of social control for a genuine socialist transition. Both an ecologically sustainable social order and substantive equality are essential for human development. Without both of these components the survival of the human species remains threatened—whether from world war or ecological collapse.[19]

Environmental concerns, in this conception, do not constitute an isolated issue. Instead they are intimately tied to the social metabolic order, which requires confronting the question of social control. Yet the capital system itself is innately "uncontrollable." It is driven inexorably, via the force of competition, to the incessant accumulation of capital, which concentrates social, economic, and political power. It imposes a particular form of rationality and interchange between human beings and nature, whereby all relationships are assessed in terms of their "productive viability" to facilitate expansion of the system.[20] The logic of capital is superimposed on everything, be it health care, education, manufacturing, or the environment. Exchange value becomes the universal measure, as owners attempt to maximize profit. The capital system is incapable of "self-sufficiency"; it must constantly be renewed, pushing outwards, revolutionizing its relations of production, devouring more labor to capture surplus value, freely appropriating nature and subsuming the world to the accumulation process.[21]

Given the distorted accountancy of capital as a system, which includes exchange value but not use value, a "universal value-equation" dominates, "obliterating substantive incommensurability everywhere." In other words, money serves as the universal medium of exchange, which extends commodity fetishism, erasing the social and natural processes—such as the time it takes for labor power to be reproduced or for trees to grow after being cut. Public wealth (the sum of use values, including natural wealth) is exploited and diminished for the sake of increasing private riches. Capital is predicated on constant growth, so it forever attempts to increase its turnover rate in order to accelerate accumulation.

Since exchange value is its exclusive focus, the social metabolic order of capital attempts to transcend whatever social or natural limits it confronts. As Mészáros puts it, "For the first time ever in history human beings have to confront a mode of social metabolic control which *can* and *must* constitute itself—in order to reach its fully developed form—as a *global* system, demolishing all obstacles that stand in the way," regardless of "how devastating the consequences."[22] Its success is solely determined by the extent to which it can amass wealth at the top of the social pyramid. Like Marx in the *Grundrisse*, Mészáros warns that capital recognizes *barriers* that can be surmounted but not *boundaries* in the sense of absolute limits. It therefore incorporates in its inner logic a tendency to overshoot all objective limits, including the conditions for life.[23]

Instead of the substantive equality necessary for universality in the social world, capitalism has produced inequality, unemployment, exploitation, human misery, war, and environmental degradation. The putative democracy offered to the world comes at the cost of disenfranchising the majority of the world's population through alienating work environments, the ever-present threat of violence for participating in political opposition, and the undermining of subsistence production and the natural infrastructure.

Mészáros stresses that the reproduction of the capital system can only be secured through ever more destructive forms that further impoverish the world's population. Increasingly, consumption and destruction are coupled within the social metabolic order of capital, as destructive forces and wastefulness, such as the military-industrial complex, are pushed to the forefront to sustain an economic system that cannot be integrated politically on the global plane. Global war, even at the expense of mutual destruction, remains a means to secure the dominant position within an international system of competition.[24] Furthermore, the profit-driven system is incapable of effectively regulating the social metabolism between human society and nature. As capitalist production intensifies its demands on nature, the scale of ecological devastation will inevitably increase—the effects of which will outlast the transformation of the system.

In "The Necessity of Social Control" in 1970 Mészáros highlighted the culminating and deepening crisis. He explained that humanity must overcome the fragmentation of society and find unity if it is to survive. Here he focused on the relation of ecological degradation to capital's extreme uncontrollability:

> Another basic contradiction of the capitalist system of control is that it cannot separate "advance" from *destruction*, nor "progress" from *waste*—however catastrophic the results. The more it unlocks the powers of productivity, the more it must unleash the powers of destruction; and the more it extends the volume of production, the more it must bury everything under mountains of suffocating waste. The concept of *economy* is radically incompatible with the "economy" of capital production which, of necessity, adds insult to injury by first using up with rapacious wastefulness the *limited resources* of our planet, and then further aggravates the outcome by *polluting and poisoning* the human environment with its mass-produced waste and effluence.
>
> Ironically, though, again, the system breaks down at the point of its supreme power; for its maximum extension inevitably generates the vital need for restraint and *conscious control* with which capital production is structurally incompatible. Thus, the establishment of the new mode of

> social control is inseparable from the realization of the principles of a *socialist economy* which centre on a *meaningful economy of productive activity*: the pivotal point of a rich human fulfillment in a society emancipated from the alienated and reified institutions of control.[25]

Joseph Schumpeter's notion of "creative destruction," itself derived from Marx, is seen by Mészáros as leading increasingly in late monopoly capitalism to a system of *destructive creation* (or destructive uncontrollability), characterized by systematic waste, chronic underutilization, environmental destruction, and both limited and potentially unlimited warfare.[26] This is the end of capitalism as a rational system. Establishment of a new system of social control, via a radical transformation of production and the human relationship to nature—resulting in a more sustainable social metabolic order—becomes an absolute necessity.

"The issue," Mészáros makes clear, "is not *whether* or *not* we produce under *some* control, but under what *kind* of control; since our present state of affairs has been produced under the 'iron-fisted control' of capital which is envisaged, by our politicians, to remain the fundamental regulating force of our life also in the future." Politics must be emancipated from the power of private capital, in order for people to gain rational *social control* over their productive lives—which includes the social metabolism with nature—and over human development. Rational planning by associated producers is thus an indispensable condition for the production and reproduction of a society of substantive equality and sustainable development.[27]

The Absolute Limits of Capital and the Law of Laws

The necessity of social control is all the more vital when we consider what Mészáros calls the "absolute limits of capital," especially in regard to the emerging ecological crisis. All social metabolic orders have "intrinsic or absolute limits which cannot be transcended" without forcing a qualitative transformation to a new mode of control.[28] In such a situation, it becomes an imperative to transition to a new social metabolic order, but just because this is necessary does not mean it will happen. Mészáros warns that even though the absolute limits of capital may be activated, capital will not come to a halt and give up its expansive thrust. It may well push onward and overshoot those absolute limits.

The reality of this situation is evident in the rapidly developing planetary environmental crisis. The *Living Planet Report 2008* indicates that the world faces a "looming ecological credit crunch." Natural resources are being consumed "faster than they can be replenished." Ecosystems are being taxed and degraded due to excessive demands and pollution, threatening to push them to the point of collapse. The loss of habitat is causing cascading extinctions throughout nature, as part of the "sixth extinction."

Recent studies have revealed that no area of the world's oceans "is unaffected by human influence." Coral reefs and continental shelves have suffered severe deterioration. Overfishing and organic pollution from agricultural runoff are driving the collapse of many aquatic ecosystems. The accumulation of carbon dioxide in the atmosphere has raised the ocean temperature and caused a drop in the pH of surface waters, making them more acidic, harming reef-building species. Scientists currently estimate that under business-as-usual emission of carbon dioxide, "the pH of the upper ocean" could produce as much "as a 150 percent increase in acidity compared with preindustrial times" with disastrous effects on ocean life. This has been dubbed "the other CO_2 problem." Human society, controlled by a rapacious system of accumulation, is in the process of ecological overshoot, exceeding the earth's "carrying capacity." The global footprint has surpassed the ability of the planet to regenerate by over 30 percent.[29]

The failure to act in the face of an environmental crisis of such scope should not come as a surprise given the union between politics and economics. Ironically, the destructive uncontrollability of capital prolongs the system's capability to grow, as it increases the prospects of expanding private riches, profiting from scarcity and degradation. As Mészáros indicates, "neither the degradation of nature nor the pain of social devastation carries any meaning at all for its system of social metabolic control when set against the absolute imperative of self-reproduction on an ever-extended scale."[30]

Surpassing the absolute ecological limits—to the point that the whole world is being run down—holds grave implications for the future of humanity. When the social metabolic order of capital confronts limits, "its destructive constituents come to the fore with a vengeance, activating the spectre of total uncontrollability in a form that foreshadows self-destruction both for this unique social reproductive system itself and for humanity in general."[31] It attempts to push ever forward—further undermining the vital conditions of existence—so long as there is a means to extend the accumulation of capital. The climate debate remains caught in the death throes of capital, as corporations, on one hand, clamor to present themselves as the solution to environmental degradation—a solution that has as its operative principle the defense of the existing social metabolic order, which must remain unchanged in all essential respects—and on the other hand, these same vested interests work to undermine even modest, utterly insufficient, political action to address climate change.[32]

For Mészáros, the overthrow of capitalist institutions is only the first step in the development of a socialist society. The logic of capital "must be eradicated from everywhere" because of how "deeply embedded" it is in every pore of society, including the "social metabolic process." A long, difficult struggle for social transformation must be undertaken to reorganize labor relations and conceptions of production, which at the same time will mend the rift between nature and society. Only the hegemonic alternative represented by labor in opposition to capital will provide the means for a transition to a new sustainable system of social metabolic reproduction. "The uncomfortable truth of the matter is that if there is no future for a radical mass movement in our time, there will be no future for humanity," because "the *extermination of humanity* is the ultimate concomitant of capital's destructive course of development."[33]

Mészáros does not limit his conception of the "absolute limits" of the system to environmental conditions. Nor is his notion of historical agency confined to labor in a narrow sense that excludes the role of other social movements (such as those based on or rooted in gender/sexual orientation, race/ethnicity/nationality, the unemployed). Thus Mészáros addresses the relation between the "activation of capital's absolute limits" in ways that encompass not only the environment but also these other issues, particularly women's emancipation. Just as the hierarchically driven, quantitative expansion built into capital's system of social metabolic reproduction has put it increasingly in conflict with the planetary "macrocosm" so is it put in conflict with the "microcosm" of the family in its various manifestations within the system. Given the unequal gender relations that are inscribed in the family, partly related to the regulation of human reproduction, and partly to the reproduction of authoritarian systems of control, the substantive equality that women's emancipation requires is negated by this reality alone.

According to Mészáros, "the economically sustainable regulation of humanity's biological reproduction is a crucial primary mediatory function of the social metabolic process." Under capital this necessary relation is completely subordinated to the management of labor and production so as to promote continued accumulation. The capital system is therefore unable to tolerate full gender equality. The "class of women," he writes, "cuts across all social class boundaries," making the demand for the "emancipation of women" the Achilles' heel of advanced monopoly capitalism. Women in this view can achieve full emancipation only through the emancipation of society in general. This, however, poses a challenge to the nuclear family, with its internal hierarchy that constitutes the micro-foundation of the system. Such an emancipatory project extends the question of substantive equality to all domains of social existence.[34]

Mészáros's argument in this respect suggests the possibility of a larger synthesis with the important work of socialist, ecofeminist Ariel Salleh, who has been focusing on the question of "meta-industrial labor"—those workers, primarily women, peasants, the indigenous, whose daily work is directed at biological growth and regeneration (including regeneration of natural systems). Salleh argues that such "meta-industrial" workers are directly connected to issues of eco-sufficiency and sustainability and offer all sorts of possibilities of overcoming the metabolic rift between humanity and nature, creating a new "metabolic fit" with respect to socio-ecological reproduction. Indeed, meta-industrial labor is directed at promoting what she calls the "metabolic value" required by all organic systems, protecting them against entropy. Such meta-industrial labor is to be seen as "rift-healing." "The material bottom line of any economy," she writes, "is a flourishing ecosystem and this can only be represented by metabolic value." In today's hierarchical world order, those responsible for this ecological bottom line are clearly women and other "carers" (caregivers, care workers), together with peasants, and indiginous peoples. All of this is connected to the issues of "substantive equality" and qualitative human development raised by Mészáros in relation to the dialectical necessity of social-ecological metabolism. For Salleh, capitalist societies metaphorically owe a vast "embodied debt" to unpaid reproductive workers engaged in the regeneration of the underlying conditions of production.[35]

Elementary Triangles: A Sustainable System of Social Metabolic Reproduction

Waiting and wishing for social change will not eliminate exploitation, social inequalities, and environmental destruction. Fortunately, the activation of capital's absolute limits, including its absolute ecological limits, coincides with new sources and strategies of mass-based revolt associated in particular with the rise of a socialism for the twenty-first century in Latin America, which has been directly influenced by Mészáros's thinking. Hugo Chávez has drawn directly on Mészáros in developing Venezuela's Bolivarian Revolution along socialist lines. As Michael Lebowitz states in *The Socialist Alternative*, "Chávez's theoretical step [in introducing the idea of a 'triangle of socialism'] can be traced to Mészáros's *Beyond Capital*," where capitalism was conceived as an organic system involving "a specific combination of production-distribution-consumption"—one in which all the elements coexist and simultaneously support one another." A socialist experiment must supersede, as Mészáros insisted, all the elements of "the totality of existing reproductive relations to go "beyond capital."

The result of this conception was a more ambitious, and at the same time more concrete, notion of socialist transition. Chávez was especially influenced by Mészáros's notion of a "communal system of production and consumption," in which the exchange of activities dominates over abstract exchange values.[36]

This more complex notion of socialism as an alternative hegemonic product aimed at sustainable development and substantive equality—requiring a new interrelated system of social metabolic reproduction, and rooted in a system of communal production and exchange—has facilitated a dialectical understanding of social-ecological metabolism. This emerging dialectic has become a defining feature of twenty-first century socialism, which is second to none in its perception of ecological imperatives. In *The Structural Crisis of Capital*, Mészáros quotes Chávez as stating: "I believe that it is not given to us to speak in terms of future centuries . . . we have no time to waste; the challenge is to save the conditions of life on this planet, to save the human species, to change the course of history, to change the world."[37]

The path to a sustainable society, in this view, necessitates social control over the social metabolic order of reproduction, encompassing all realms of productive life, including what is produced and how it is produced, as well as social relations with nature. Marx argued that a society of associated producers must live within the "metabolism prescribed by the natural laws of life itself" to sustain the vital conditions of existence for present and future generations. This conception is not simply a question of sustaining human conditions, since a metabolic approach means that ecosystems need to continue to function and provide the various ecological services that enrich the world and support other life forms.[38]

There has arisen a vital synthesis between Marx and Mészáros in formulating a conception of transition to a sustainable system of social metabolic reproduction. Both substantive equality and ecological sustainability are the cornerstones of a society freed from the dictates and logic of capital. Substantive equality (what Simón Bolívar called "the law of laws") helps overcome the divisions, the social isolation, and the alienation that characterize capitalist relations.[39] Ecological sustainability involves transcending the alienation from nature, which is the precondition for the capitalist system of production and exploitation.

Influenced by Marx's conception of a society of associated producers and drawing directly on Mészáros's theory of transition, Hugo Chávez has proposed a new socialism for the twenty-first century rooted in the "elementary triangle of socialism." This triangle consists of: (1) social ownership, (2) social production organized by workers, and (3) satisfaction of communal needs. Social control serves as the root basis for this transformation to a socialist metabolic order. If socialism fails to embody all sides of the triangle simultaneously, it will not take hold and will cease to be sustainable.[40]

It is clear that this *elementary triangle of socialism* is dialectically interconnected at a more fundamental level with what could be called the *elementary triangle of ecology*, as delineated by the natural laws of life: (1) social use, not ownership, of nature, (2) rational regulation by the associated producers of the metabolism between human beings and nature, and (3) the satisfaction of communal needs—not only of present but also future generations. Marx insisted that human development must be rooted in sustainable human relations with the material world, demanding constant vigilance and a scientifically informed public.[41] As a result, the two triangles must become one, allowing "an entire society . . . to bequeath [the earth] in an improved state to succeeding generations."[42]

In gaining social control over the social metabolic order, Mészáros emphasizes, we must eradicate the capital relation, constructing an entirely new foundation for society. This radical reorientation toward substantive equality is particularly evident, as noted, in the struggles today associated with the Bolivarian revolution in Venezuela and elsewhere in Latin America. In Venezuela, a historic transformation is under way, as a nation and its people work to transition to socialism. This is a process whereby the logic of capital must be uprooted, and the logic of a sustainable, human society sown. Major strides have been taken to establish communal councils, to encourage cooperatives, to create worker councils, to increase the education of workers, to train workers in co-management and self-management, and to extend social control over production. These steps are part of an effort to empower and invest people in the social transformation, which, as Chávez explains, also facilitates "the construction of the new man, of the new woman, of the new society."

To some extent, the fact that Venezuela's economy is heavily dependent on fossil fuels poses a contradiction in this respect, though one not internal to Venezuela itself but a product of the whole nature of development and energy-use in the capitalist world economy. Given the historical conditions of Venezuela in this respect, the question becomes not so much the export of oil but of how the proceeds are being used to transform the economy and society in a sustainable direction. Oil revenues have funded many projects, including programs to increase health care and education, within Venezuela. As part of the revolutionary process, an attempt is being made to diversify internal production to reduce the need to import goods to meet human needs. Here production is being focused on "stimulating the full development of human beings."

These changes may open up more revolutionary possibilities as a new society is created. In January 2010, Chávez announced that Venezuela must move beyond the oil-rentier development model as part of its transition. What this will mean in practice only time will tell. Nevertheless, peak oil may force a transformation to a less resource-extractive society. The more Venezuela has moved toward food self-sufficiency and ecological sustainability the easier such a transformation will be.[43]

Similar contradictions and initiatives are evident in Ecuador, which under its current president, Rafael Correa, has embraced the cause of a socialism for the twenty-first century and has joined Venezuela, Cuba, Bolivia, and other countries in the Bolivarian Alliance for the Peoples of Our America (ALBA). Ecuador has been the site of one of the highest rates of deforestation in South America as well as uncontrolled oil extraction. For the last seventeen years, Chevron and some 30,000 Ecuadorians have been involved in a legal dispute over water and land contaminated by oil spills and toxic open waste pits. In opposition to the capital system's unlimited extractive economy, Ecuador's new constitution introduces the Rights of Nature as a constitutional principle integrated with the concept of the Good Life (Sumak Kawsay). This makes Ecuador the first country in the world to recognize the rights of nature and of ecosystems to survive and flourish, allowing citizens to sue on nature's behalf if these rights are infringed.

Although Ecuador is an oil exporter, with oil currently accounting for 60 percent of exports, Correa has come out in favor of what is called the Daly-Correa tax on exports of oil (named after Correa and ecological economist Herman Daly) to help fund energy and nature conservation. The Ecuadorian government has also proposed, in what is called the ITT Yasuní Initiative, leaving 20 percent of Ecuador's total oil reserves—within the Ishpingo Tambococha-Tiputini (ITT) corridor (a 675-square-mile area)—in the ground forever. The goal is to protect the Yasuní National Park area, one of the world's greatest sites of biological diversity.

The UN-backed plan, to be administered by the UN Development Programme, which signed an agreement with Ecuador in August 2010, would result in some 846 billion barrels of crude oil being permanently left in the ground, preventing carbon emissions in excess of the total annual emissions of France. Ecuador is asking for rich countries to provide some $3.6 billion in compensation, half of the market value of the oil to remain in the ground. The money received would be used to develop alternative energy in order to lessen Ecuador's oil dependence, and to introduce projects/measures to lessen illegal logging in the Ecuadorian Amazon. The plan would give countries that contribute to the compensation fund what are known as Yasuní Guarantee Certificates (CGYs in Spanish) based on the value of the non-emitted carbon dioxide their contribution has secured. If a future Ecuadorian government goes back on the commitment not to drill for oil in the area, the value of these certificates would have to be paid back with interest.

The area to be protected in the heart of the Ecuadorian Amazon is one of the most ecologically diverse areas of the world. One hectare contains more tree species than all of the United States and Canada combined, as well as more than 500 bird species, 200 mammal species, 105 amphibian species, and countless plants and insects. The Yasuní National Park is the ancestral home of two of the world's last remaining "uncontacted" indigenous tribes, the Taromenane and the Tagaeri. The *Independent* (UK) has called the ITT Yasuní Initiative "the world's first genuinely green energy deal."[44]

Significant attempts to alter the human metabolism with nature are being made through agrarian reform as part of the Bolivarian revolutionary process. Throughout much of the twentieth century, Venezuela's agricultural sector was dismantled, and the rural population migrated to cities. The nation became dependent on food imports. As part of the effort to establish a social economy—which is focused on use values—and to pursue human development, the Bolivarian Revolution has committed itself to pursuing "food sovereignty." Under this framework, small farmers and collectives rather than agribusiness have control over food production and distribution. This change helps reduce alienation from nature. Education has become integral to the production process, as the farmers and agricultural centers are increasingly concerned with the natural conditions under which food is produced. Agroecological approaches are being studied and applied, in order to build up the soil and to work within natural cycles. Farmers are planting diverse traditional crops, saving seeds, and collecting compost. The government is supporting these efforts by extending credit to those who use them. Like Cuba, Venezuela has created research facilities to develop "biological pest control and fertilizers" to eliminate the use of pesticides. The Law for Integrated Agricultural Health (2008) mandates that the use of "toxic agrochemicals" be phased out, in favor of agroecological practices.[45] Here the elementary triangles of socialism and ecology intersect as the revolutionary process continues to take root.

Under the presidency of Evo Morales, a socialist and the first indigenous Bolivian head of state, Bolivia, like Venezuela, has pushed forward an agenda aimed at ecological and social justice. Morales rose to power partly as a result of the Cochabamba water wars, in which the poor and indigenous populations rose up against water privatization. With the rise of Morales, Bolivia has come to play a leading role within the world in opposing the ecological depredations of capitalism and promoting radical ecological change. Not only did Bolivia host the World People's Conference on Climate Change and the Rights of Mother Earth in April 2010, it also played the primary role in this context in issuing the *Cochabama Protocol* or *The People's Agreement on Climate Change and the Rights of Mother Earth*—the leading third world revolutionary strategy for addressing the global ecological crisis. According to the *Cochabamba Protocol*, "Humanity confronts a great dilemma: to continue on the path of capitalism, depredation and death, or to choose the path of harmony with nature and respect for life."[46]

All of this suggests that the struggle for socialism in the twentieth century arises out of a dual struggle for substantive equality and ecological sustainability, as mutually dependent conditions of revolutionary change. In Mészáros's conception, the creation of a more ecological relation for humanity is not a separate problem—but an indispensable, even defining (though not all-encompassing) part of the struggle to create a qualitatively new social order dedicated to the realization of genuine human needs. Ecological struggle in the abstract in this view is meaningless, since it cannot be achieved except as part of a wider social revolt that encompasses the totality of human relations: not just those with nature directly. As he writes in *The Challenge and Burden of Historical Time*, "Ecology . . . is an important but subordinate aspect of the necessary *qualitative redefinition* of utilizing the produced goods and services, without which the advocacy of humanity's permanently sustainable ecology—again, an absolute must—can be nothing more than a pious hope."[47] Such a qualitative redefinition relates of course to the creation of a culture of substantive equality.[48]

In this universal, dialectical view, the problem of constructing a viable system of social and ecological metabolism becomes—in contradistinction to the *Limits to Growth* argument, which targets the abstract commitment to "growth" rather than the capital system itself—a central aspect of a wide-ranging revolutionary process. This process demands for its completion *social* control: wresting the determining power away from the agency of *capital* and placing it back in the sovereign population. It is a matter of "putting to humanly commendable and rewarding use the attained *potential* of productivity, in a world of now criminally wasted material and human resources."[49] Repairing the rift in the ecological metabolism requires that the rift in the social metabolism be overcome.

18. Why Ecological Revolution?

It is now universally recognized within science that humanity is confronting the prospect—if we do not soon change course—of a planetary ecological collapse. Not only is the global ecological crisis becoming more and more severe, with the time in which to address it fast running out, but the dominant environmental strategies are also forms of denial, demonstrably doomed to fail, judging by their own limited objectives. This tragic failure can be attributed to the refusal of the powers that be to address the roots of the ecological problem in capitalist production and the resulting necessity of ecological and social revolution.

The term "crisis," attached to the global ecological problem, although unavoidable, is somewhat misleading, given its dominant economic associations. Since 2008, we have been living through a world economic crisis—the worst economic downturn since the 1930s. This has been a source of untold suffering for hundreds of millions, indeed billions, of people. But insofar as it is related to the business cycle and not to long-term factors, expectations are that it is temporary and will end, to be followed by a period of economic recovery and renewed growth—until the advent of the next crisis. Capitalism is, in this sense, a crisis-ridden, cyclical economic system. Even if we were to go further, to conclude that the present crisis of accumulation is part of a long-term economic stagnation of the system—that is, a slowdown of the trend-rate of growth beyond the mere business cycle—we would still see this as a partial, historically limited calamity, raising, at most, the question of the future of the present system of production.[1]

When we speak today of the world ecological crisis, however, we are referring to something that could turn out to be *final*, that is, there is a high probability, if we do not quickly change course, of a *terminal crisis*—a death of the whole period of human dominance of the planet. Human actions are generating environmental changes that threaten the extermination of most species on Earth, along with civilization, and conceivably our own species as well.

What makes the current ecological situation so serious is that climate change, arising from human-generated increases in greenhouse gas emissions, is not occurring gradually and in a linear process, but is undergoing a dangerous acceleration, pointing to sudden shifts in the state of the earth system. We can therefore speak, to quote James Hansen, of "tipping points . . . fed by amplifying feedbacks."[2] Four amplifying feedbacks are significant at present: (1) rapid melting of Arctic sea ice, with the resulting reduction of the earth's albedo (reflection of solar radiation) due to the replacement of bright reflective ice with darker blue sea water, leading to greater absorption of solar energy and increasing global average temperatures; (2) melting of the frozen tundra in northern regions, releasing methane (a much more potent greenhouse gas than carbon dioxide) that is trapped beneath the surface, causing accelerated warming; (3) a drop in the efficiency of the carbon absorption of the world's oceans since the 1980s, and particularly since 2000, due to growing ocean acidification (from past carbon absorption) and other factors, potentially contributing to a faster carbon buildup in the atmosphere and enhanced warming; (4) extinction of species due to changing climate zones, leading to the collapse of ecosystems dependent on these species, and the death of still more species.[3]

Due to this acceleration of climate change, the time line in which to act before calamities hit, and before climate change increasingly escapes our control, is extremely short. In October 2009, Luc Gnacadja, executive secretary of the United Nations Convention to

Combat Desertification, reported that based on current trends close to 70 percent of the land surface of the earth could be drought-affected by 2025, compared to nearly 40 percent today.[4] The United Nations Intergovernmental Panel on Climate Change (IPCC) has warned that glaciers are melting throughout the world and could recede substantially this century. Rivers fed by the Himalayan glaciers currently supply water to countries with around 3 billion people. Their melting will give rise to enormous floods, followed by acute water shortages.[5]

Many of the planetary dangers associated with current global warming trends are by now well known: rising sea levels engulfing islands and low-lying coastal regions throughout the globe; loss of tropical forests; destruction of coral reefs; a "sixth extinction" rivaling the great die-downs in the history of the planet; massive crop losses; extreme weather events; spreading hunger and disease. But these dangers are heightened by the fact that climate change is not the entirety of the world ecological crisis. For example, independently of climate change, tropical forests are being cleared as a direct result of the search for profits. Soil destruction is occurring, due to current agribusiness practices. Toxic wastes are being diffused throughout the environment. Nitrogen runoff from the overuse of fertilizer is affecting lakes, rivers, and ocean regions, contributing to oxygen-poor "dead zones."

Since the whole earth is affected by the vast scale of human impact on the environment in complex and unpredictable ways, even more serious catastrophes could conceivably be set in motion. One growing area of concern is ocean acidification due to rising carbon dioxide emissions. As carbon dioxide dissolves, it turns into carbonic acid, making the oceans more acidic. Because carbon dioxide dissolves more readily in cold than in warm water, the cold waters of the Arctic are becoming acidic at an unprecedented rate. Within a decade, the waters near the North Pole could become so corrosive as to dissolve the living shells of shellfish, affecting the entire ocean food chain. At the same time, ocean acidification appears to be reducing the carbon uptake of the oceans, possibly speeding up global warming.[6]

There are endless predictive uncertainties in all of this. Nevertheless, evidence is mounting that the continuation of current trends is unsustainable, even in the short-term. The only rational answer, then, is a radical change of course. Moreover, given certain imminent tipping points, there is no time to be lost. Catastrophic changes in the earth system could be set irreversibly in motion within a few decades, at most.

The IPCC, in its 2007 report, indicated that an atmospheric carbon dioxide level of 450 parts per million (ppm) should not be exceeded, and implied that this was the fail-safe point for carbon stabilization. But these findings are already out of date. "What science has revealed in the past few years," Hansen states, "is that the safe level of carbon dioxide in the long run is no more than 350 ppm," as compared with 390 ppm in 2010. This means that carbon emissions have to be reduced faster and more drastically than originally thought, to bring the overall carbon concentration in the atmosphere down. The reality is that "if we burn all the fossil fuels, or even half of remaining reserves, we will send the planet toward the ice-free state with sea level about 250 feet higher than today. It would take time for complete ice sheet disintegration to occur, but a chaotic situation would be created with changes occurring out of control of future generations." More than eighty of the world's poorest and most climate-vulnerable countries have now declared that carbon dioxide atmospheric concentration levels must be reduced below 350 ppm, and that the rise in global average temperature by century's end must not exceed 1.5°C.[7]

Strategies of Denial

The central issue we have to confront is devising social strategies to address the world ecological crisis. Not only do the solutions have to be large enough to deal with the problem, but also all of this must take place on a world scale in a generation or so. The speed and scale of change necessary means that what is required is an ecological revolution that would also need to be a social revolution. However, rather than addressing the real roots of the crisis and drawing the appropriate conclusions, the dominant response is to avoid all questions about the nature of our society and turn to technological fixes or market mechanisms of one sort or another. In this respect, there is a certain continuity of thought between those who deny the climate change problem altogether, and those who, while acknowledging the severity of the problem at one level, nevertheless deny that it requires a revolution in our social system.

We are increasingly led to believe that the answers to climate change are primarily to be found in new energy technology, specifically increased energy and carbon efficiencies in both production and consumption. Technology in this sense, however, is often viewed abstractly as a *deus ex machina*, separated from both the laws of physics (such as entropy or the second law of thermodynamics) and from the way technology is embedded in historically specific conditions. With respect to historical conditions, it is worth noting that, under the present economic system, increases in energy efficiency normally lead to increases in the scale of economic output, effectively negating any gains from the standpoint of resource use or carbon efficiency—a problem known as the Jevons Paradox. As William Stanley Jevons observed in the nineteenth century, every new steam engine was more efficient in the use of coal than the one before, which did not prevent coal burning from increasing overall, since the efficiency gains only led to the expansion of the number of steam engines and of growth in general. This relation between efficiency and scale has proven true for capitalist economies up to the present day.[8]

Technological fetishism with regard to environmental issues is usually coupled with a form of market fetishism. So widespread has this become that even a militant ecologist like Bill McKibben, author of *The End of Nature*, has stated: "There is only one lever even possibly big enough to make our system move as fast as it needs to, and that's the force of markets."[9]

Green-market fetishism is most evident in what is called "cap and trade"—a catch phrase for the creation, via governments, of artificial markets in carbon trading and so-called offsets. The important thing to know about cap and trade is that it is a proven failure. Although enacted in Europe as part of the implementation of the Kyoto

Protocol, it has failed where it was supposed to count: in reducing emissions. Carbon-trading schemes have been shown to be full of holes. Offsets allow all sorts of dubious forms of trading that have no effect on emissions. Indeed, the only area in which carbon trading schemes have actually been effective is in promoting profits for speculators and corporations, which are therefore frequently supportive of them. Recently, Friends of the Earth released a report titled *Subprime Carbon?* which pointed to the emergence, under cap-and-trade agreements, of what could eventually turn out to be the world's largest financial derivatives market in the form of carbon trading. All of this has caused Hansen to refer to cap and trade as "the temple of doom," locking in "disasters for our children and grandchildren."[10]

The whole huge masquerade associated with the dominant response to global warming was dramatically revealed in the comprehensive climate bill passed by the U.S. House of Representatives in late June 2009 (only to be killed by the U.S. Senate a little over a year later). The climate bill was ostensibly aimed at reducing greenhouse gas emissions by 17 percent relative to 2005 levels by 2020, which would have translated into 4 to 5 percent less U.S. global warming pollution than in 1990. This fell far short of the target level of a 6 to 8 percent cut (relative to 1990) for wealthy countries that the Kyoto accord set for 2012, which was supposed to have been only a minor, first step in dealing with global warming at a time when the problem was seen as much less severe and action less pressing.

But the small print in the House climate bill made achieving even this meager target unrealistic. The coal industry was given until 2025 to comply with the bill's pollution reduction mandates, with possible extensions afterward. As Hansen exclaimed, the bill actually built "in [the] approval of new coal-fired power plants!" Agribusiness, which accounts for a quarter of U.S. greenhouse gas emissions, was to be entirely exempt from the mandated reductions. The cap-and-trade provisions of the legislation would have given annual carbon dioxide emission allowances to some 7,400 facilities across the United States, most of them handed out for free. These pollution allowances were to increase up through 2016, and companies would have been permitted to "bank" them indefinitely for future use. Corporations exceeding their allowances could have fulfilled their entire set of obligations by buying offsets associated with pollution control projects until 2027.

If all of this were not bad enough, it was understood from the start that the Senate version of the bill, slated to emerge in the following year, would inevitably be weaker than the House version, giving even more concessions to corporations. Hence, the final unified legislation, if it were eventually to wind its way through both houses of Congress, was destined, as Hansen put it, to be "worse than nothing."

As it turned out, faced with the concerted opposition of corporations, the Senate found itself unable to act at all. Even the outright sell-out to business as usual that constituted the House bill was regarded as too anti-business once it got to the Senate. After numerous delays, Senate Democrats—who had not been able to get the backing of one Republican and were faced with the prospect of numerous desertions from their own ranks—simply gave up the fight. The climate legislation was pronounced dead in July 2010 before ever having reached the Senate floor.

This U.S. failure mirrored the collapse in December 2009 of the world climate negotiations in Copenhagen. There Washington had led the way in blocking all but the most limited, voluntary agreements, insisting on only market-based approaches, such as cap and trade, and proposing meager levels of emissions reductions. In the end, the negotiations simply collapsed. The so-called Copenhagen Accord, agreed to by the United States and four leading emerging economies—China, Brazil, India, and South Africa—was a mere last-ditch attempt at diplomatic face-saving, with no positive significance.[11]

Recognizing that world powers are playing the role of Nero as Rome burns, James Lovelock, the earth system scientist famous for his Gaia hypothesis, argues that massive climate change and the destruction of human civilization as we know it may now be irreversible. Nevertheless, he proposes as "solutions" either a massive building of nuclear power plants all over the world (closing his eyes to the enormous dangers accompanying such a course)—or geoengineering our way out of the problem by using the world's fleet of aircraft to inject huge quantities of sulfur dioxide into the stratosphere to block a portion of the incoming sunlight, reducing the solar energy reaching the earth. Another common geoengineering proposal includes dumping iron filings throughout the ocean to increase its carbon-absorbing properties.

Rational scientists recognize that interventions in the earth system on the scale envisioned by geoengineering schemes (for example, blocking sunlight) have their own massive, unforeseen consequences. Nor could such schemes solve the crisis. The dumping of massive quantities of sulfur dioxide into the stratosphere would, even if effective, have to be done again and again, on an increasing scale, if the underlying problem of cutting greenhouse gas emissions were not dealt with. Moreover, it could not possibly solve other problems associated with massive carbon dioxide emissions, such as the acidification of the oceans.[12]

The dominant approach to the world ecological crisis, focusing on technological fixes and market mechanisms, is thus a kind of denial; one that serves the vested interests of those who have the most to lose from a change in economic arrangements. Al Gore exemplifies the dominant form of denial in his new book, *Our Choice: A Plan to Solve the Climate Crisis*. For Gore, the answer is the creation of a "sustainable capitalism." He is not, however, altogether blind to the faults of the present system. He describes climate change as the "greatest market failure in history," and decries the "short-term" perspective of present-day capitalism, its "market triumphalism," and the "fundamental flaws" in its relation to the environment. Yet, in defiance of all this, he assures his readers that the "strengths of capitalism" can be harnessed to a new system of "sustainable development."[13]

The System of Unsustainable Development

In reality, capitalism can be defined as *a system of unsustainable development*. To understand why this is so, it is useful to turn to Karl Marx, the core of whose entire intellectual corpus might be interpreted as a critique of the political economy of unsustainable development and its human and natural consequences.

Capitalism, Marx explains, is a system of generalized commodity production. There were other societies prior to capitalism in which commodity markets played important roles, but it is only in capitalism that a system emerges that is centered entirely on the production of commodities. A "commodity" is a good produced to be sold and exchanged for profit in the market. We call it a "good" because it is has a use value—that is, it normally satisfies some use, otherwise there would be no need for it. But it is the exchange value, the money income and the profit the commodity generates, that is the exclusive concern of the capitalist.

What Marx called "simple commodity production" is an idealized economic formation—often assumed to describe the society wherein we live—in which the structure of exchange is such that a commodity embodying a certain use value is exchanged for money (acting as a mere means of exchange), which is, in turn, exchanged for another commodity (use value) at the end. The exchange process from beginning to end can be designated by the shorthand C-M-C. In such a process, exchange is simply a modified form of barter, with money merely facilitating exchange. The goal of exchange is concrete use values, embodying qualitative properties. Such use values are normally consumed—thereby bringing a given exchange process to an end.

Marx, however, insisted that a capitalist economy, in reality, works altogether differently, with exchange taking the form of M-C-M′. Here money capital (M) is used to purchase commodities (labor power and means of production) to produce a commodity that can be sold for more money, M′ (that is, M + Δm or surplus value) at the end. This process, once set in motion, never stops of its own accord, since it has no natural end. Rather, the surplus value (profit) is reinvested in the next round, with the object of generating M″; and, in the following round, the returns are again reinvested with the goal of obtaining M‴, and so on, *ad infinitum.*[14]

For Marx, therefore, capital is *self-expanding value*, driven incessantly to ever larger levels of accumulation, knowing no bounds. "Capital," he wrote, "is the endless and limitless drive to go beyond its limiting barrier. Every boundary is and has to be a [mere] barrier for it [and thus capable of being surmounted]. Else it would cease to be capital—money as self-reproductive." It thus converts all of nature and nature's laws as well as all that is distinctly human into a mere means of its own self-expansion. The result is a system, fixated on the exponential growth of profits and accumulation. "Accumulate, accumulate! That is Moses and the prophets!"[15]

Any attempt to explain where surplus value (or profits) comes from must penetrate beneath the exchange process and enter the realm of labor and production. Here Marx argues that value added in the working day can be divided into two parts: (1) the part that reproduces the value of labor power (the wages of the workers) and thus constitutes necessary labor; and (2) the labor expended in the remaining part of the working day, which can be regarded as surplus labor, and which generates surplus value (or gross profits) for the capitalist. Profits are thus to be regarded as residual, consisting essentially of what is left over after wages are paid out—something that every businessperson instinctively understands. The ratio of surplus (unpaid) labor to necessary (paid) labor in the working day is, for Marx, the rate of exploitation.

The logic of this process is that the increase in surplus value appropriated depends on the effective exploitation of human labor power. This can be achieved in two ways: (1) either workers are compelled to work longer hours for the same pay, thereby increasing the surplus portion of the working day simply by adding to the total working time (Marx calls this "absolute surplus value"); or (2) the value of labor power—the value equivalent of workers' wages—is generated in less time (as a result of increased productivity, etc.), thereby augmenting the surplus portion of the working day to that extent (Marx calls this "relative surplus value").

In its unrelenting search for greater (relative) surplus value, capitalism is dependent on the revolutionization of the means of production with the aim of increasing productivity and reducing the paid portion of the working day. This leads inexorably to additional revolutions in production, additional increases in productivity, in what constitutes an endless treadmill of production/accumulation. The logic of accumulation concentrates more and more of the wealth and power of society in fewer and fewer hands, and generates an enormous industrial reserve army of the unemployed.

This is all accompanied by the further alienation of labor, robbing human beings of their creative potential, and often of the environmental conditions essential for their physical reproduction. "The factory system," Marx wrote, "is turned into systematic robbery of what is necessary for the life of the worker while he is at work, i.e., space, light, air and protection against the dangerous or the unhealthy contaminants of the production process."[16]

For classical political economists, beginning with the physiocrats and Adam Smith, nature was explicitly designated as a "free gift" to capital. It did not directly enter into the determination of exchange value (value), which constituted the basis of the accumulation of private capital. Nevertheless, classical political economists did see nature as constituting public wealth, since this was identified with use values, and included not only what was scarce, as in the case of exchange values, but also what was naturally abundant—air, water, etc. Out of these distinctions arose what came to be known as the Lauderdale Paradox, associated with the ideas of James Maitland, the eighth Earl of Lauderdale, who observed in 1804 that private riches (exchange values) could be expanded by destroying public wealth (use values)—that is, by generating scarcity in what was formerly abundant. This meant that individual riches could be augmented by landowners monopolizing the water of wells and charging a price for what had previously been free—or by burning crops (the produce of the earth) to generate scarcity and thus exchange value. Even the air itself, if it became scarce enough, could expand private riches, once it was possible to put a price on it. Lauderdale saw such artificial creation of scarcity as a way in which those with private monopolies of land and resources robbed society of its real wealth.[17]

Marx (following Ricardo) strongly embraced the Lauderdale Paradox, and its criticism of the inverse relation between private riches and public wealth. Nature, under the system of generalized commodity production, was, Marx insisted, reduced to being merely a *free gift to capital* and was thus robbed. Indeed, the fact that part of the working day was unpaid and went to the surplus of the capitalist meant that an analogous situation pertained to human labor power, itself a "natural force." The worker was allowed to "work for his own life, i.e. *to live*, only insofar as he works for a certain time gratis for the capitalist . . . [so that] the whole capitalist system of production turns on the prolongation of this gratis labour by extending the working day or by developing the productivity, i.e., the greater intensity of labour power, etc." Both nature and the unpaid labor of the worker were then to be conceived in analogous ways as free gifts to capital.[18]

Given the nature of this classical critique, developed to its furthest extent by Marx, it is hardly surprising that later neoclassical economists, exercising their primary role as apologists for the system, were to reject both the classical theory of value and the Lauderdale Paradox. The new marginalist economic orthodoxy that emerged in the late nineteenth century erased all formal distinctions within economics between use value and exchange value, between wealth and value. Nature's contribution to wealth was simply defined out of existence within the prevailing economic view. However, a minority of heterodox economists, including such figures as Henry George, Thorstein Veblen, and Frederick Soddy, were to insist that this rejection of nature's contribution to wealth only served to encourage the squandering of common resources characteristic of the system. "In a sort of parody of an accountant's nightmare," John Maynard Keynes was to write of the financially driven capitalist system, "we are capable of shutting off the sun and the stars because they do not pay a dividend."[19]

For Marx, capitalism's robbing of nature could be seen concretely in its creation of a rift in the human-earth metabolism, whereby the reproduction of natural conditions was undermined. He defined the labor process in ecological terms as the "metabolic interaction" between human beings and nature. With the development of industrial agriculture under capitalism, a rift was generated in the nature-given metabolism between human beings and the earth. The shipment of food and fiber hundreds, and sometimes thousands, of miles to the cities meant the removal of soil nutrients, such as nitrogen, phosphorus, and potassium, which ended up contributing to the pollution of the cities, while the soil itself was robbed of its "constituent elements." This created a rupture in "the eternal natural condition for the lasting fertility of the soil," requiring the "systematic restoration" of this metabolism. Yet even though this had been demonstrated with the full force of natural science (for example, in Justus von Liebig's chemistry), the rational application of scientific principles in this area was impossible for capitalism. Consequently, capitalist production simultaneously undermined "the original sources of all wealth—the soil and the worker."[20]

Marx's critique of capitalism as an unsustainable system of production was ultimately rooted in its "preconditions," that is, the historical bases under which capitalism as a mode of production became possible. These were to be found in "primitive accumulation," or the expropriation of the commons (of all customary rights to the land), and hence the expropriation of the workers themselves—of their means of subsistence. It was this expropriation that would help lay the grounds for industrial capitalism in particular. The turning of the land into private property, a mere means of accumulation, was at the same time the basis for the destruction of the metabolism between human beings and the earth.[21]

This was carried out on an even greater and more devastating scale in relation to the pillage of the third world. Here, trade in human slavery went hand in hand with the seizure of the land and resources of the entire globe as mere plunder to feed the industrial mills of England and elsewhere. Whole continents (or at least those portions that European colonialism was able to penetrate) were devastated. Nor is this process yet complete, with de-peasantization of the periphery by expanding agribusiness constituting one of the chief forms of social and ecological destruction in the present day.[22]

Marx's whole critique pointed to the reality of capitalism as a system of unsustainable development, rooted in the unceasing exploitation and pillage of human and natural agents. As he put it: "*Après moi le déluge!* is the watchword of every capitalist and of every capitalist nation. Capital therefore takes no account of the health and the length of life of the worker [or the human-nature metabolism], unless society forces it to do so."[23]

He wryly observed in *Capital* that, when the Germans improved the windmill (in the form to be taken over by the Dutch), one of the first concerns, vainly fought over by the emperor Frederick I, the nobility, and the clergy, was who was "the 'owner' of the wind." Nowadays, this observation on early attempts to commodify the air takes on even greater irony—at a time when markets, in what Gore himself refers to as "subprime carbon assets," are helping to generate a speculative bubble with respect to earth's atmosphere.[24]

Toward Ecological Revolution

If the foregoing argument is correct, humanity is facing an unprecedented challenge. On the one hand, we are confronting the question of a terminal crisis, threatening most life on the planet, civilization, and the very existence of future generations. On the other hand, attempts to solve this through technological fixes, market magic, and the idea of a "sustainable capitalism" are mere forms of ecological denial, since they ignore the inherent destructiveness of the current system of unsustainable development—capitalism. This suggests that the only rational answer lies in an ecological revolution, which would also have to be a social revolution, aimed at the creation of a just and sustainable society.

In addressing the question of an ecological revolution in the present dire situation, both short-term and long-term strategies are necessary, and should complement each other. One short-term strategy, directed mainly at the industrialized world, has been presented by Hansen. He starts with what he calls a "geophysical fact": most of the remaining fossil fuel, particularly coal, must stay in the ground, and carbon emissions have to be reduced as quickly as possible to near zero. He proposes three measures: (1) coal burning (except where carbon is sequestered—right now not technologically feasible) must cease; (2) the price of fossil fuel consumption should be steadily increased by imposing a progressively rising tax at the point of production: well head, mine shaft, or point of entry—redistributing 100 percent of the revenue, on a monthly basis, directly to the population as dividends; (3) a massive, global campaign to end deforestation and initiate large-scale reforestation needs to be introduced. A carbon tax, he argues, if it were to benefit the people directly—the majority of whom have below average per-capita carbon footprints and would experience net gains from the carbon dividends once their added energy costs were subtracted—would create massive support for change. It would help to mobilize the population, particularly those at the bottom of society, in favor of a climate revolution. Hansen's "fee and dividend" proposal is explicitly designed not to feed the profits of vested interests. Any revenue from the carbon tax, in this plan, has to be democratically structured so as to redistribute income and wealth to those with smaller carbon footprints (the poor), and away from those with the larger carbon footprints (the rich).[25]

Hansen has emerged as a leading figure in the climate struggle, not only as a result of his scientific contributions, but also due to his recognition that at the root of the problem is a system of economic power, and his increasingly radical defiance of the powers that be. Thus he declares: "The trains carrying coal to power plants are death trains. Coal-fired plants are factories of death." He criticizes those such as Gore, who have given in to cap and trade, locking in failure. Arguing that the unwillingness and inability of the authorities to act means that desperate measures are necessary, he is calling for mass "civil resistance." In June 2009, he was arrested, along with thirty-one others, in the exercise of civil resistance against mountaintop-removal coal mining.[26]

In strategizing an immediate response to the climate problem, it is crucial to recognize that the state, through government regulation and spending programs, could intervene directly in the climate crisis. Carbon dioxide could be considered an air pollutant to be regulated by law. Electrical utilities could be mandated to obtain their energy increasingly from renewable sources. Solar panels could be included as a mandatory part of the building code. The state could put its resources behind major investments in public environmental infrastructure and planning, including reducing dependence on the automobile through massive funding of public transportation (for example, intercity trains and light rail) and the necessary accompanying changes in urban development and infrastructure.

Globally, the struggle has to take into account the reality of economic and ecological imperialism. The allowable carbon-concentration limits of the atmosphere have already been taken up as a result of the carbon emissions from the rich states at the center of the world system. The economic and social development of poor countries is, therefore, now being further limited by the pressing need to impose restrictions on carbon emissions for the sake of the planet as a whole—despite the fact that underdeveloped economies had little or no role in the creation of the problem. The global South is likely to experience the effects of climate change much earlier and more severely than the North, and has fewer economic resources with which to adapt. All of this means that a non-imperialistic and more sustainable world solution depends initially on what is called "contraction and convergence"—a drastic *contraction* in greenhouse gas emissions overall (especially in the rich countries), coupled with the *convergence* of per-capita emissions in all countries at levels that are sustainable for the planet.[27] Since science suggests that even low greenhouse gas emissions may be unsustainable over the long run, strategies have to be developed to make it economically feasible for countries in the periphery to introduce solar and renewable technologies—reinforcing those necessary radical changes in social relations that will allow them to stabilize and reduce their emissions.

For the anti-imperialist movement, a major task should be creating stepped-up opposition to military spending (amounting to a trillion dollars in the United States in 2007) and ending government subsidies to global agribusiness. The goal should be to shift those monies into environmental defense and meeting the social needs of the poorest countries, as suggested in 2006 by the *Bamako Appeal*—the third world strategy for world economic and social reform.[28] It must be firmly established as a principle of world justice that the wealthy countries owe an enormous ecological debt to poorer countries, due to the robbing by the imperial powers of the global commons and the pillage of the periphery at every stage of world capitalist development.

Already, the main force for ecological revolution stems from movements in the global South, marked by the growth of the Vía Campesina movement, socialist organizations like Brazil's MST, and ongoing revolutions in Latin America (the ALBA countries) and Asia (Nepal). Cuba has been applying permaculture design techniques that mimic energy-maximizing natural systems to its agriculture since the 1990s, generating a revolution in food production. Venezuela, although for historic reasons an oil power economically dependent on the sale of petroleum, has made extraordinary achievements in recent years by moving toward a society directed at collective needs, including dramatic achievements in food sovereignty.[29] Ecuador has placed the "Rights of Mother Earth" in its constitution and is proposing leaving 20 percent of its oil in the ground. Bolivia has taken a leading role in promoting socialist, third world, and indigenous ecological strategies.

Reaching back into history, it is worth recalling that the proletariat in Marxian theory was the revolutionary agent because this class had nothing to lose, and thus came to represent the universal interest in abolishing, not only its own oppression, but oppression itself. As Marx put it, "The living conditions of the proletariat represent the focal point of all inhuman conditions in contemporary society. . . . However, it [the proletariat] cannot emancipate itself without abolishing the conditions which give it life, and it cannot abolish these conditions without abolishing all those inhuman conditions of social life which are summed up in its own situation."[30]

Later Marxist theorists were to argue that, with the growth of monopoly capitalism and imperialism, the "focal point of inhuman conditions" had shifted from the center to the periphery of the world system. Paul Sweezy contended that although the objective conditions that Marx associated with the proletariat did not match those of better-off workers in the United States and Europe in the 1960s, they did correspond to the harsh, inhuman conditions imposed on "the masses of the much more numerous and populous underdeveloped dependencies of the global capitalist system." This helped explain the pattern of socialist revolutions following the Second World War, as exemplified by Vietnam, China, and Cuba.[31]

Looking at this today, it is conceivable that the main historic agent and initiator of a new epoch of ecological revolution is to be found in the third world masses most directly in line to be hit first by the impending disasters. Today the ecological front line is arguably to be found in the Ganges-Brahmaputra Delta and the low-lying fertile coast area of the Indian Ocean and China Seas—the state of Kerala in India, Thailand, Vietnam, and Indonesia. The inhabitants of these places, as in the case of Marx's proletariat, have nothing to lose from the radical changes necessary to avert (or adapt to) disaster. In fact, with the universal spread of capitalist social relations and the commodity form, the world proletariat and the masses most exposed to sea-level rise—for example, in the low-lying delta of the Pearl River and the Guangdong industrial region from Shenzhen to Guangzhou—sometimes overlap. This, then, potentially constitutes the global epicenter of a new environmental proletariat.[32]

The planetary crisis we are now caught up in, however, requires a world uprising transcending all geographical boundaries. This means that ecological and social revolutions in the third world have to be accompanied by, or inspire, universal revolts against imperialism, the destruction of the planet, and the treadmill of accumulation. The recognition that the weight of environmental disaster is such that it would cross all class lines and all nations and positions, abolishing time itself by breaking what Marx called "the chain of successive generations," could lead to a radical rejection of the engine of destruction in which we live, and put into motion a new conception of global humanity and earth metabolism. As always, real change will have to come from those most alienated from the existing systems of power and wealth. The most hopeful development within the advanced capitalist world at present is the meteoric rise of the youth-based climate justice movement, which is emerging as a considerable force in direct action mobilization and in challenging the current climate negotiations.[33]

What is clear is that the long-term strategy for ecological revolution throughout the globe involves the building of a society of substantive equality—the struggle for socialism. Not only are the two inseparable, but they also provide essential content for each other. There can be no true ecological revolution that is not socialist; no true socialist revolution that is not ecological. This means recapturing Marx's own vision of socialism/communism, which he defined as a society where "the associated producers govern the human metabolism with nature in a rational way, bringing it under their collective control . . . accomplishing it with the least expenditure of energy and in conditions most worthy and appropriate for their human nature."[34]

One way to understand this interdependent relation between ecology and socialism is in terms of what Hugo Chávez has called "the elementary triangle of socialism" (derived from Marx) consisting of: (1) social ownership; (2) social production organized by workers; and (3) satisfaction of communal needs. All three components of the elementary triangle of socialism are necessary if socialism is to be sustained. Complementing and deepening this is what could be called "the elementary triangle of ecology" (derived even more directly from Marx): (1) social use, not ownership, of nature; (2) rational regulation by the associated producers of the metabolic relation between humanity and nature; and (3) satisfaction of communal needs—not only of present but also future generations (and life itself).[35]

As Lewis Mumford explained in 1944, in his *Condition of Man*, the needed ecological transformation required the promotion of "basic communism," applying "to the whole community the standards of the household," distributing benefits "according to need, not ability or productive contribution." This meant focusing first and foremost on "education, recreation, hospital services, public hygiene, art, food production, the rural and urban environments, and, in general, collective needs." The idea of "basic communism" drew on Marx's principle of substantive equality in the *Critique of the Gotha Programme*: "From each according to his ability, to each according to his needs!" But Mumford also associated this idea with John Stuart Mill's vision, in his most socialist phase, of a "stationary state"—

viewed, in this case, as a system of economic production no longer geared to the accumulation of capital, in which the emphasis of society would be on collective development and the quality of life.[36] For Mumford, this demanded a new "organic person"—who would emerge from the struggle itself.

An essential element of such an ecological and socialist revolution for the twenty-first century is a truly radical conception of sustainability, as articulated by Marx: "From the standpoint of a higher socio-economic formation, the private property of particular individuals in the earth will appear just as absurd as the private property of one man in other men [slavery]. Even an entire society, a nation, or all simultaneously existing societies taken together, are not the owners of the earth. They are simply its possessors, its beneficiaries, and have to bequeath it in an improved state to succeeding generations as *boni patres familias* [good heads of the household]."[37]

Such a vision of a sustainable, egalitarian society must define the present social struggle, not only because it is ecologically necessary for human survival but also because it is historically necessary for the development of human freedom. Today we face the challenge of forging a new organic revolution in which the struggles for human equality and for the earth are becoming one. There is only one future: that of sustainable human development.[38]

Notes

PREFACE

1. Derrick Jensen and Aric McBay, *What We Leave Behind* (New York: Seven Stories Press, 2009), 201-05.

INTRODUCTION: A RIFT IN EARTH AND TIME

1. Karl Marx and Frederick Engels, *Collected Works* (New York: International Publishers, 1975), vol. 5, 40.
2. James Hansen, *Storms of My Grandchildren* (New York: Bloomsbury, 2009), ix.
3. Paul J. Crutzen, "Geology of Mankind: The Anthropocene," *Nature* 415 (2002), 22–23; Will Steffan, Paul J. Crutzen, and John R. McNeil, "The Anthropocene: Are Humans Now Overwhelming the Great Forces of Nature," *Ambio* 36/8 (December 2007), 614–21.
4. Johan Rockström, Will Steffen, Kevin Noone, Åsa Persson, F. Stuart Chapin, III, Eric F. Lambin, Timothy M. Lenton, Marten Scheffer, Carl Folke, Hans Joachim Schellnhuber, Björn Nykvist, Cynthia A. de Wit, Terry Hughes, Sander van der Leeuw, Henning Rodhe, Sverker Sörlin, Peter K. Snyder, Robert Constanza, Uno Svedin, Malin Falkenmark, Louise Karlberg, Robert W. Corell, Victoria J. Fabry, James Hansen, Brian Walker, Diana Liverman, Katherine Richardson, Paul Crutzen, and Jonathan A. Foley, "A Safe Operating Space for Humanity," *Nature* 461/24 (September

2009), 472–75, and "Planetary Boundaries," *Ecology and Society* 14/2 (2009), http://ecologyandsociety.org; Peter M. Vitousek, Harold A. Mooney, Jane Lubchenko, and Jerry A. Melilo, "Human Domination of Earth's Ecosystems," *Science* 277 (1997): 494–99; Peter M. Vitousek, "Beyond Global Warming: Ecology and Global Change," *Ecology* 75/7 (1994): 1861–76; Benjamin S. Halpern, Shaun Walbridge, Kimberly A. Selkoe, Carrie V. Kappel, Fiorenza Micheli, Caterina D'Agrosa, John F. Bruno, Kenneth S. Casey, Colin Ebert, Helen E. Fox, Rod Fujita, Dennis Heinemann, Hunter S. Lenihan, Elizabeth M. P. Madin, Matthew T. Perry, Elizabeth R. Selig, Mark Spalding, Robert Steneck, and Reg Watson, "A Global Map of Human Impact on Marine Ecosystems," *Science* 319 (2008): 948–52; Intergovernmental Panel on Climate Change, *Climate Change 2007: The Physical Science Basis* (Cambridge: Cambridge University Press, 2007); Millennium Ecosystem Assessment, *Ecosystems and Human Well-Being* (2005), available at http://millenniumassessment.org/en/Index.aspx.

5. Rockström et al., "A Safe Operating Space for Humanity," and "Planetary Boundaries"; Jonathan Foley, "Boundaries for a Healthy Planet," *Scientific American* 302/4 (April 2010), 54–57.
6. Rockström et al., "A Safe Operating System for Humanity," and "Planetary Boundaries"; "Dobson Unit," http://theozonehole.com.
7. World Wildlife Fund, *Living Planet Report, 2008*, http://wwf.panda.org, 1–3, 22–23; Global Footprint Network, "Earth Overshoot Day 2009," http://www.footprintnetwork.org. Also see William Catton, *Overshoot: The Ecological Basis of Revolutionary Change* (Chicago: University of Illinois Press, 1980).
8. See Arthur P. J. Mol and David Sonnenfeld, eds., *Ecological Modernisation Around the World* (London: Frank Cass, 2000), 22–25; Arthur P. J. Mol, David Sonnenfeld, and Gert Spaargaren, *The Ecological Modernisation Reader* (London: Routledge, 2009); Worldwatch, *The State of the World, 2010* (New York: W. W. Norton, 2010).
9. Thomas Friedman, *Hot, Flat and Crowded 2.0: Why We Need a Green Revolution—And How It Can Renew America* (New York: Picador, 2009); Fred Krupp and Miriam Horn, *Earth: The Sequel* (New York: W. W. Norton, 2009); Ted Nordhaus and Michael Shellenberger, *Break Through* (Boston: Houghton Mifflin, 2007); Newt Gingrich and Terry Maple, *A Contract with the Earth* (New York: Penguin, 2007).
10. Ulrich Beck, "World Risk Society as Cosmopolitan Society," in *Human Footprints on the Global Environment*, ed. Eugene A. Rosa, Andreas Diekmann, Thomas Dietz, and Carlo C. Jagger (Cambridge, Mass.: MIT Press, 2010), 73; Arthur P. J. Mol, *Globalization and Economic Reform* (Cambridge, Mass.: MIT Press, 2001), 206–7.

11. J. D. Bernal, *Science in History* (London: Watts and Co., 1954), 696–700; Plato, *Republic* (New York: Oxford University Press, 1945), 106 (III.414).
12. Karl Marx, *Capital*, vol. 1 (London: Penguin, 1976), 97.
13. Bernal, *Science in History*, 702.
14. Joan Robinson, "Review of Money, Trade and Economic Growth by J. G. Johnson," *Economic Journal* 72/287 (September 1962): 690–92; Lynn Turgeon, *Bastard Keynesianism* (Westport, Conn.: Greenwood Press, 1996).
15. Richard York and Brett Clark, "Critical Materialism: Science, Technology, and Environmental Sustainability," *Sociological Inquiry* 80/3 (2010): 475–99.
16. C. Wright Mills, *The Sociological Imagination* (New York: Oxford University Press, 1959), 85–87, and *The Marxists* (New York: Dell Publishing, 1962), 10–11, 18, 28–29. See also John Bellamy Foster, "Liberal Practicality and the U.S. Left," in *The Retreat of the Intellectuals: Socialist Register, 1990*, ed. Ralph Miliband, Leo Panitch, and John Saville, (London: Merlin Press, 1990), 265–89.
17. G. W. F. Hegel, *Natural Law* (Philadelphia: University of Pennsylvania Press, 1975), 62.
18. Quoted in Jeffrey A. Coyne, "Intelligent Design: The Faith that Dare Not Speak its Name," in *Intelligent Thought*, ed. John Brockman (New York: Vintage, 2006), 11.
19. Inspired by Georg Lukács's argument in *History and Class Consciousness*, Paul Sweezy introduced the phrase "the present as history" to define what distinguished the Marxian approach to political economy from its mainstream counterpart, with its assumption of the present as *non-history*. See Paul M. Sweezy, *The Present as History* (New York: Monthly Review Press, 1953), and *The Theory of Capitalist Development* (New York: Monthly Review Press, 1970), 20–22.
20. See Ellen Meiksins Wood, *Democracy Against Capitalism* (Cambridge: Cambridge University Press, 1995).
21. Francis Fukuyama, *The End of History and the Last Man* (New York: Free Press, 1992).
22. John Stuart Mill, *A System of Logic* (London: Longmans, Green and Co., 1889), 595.
23. On the role of the principle of historical specificity of historical materialism see Karl Korsch, *Karl Marx* (New York: Russell and Russell, 1938), 24–44. See also Mills, *The Sociological Imagination*, 143–64.
24. Fred Block, *Postindustrial Possibilities* (Berkeley: University of California Press, 1990), 44. See also his discussion of background conditions, 29–33.
25. Much of the discussion of "background conditions," "social gravity," and historicity in this and the following two paragraphs has been adapted from Richard York and Brett Clark, "Marxism, Positivism, and Scientific

Sociology: Social Gravity and Historicity," *The Sociological Quarterly* 47 (2006): 425–50, "Science and History: A Reply to Turner," *The Sociological Quarterly* 47 (2006): 465–70.

26. "Greenspan Testimony on Sources of Financial Crisis," *Wall Street Journal*, October 23, 2008; "Greenspan Says He Was Wrong on Regulation," *Washington Post*, October 24, 2008.
27. Jonathan H. Turner, "Explaining the Social World: Historicism versus Positivism," *The Sociological Quarterly* 47 (2006): 453.
28. Jacob Burckhardt, *Reflections on History* (Indianapolis: Liberty Press, 1979), 213, 224.
29. On "crackpot realism" see C. Wright Mills, *The Causes of World War Three* (New York: Simon and Schuster, 1958), 87.
30. Karl Marx and Frederick Engels, *The Communist Manifesto* (New York: Monthly Review Press, 1964), 11. See also Robert W. McChesney and John Bellamy Foster, "Capitalism, the Absurd System," *Monthly Review* 62/2 (June 2010): 1–16.
31. Our purpose here is to place in broader environmental terms Marx's famous proposition that history is made by human beings, but under preexisting conditions not of their choosing. See Karl Marx, *The Eighteenth Brumaire of Louis Bonaparte* (New York: International Publishers, 1963), 15.
32. Marx and Engels, *The Communist Manifesto*, 10.
33. Paul M. Sweezy, "Capitalism and the Environment," *Monthly Review* 41/2 (June 1989): 1–10.
34. Joseph A. Schumpeter, *Capitalism, Socialism and Democracy* (New York: Harper and Row, 1950), 81–86.
35. On the financialization tendency of today's monopoly-finance capital see John Bellamy Foster and Fred Magdoff, *The Great Financial Crisis* (New York: Monthly Review Press, 2009).
36. Joseph A. Schumpeter, *Essays* (Reading, Mass.: Addison-Wesley, 1951), 293.
37. Al Gore argues for a "green" or "sustainable capitalism" today in his *Our Choice* (New York: Melcher Media, 2009), 346. See also Jonathan Porritt, *Capitalism: As If the World Matters* (London: Earthscan, 2007).
38. Max Horkheimer and Theodor W. Adorno, *Dialectic of Enlightenment* (Stanford: Stanford University Press, 2002), 230.
39. Mol, *Globalization and Environmental Reform*, 1; Arthur P. J. Mol and Martin Jänicke, "The Origins and Theoretical Foundations of Ecological Modernisation Theory," in *The Ecological Modernisation Reader*, ed. Arthur P. J. Mol, David A. Sonnenfeld, and Gert Spaargaren (London: Routledge, 2009), 24; Kenneth A. Gould, David N. Pellow, and Allan Schnaiberg, *The Treadmill of Production* (Boulder, Colo.: Paradigm Publishers, 2008). Although arising out of a radical perspective, one of the

characteristics of the "treadmill" argument as represented in the last book referred to here is the very little attention devoted to the realities of capitalism itself, which is scarcely mentioned by name in the analysis. This contrasts sharply with the classic work in this tradition: Alan Schnaiberg, *The Environment: From Surplus to Scarcity* (New York: Oxford University Press, 1980).

40. John Kenneth Galbraith, *The Economics of Innocent Fraud* (Boston: Houghton Mifflin, 2004), 2–7.
41. See Georg Lukács, *History and Class Consciousness* (London: Merlin Press, 1968), 83–92; Horkheimer and Adorno, *Dialectic of Enlightenment*, 230.
42. Alfred North Whitehead, *Science and the Modern World* (New York: Free Press, 1925), 51. On the application of Whitehead's principle to the ecological critique of economics, see Herman E. Daly and John B. Cobb, Jr., *For the Common Good* (Boson: Beacon Press, 1994), 25–117. James K. Galbraith notes in *The Predatory State* (New York: Free Press, 2008), 19: "When you come down to it, the word market is a *negation*. It is a word to be applied to the context of any transaction so long as that transaction is not dictated by the state."
43. Max Weber, *Critique of Stammler* (New York: The Free Press, 1977), 95–97; see also Raymond Williams, *Problems in Materialism and Culture* (London: Verso, 1980), 67–85.
44. Fredric Jameson, *Valences of the Dialectic* (London: Verso, 2009), 23.
45. Roy Bhaskar, *The Possibility of Naturalism* (New York: Routledge, 1979), 2. Our own commitment to critical materialism (critical realism) is in evidence in John Bellamy Foster, Brett Clark, and Richard York, *The Critique of Intelligent Design: Materialism versus Creationism from Antiquity to the Present* (New York: Monthly Review Press, 2008).
46. C. P. Snow, *The Two Cultures* (Cambridge: Cambridge University Press, 1998).
47. Theodor W. Adorno, *Introduction to Sociology* (Stanford: Stanford University Press, 2000), 149.
48. See Riley E. Dunlap, William Michelson, and Glen Stalker, "Environmental Sociology: An Introduction," in Dunlap and Michelson, *Handbook of Environmental Sociology* (Westport, Conn.: Greenwood Press, 2002), 20–21; William R. Catton, Jr. and Riley E. Dunlap, "Paradigms, Theories and the Primacy of the HEP–NEP Distinction," *The American Sociologist* 13 (1978): 256–59; Riley E. Dunlap and William R. Catton, Jr., "Environmental Sociology," *Annual Review of Sociology* 5 (1979): 243–73.
49. Adorno, *Introduction to Sociology*, 149.
50. Keith Tester, *Animals and Society* (New York: Routledge, 1991). For a critical

view, see Raymond Murphy, *Rationality and Nature* (Boulder, Colo.: Westview Press, 1994). On the retreat from history represented by certain influential forms of contemporary postmodernism, see John Bellamy Foster, "In Defense of History," in *In Defense of History*, ed. Ellen Meiksins Wood and John Bellamy Foster (New York: Monthly Review Press, 1997), 184–94. On the relation to the sociology of science, see chapter 14 in this book.

51. See Ted Benton, "Biology and Social Theory in the Environmental Debate," in *Social Theory and the Global Environment*, ed. Michael Redclift and Ted Benton (New York: Routledge, 1994), 28–50. See also Riley E. Dunlap and William R. Catton, Jr., "Struggling with Human Exemptionalism," *American Sociologist* 25/1 (1994): 5–30.
52. John Bellamy Foster, *Marx's Ecology* (New York: Monthly Review Press, 2000), 242.
53. G. W. F. Hegel, *The Phenomenology of Spirit* (New York: Oxford University Press, 1977), 11.
54. See David R. Keller and Frank B. Golley, eds., *The Philosophy of Ecology: From Science to Synthesis* (Athens: University of Georgia Press, 2000). On the crucial concept of coevolution see Richard B. Norgaard, *Development Betrayed* (New York: Routledge, 1994).
55. Peter M. Vitousek, "Beyond Global Warming: Ecology and Global Change," *Ecology* 75/7 (1994): 1861–1876.
56. This awareness no doubt accounts for the recent popularity of so-called big history. See, for example, David Christian, *Maps of Time* (Berkeley: University of California Press, 2004).
57. See the discussion in John Bellamy Foster, *The Vulnerable Planet* (New York: Monthly Review Press, 1999), 28–30. On the larger problem of the "acceleration of history" and its relation to the acceleration of nature see 143–49.
58. James Hansen, *Storms of My Grandchildren* (New York: Bloomsbury, 2009), 146.
59. Martin J. S. Rudwick, *Bursting the Limits of Time: The Reconstruction of Geohistory in the Age of Revolution* (Chicago: University of Chicago Press, 2005), 102, 296.
60. Stephen J. Gould, *The Lying Stones of Marrakech* (New York: Three Rivers Press, 2000), 147–68.
61. See Erik Assadourian, "The Rise and Fall of Consumer Cultures," *State of the World, 2010* (New York: W.W. Norton, 2010), 3–20.
62. See chapter 1 of this book.
63. Richard Leakey and Roger Lewin, *The Sixth Extinction: Patterns of Life and the Future of Humankind* (New York: Anchor, 1996); Niles Eldredge, *The Miner's Canary* (Princeton, N.J.: Princeton University Press, 1991); Franz J. Broswimmer, *Ecocide* (London: Pluto Press, 2002).

64. Frederick Buell, *From Apocalypse to Way of Life* (New York: Routledge, 2003).
65. This is the description of human nature under conditions of the state of nature (a war of all against all) as depicted in Thomas Hobbes's *Leviathan* (London: Penguin, 1968), 186 (italics added). But it applies equally to the final results to be expected of such a state of human relations if raised to the level of the planet. "If we continue to act on the assumption that the only thing that matters is personal greed and personal gain," Noam Chomsky has written, "the [ecological] commons will be destroyed." Noam Chomsky (interview) in Bill Moyers, *A World of Ideas* (New York; Doubleday, 1989), 58.
66. Robert Heilbroner, *The Nature and Logic of Capitalism* (New York: Norton, 1985), 33–52; Paul Burkett, "Natural Capital, Ecological Economics, and Marxism," *International Papers in Political Economy* 10/3 (2003): 1–61.
67. Marx, *Capital*, vol. 1, 799.
68. István Mészáros, *Beyond Capital* (New York: Monthly Review Press, 1995), 107.
69. Heilbroner, *The Nature and Logic of Capital*, 36.
70. Marx, *Capital*, vol. 1, 769.
71. E. F. Schumacher, *Small Is Beautiful* (New York: Harper and Row, 1973), 255.
72. G. W. F. Hegel, *Science of Logic* (London: George Allen and Unwin, 1969), 132–135. See also Michael Inwood, *A Hegel Dictionary* (Oxford: Blackwell, 1992), 177–80.
73. Karl Marx, *Grundrisse* (London: Penguin, 1973), 334–35. See also the discussion in chapter 13 of this book.
74. See the detailed treatment of *The Stern Review* in chapter 3 of this book.
75. David Pearce, *Blueprint 3: Measuring Sustainable Development* (London: Earthscan, 1993), 8.
76. Arthur P. J. Mol and David A. Sonnenfeld, "Ecological Modernisation Theory in Debate," in Mol and Sonnenfeld, ed., *Ecological Modernisation Around the World*, 22. Mol and Hajer quoted in Oluf Langhelle, "Why Ecological Modernisation and Sustainable Development Should Not Be Conflated," in Mol, Sonnenfeld, and Spaargaren, *The Ecological Modernisation Reader*, 402–5, 414.
77. Langhelle, "Why Ecological Modernisation," 403–5.
78. Gert Spaargaren, Arthur P.J. Mol, and David Sonnenfeld, "Ecological Modernisation," in Mol, Sonnenfeld, and Spaaargaren, *The Ecological Modernisation Reader*, 509.
79. Herman E. Daly, *Steady-State Economics* (Washington, D.C.: Island Press, 1991), 107.

80. Charles Leadbeater, *The Weightless Society* (New York: Texere, 2000); see the discussion of "the myth of dematerialization" in John Bellamy Foster, *Ecology Against Capitalism* (New York: Monthly Review Press, 2002), 22–24.
81. Maarten A. Hajer, "Ecological Modernisation as Cultural Politics," in Mol, Sonnenfeld, and Spaargaren, *The Ecological Modernisation Reader*, 83, 94–96.
82. Huber simply claims that the concerns regarding entropy (the second law of thermodynamics) and energy and the limits that they place on the scale of human production raised by Nicholas Georgescu-Roegen "are wrong and in no way helpful." See Huber, "Ecological Modernization," in Mol, Sonnenfeld, and Spaargaren, *The Ecological Modernisation Reader*, 44; Nicholas Georgescu-Roegen, *Energy and Economic Myths* (New York: Pergamon, 1976), 3–36.
83. Nicholas Georgescu-Roegen, *The Entropy Law and the Economic Process* (Cambridge, Mass.: Harvard University Press, 1971), 19.
84. See chapter 7 in this book.
85. Juliet Schor, *Plenitude* (New York: Penguin, 2010), 89.
86. Ibid., 87.
87. Donella Meadows, Jorgen Randers, and Dennis Meadows, *Limits to Growth: The 30-Year Update* (White River Junction, Vt.: Chelsea Green, 2004), 223–24.
88. See the powerful critique in Marilyn Waring, *Counting for Nothing* (Toronto: University of Toronto Press, 1999). On the inbuilt tendency toward waste under monopoly capitalism, see also Paul A. Baran and Paul M. Sweezy, *Monopoly Capital* (New York: Monthly Review Press, 1966).
89. Charles Lindbloom, *The Market System* (New Haven: Yale University Press, 2001), 146–77.
90. Lester Thurow, *The Future of Capitalism* (London: Penguin Books, 1996), 303.
91. On the concept of ecological debt see John Bellamy Foster, *The Ecological Revolution* (New York: Monthly Review Press, 2009), 242–47.
92. Marx, *Capital*, vol. 1, 636–38; Karl Marx, *Capital*, vol. 3 (London: Penguin, 1981), 949.
93. See the discussion of the problem of the old growth forest in Foster, *Ecology Against Capitalism*, 106–12.
94. John Bellamy Foster, *The Vulnerable Planet* (New York: Monthly Review Press, 1994), 110–12.
95. See chapter 2 in this book.
96. L. S. Stavrianos, *The Global Rift* (New York: William Morrow, 1981).
97. Rhodes quoted in Sarah Gertrude Millin, *Rhodes* (London: Chatto and Windhurst, 1952), 138.
98. Maude Barlow, *Blue Covenant* (New York: New Press, 2007), 103–5.
99. *Cochabamba Protocol*, http://pwccc.wordpress.com/2010/04/28/peoples-agreement.

100. "To Save Planet, End Capitalism, Morales Says," *Green Left*, May 9, 2009, http://www.greenleft.org.
101. Epicurus, *The Epicurus Reader* (Indianapolis: Hackett, 1994), 39.
102. Elmar Altvater, *The Future of the Market* (London: Verso, 1993), 203. The leading exponent of a steady-state economics is Herman Daly. His views of the changes that need to be wrought to generate such an economy are far-reaching, as shown most dramatically in his work together with John Cobb. Nevertheless, Daly attempts to avoid the contradiction raised by Altvater above and stops short of the conclusion that a decisive move in this direction must necessarily mean a break with capitalism. See especially Herman E. Daly, *Beyond Growth* (Boston: Beacon Press, 1996); and Daly and Cobb, *For the Common Good*.
103. Schumacher, *Small Is Beautiful*, 254–55.

CHAPTER ONE: THE PARADOX OF WEALTH

This chapter is adapted and revised for this book from the following article: John Bellamy Foster and Brett Clark, "The Paradox of Wealth: Capitalism and Ecological Degradation," *Monthly Review* 61/6 (November 2009): 1–18.

1. See the discussion of William Nordhaus's stance on climate change in chapter 3 of this book.
2. John Maynard Keynes, *The General Theory of Employment, Interest and Money* (London: Macmillan, 1973), 32.
3. James Maitland, Earl of Lauderdale, *An Inquiry into the Nature and Origin of Public Wealth and into the Means and Causes of Its Increase* (Edinburgh: Archibald Constable and Co., 1819), 37–59, and *Lauderdale's Notes on Adam Smith*, ed. Chuhei Sugiyama (New York: Routledge, 1996), 140–41. Lauderdale was closest to Malthus in classical political economy, but generally rejected classical value theory, emphasizing the three factors of production (land, labor, and capital). Marx, who took Ricardo as his measure of bourgeois political economy, therefore had little genuine interest in Lauderdale as a theorist, apart from the latter's sense of the contradiction between use value and exchange value. Still, Lauderdale's devastating critique of the pursuit of private riches at the expense of public wealth earns him a position as one of the great dissident voices in the history of economics.
4. Robert Brown, *The Nature of Social Laws* (Cambridge: Cambridge University Press, 1984), 63–64.
5. Karl Marx, *A Contribution to the Critique of Political Economy* (Moscow: Progress Publishers, 1970), 27. In this chapter, for simplicity's sake, we do not explicitly address Marx's distinction between *exchange value* and its

basis in *value* (abstract labor), treating them as basically synonymous within the limits of our discussion.

6. David Ricardo, *On the Principles of Political Economy and Taxation,* vol. 1, *Works and Correspondence of David Ricardo* (Cambridge: Cambridge University Press, 1951), 276–87; George E. Foy, "Public Wealth and Private Riches," *Journal of Interdisciplinary Economics* 3 (1989): 3–10.
7. Jean Baptiste Say, *Letters to Thomas Robert Malthus on Political Economy and Stagnation of Commerce* (London: G. Harding's Bookshop, Ltd., 1936), 68–75.
8. Karl Marx, *Capital*, vol. 1 (London: Penguin, 1976), 98.
9. John Stuart Mill, *Principles of Political Economy with Some of Their Applications to Social Philosophy* (New York: Longmans, Green, and Co., 1904), 4, 6.
10. Mill appears to break out of these limits only briefly in his book, in his famous discussion of the stationary state. See Mill, *Principles of Political Economy*, 452–55.
11. Karl Marx, *The Poverty of Philosophy* (New York: International Publishers, 1964), 35–36, and *Theories of Surplus Value* (Moscow: Progress Publishers, 1968), part 2, 245; Karl Marx and Frederick Engels, *Selected Correspondence* (Moscow: Progress Publishers, 1975), 180–81 (Marx to Engels, August 24, 1867).
12. See Paul Burkett and John Bellamy Foster, "Metabolism, Energy, and Entropy in Marx's Critique of Political Economy," *Theory and Society* 35/1 (February 2006): 109–56, and "The Podolinsky Myth," *Historical Materialism* 16 (2008): 115–61; John Bellamy Foster and Paul Burkett, "Classical Marxism and the Second Law of Thermodynamics," *Organization & Environment* 21/1 (March 2008): 1–35.
13. Karl Marx, *Early Writings* (New York: Vintage, 1974), 359–60.
14. Karl Marx, *Capital*, vol. 3 (London: Penguin, 1981), 911, 959, and *Theories of Surplus Value*, part 2, 245. For a discussion of the metabolic rift, see John Bellamy Foster, *The Ecological Revolution* (New York: Monthly Review Press, 2009), 161–200.
15. Karl Marx, *Dispatches for the New York Tribune* (London: Penguin, 2007), 128–29; Herbert Spencer, *Social Statics* (New York; D. Appleton and Co. 1865), 13–44. Herbert Spencer was to recant these views beginning in 1892, which led Henry George to polemicize against him in *A Perplexed Philosopher* (New York: Charles L. Webster & Co., 1892). See also George R. Geiger, *The Philosophy of Henry George* (New York: Macmillan, 1933), 285–335.
16. Marx, *Early Writings*, 239; Thomas Müntzer, *Collected Works* (Edinburgh: T & T Clark, 1988), 335.
17. E. F. Schumacher, *Small Is Beautiful* (New York: Harper and Row, 1973), 15; Luiz C. Barbosa, "Theories in Environmental Sociology," in *Twenty*

Lessons in Environmental Sociology, ed. Kenneth A. Gould and Tammy Lewis (Oxford: Oxford University Press, 2009), 28; Jean-Paul Deléage, "Eco-Marxist Critique of Political Economy," in *Is Capitalism Sustainable?*, ed. Martin O'Connor (New York: Guilford, 1994), 48. Mathew Humphrey, *Preservation Versus the People?* (Oxford: Oxford University Press, 2002), 131–41.

18. Thomas Malthus, *Pamphlets* (New York: Augustus M. Kelley, 1970), 185; Ricardo, *Principles of Political Economy*, 76, 287; Paul Burkett, *Marxism and Ecological Economics* (Boston: Brill, 2006), 25–27, 31, 36.
19. Marx and Engels, *Collected Works* (New York: International Publishers, 1975), vol. 34, 443–507.
20. Campbell McConnell, *Economics* (New York: McGraw Hill, 1987), 20, 672; Alfred Marshall, *Principles of Economics* (London: Macmillan and Co., 1895), chap. 2.
21. Nick Hanley, Jason F. Shogren, and Ben White, *Introduction to Environmental Economics* (Oxford: Oxford University Press, 2001), 135.
22. Karl Marx, *Critique of the Gotha Programme* (New York: International Publishers, 1938), 3; Marx, *Capital*, vol. 1, 134.
23. Paul Burkett, *Marx and Nature* (New York: St. Martin's Press, 1999), 99.
24. Karl Marx, *Critique of the Gotha Programme*, 3, and *Capital*, vol. 1, 133–34, 381, 751–52; Paul Burkett, *Marx and Nature*, 99.
25. On Marx's use of the vampire metaphor, see Mark Neocleous, "The Political Economy of the Dead: Marx's Vampires," *History of Political Thought* 24/4 (Winter 2003), 668–84.
26. Carl Menger, *Principles of Political Economy* (Auburn, Ala.: Ludwig von Mises Institute, 2007), 110–11. For related views see Eugen Böhm-Bawerk, *Capital and Interest* (South Holland, Ill.: Libertarian Press, 1959), 127–34.
27. Henry George, *Progress and Poverty* (New York: Modern Library, no copyright, published 1879), 39–40.
28. Henry George, *Complete Works* (New York: Doubleday, 1904), vol. 6, 121–28, 158, 212–25, 242, 272–76, 292, and *A Perplexed Philosopher*, 51–61; Marx, *Capital*, vol. 1, 323.
29. Thorstein Veblen, *Absentee Ownership* (New York: Augustus M. Kelley, 1923), 168–70.
30. Frederick Soddy, *Wealth, Virtual Wealth, and Debt* (London: Allen and Unwin, 1933), 73–74.
31. Frederick Soddy, *Cartesian Economics* (London: Hendersons, 1922), 15–16; Soddy, *Matter and Energy* (New York: Henry Holt, 1912), 34–36.
32. Soddy, *Wealth, Virtual Wealth, and Debt*, 63–64.
33. K. William Kapp, *The Social Costs of Private Enterprise* (New York: Schocken, 1971), 8, 29, 34–36, 231.

34. Herman E. Daly, "The Return of the Lauderdale Paradox," *Ecological Economics* 25 (1998): 21–23, and *Ecological Economics and Sustainable Development* (Cheltenham, UK: Edward Elgar, 2007), 105–6; Herman E. Daly and John B. Cobb, Jr., *For the Common Good* (Boston: Beacon Press, 1994), 147–48.
35. Nordhaus quoted in Leslie Roberts, "Academy Panel Split on Greenhouse Adaptation," *Science* 253 (September 13, 1991): 106; Wilfred Beckerman, *Small Is Stupid* (London: Duckworth, 1995), 91; "The Environment as a Commodity," *Nature* 357 (June 4, 1992): 371–72; Thomas C. Shelling, "The Cost of Combating Global Warming," *Foreign Affairs* (November/December 1997): 8–9; Daly, *Ecological Economics and Sustainable Development*, 188–90.
36. Fred Magdoff and Brian Tokar, "Agriculture and Food in Crisis," *Monthly Review* 61/3 (July–August 2009), 1–3.
37. William D. Nordhaus, "Reflections on the Economics of Climate Change," *Journal of Economic Perspectives* 7/4 (Fall 1993): 22–23.
38. The argument that such a feedback mechanism exists is known in Marxist ecological analysis as the "second contradiction of capitalism." See James O'Connor, *Natural Causes* (New York: Guilford, 1998). For a critique see Foster, *The Ecological Revolution*, 201–12.
39. Maude Barlow and Tony Clarke, *Blue Gold* (New York: New Press, 2002), 88, 93, 105.
40. Fred Magdoff, "World Food Crisis," *Monthly Review* 60/1 (May 2008): 1–15.
41. Lauderdale, *Inquiry into the Nature and Origin of Public Wealth*, 41–42.
42. On green accounting, see Andrew John Brennan, "Theoretical Foundations of Sustainable Economic Welfare Indicators," *Ecological Economics* 67 (2008): 1–19; Daly and Cobb, *For the Common Good*, 443–507.
43. Burkett, *Marx and Nature*, 82–84.
44. Marx, *A Contribution to a Critique of Political Economy*, 36, and *Capital*, vol. 1, 638.

CHAPTER TWO: RIFTS AND SHIFTS

This chapter is adapted and revised for this book from the following article: Brett Clark and Richard York, "Rifts and Shifts: Getting to the Roots of Environmental Crises," *Monthly Review* 60/6 (November 2008): 13–24.

1. James Hansen, "Tipping Point," in *State of the World 2008*, ed. E. Fearn and K. H. Redford (Washington, D.C.: Island Press, 2008), 7–8.
2. Arthur P. J. Mol, *Globalization and Environmental Reform* (Cambridge: MIT Press, 2001); Charles Leadbeater, *The Weightless Society* (New York: Texere, 2000).

3. John Bellamy Foster, *Marx's Ecology* (New York: Monthly Review Press, 2000), 158; Karl Marx, *Capital*, vol. 1 (New York: Vintage, 1976), 637–38.
4. István Mészáros, *Beyond Capital* (New York: Monthly Press, 1995), 40–45, 170–71; István Mészáros, "The Necessity of Planning," *Monthly Review* 58/5 (2006): 27–35.
5. Karl Marx, *Grundrisse* (New York: Penguin Books, 1993), 409–10; Paul Sweezy, "Capitalism and the Environment," *Monthly Review* 56/5 (2004): 86–93; Paul Burkett, *Marx and Nature* (New York: St. Martin's Press, 1999).
6. Justus von Liebig, *Letters on Modern Agriculture* (London: Walton & Maberly, 1859).
7. Karl Marx, *Capital*, vol. 3 (New York: Penguin Books, 1991), 949.
8. John Bellamy Foster and Brett Clark, "Ecological Imperialism," in *Socialist Register 2004*, ed. Leo Panitch and Colin Leys (London: Merlin Press, 2003); Jason W. Moore, "*The Modern World-System* as Environmental History?," *Theory and Society* 32/3 (2003): 307–77; Marx, *Capital*, vol. 1, 637.
9. Karl Marx, "Wage Labour and Capital," in *The Marx-Engels Reader*, ed. Robert Tucker (New York: W.W. Norton, 1978), 213.
10. Foster and Clark, "Ecological Imperialism"; Jimmy M. Skaggs, *The Great Guano Rush* (New York: St. Martin's Griffin, 1994).
11. Marx, *Grundrisse*, 527.
12. Marx, *Capital*, vol. 3, 950.
13. Marx, *Capital*, vol. 1, 638.
14. Karl Kautsky, *The Agrarian Question* (Winchester, Mass.: Swan, 1988), 215.
15. John Bellamy Foster and Fred Magdoff, "Liebig, Marx, and the Depletion of Soil Fertility," in *Hungry For Profit*, ed. Fred Magdoff, John Bellamy Foster, and Frederick Buttel (New York: Monthly Review Press, 2000); Fred Magdoff, Les Lanyon, and Bill Liebhardt, "Nutrient Cycling, Transformation and Flows," *Advances in Agronomy* 60 (1997): 1–73; Philip Mancus, "Nitrogen Fertilizer Dependency and its Contradictions," *Rural Sociology* 72/2 (2007): 269–88.
16. Vaclav Smil, *Energy in World History* (Boulder, Colo.: Westview, 1994); Michael Williams, *Deforesting the Earth* (Chicago: University of Chicago Press, 2003); Brett Clark and Richard York, "Carbon Metabolism," *Theory and Society* 34/4 (2005): 391–428.
17. See Joseph Fargione, Jason Hill, David Tilman, Stephen Polasky, and Peter Hawthorne, "Land Clearing and the Biofuel Carbon Debt," *Science* 319 (2008): 1235–38; Timothy Searchinger, Ralph Heimlich, R. A. Houghton, Fengxia Dong, Amani Elobeid, Jacinto Fabiosa, Simla Tokgoz, Dermot Hayes, and Tun-Hsiang Yu, "Use of U.S. Croplands for Biofuels Increase Greenhouse Gases through Emissions from Land-Use Change," *Science*

319 (2008): 1238–40; Fred Magdoff, "The Political Economy and Ecology of Biofuels," *Monthly Review* 60/3 (2008): 34–50.

18. Paul J. Crutzen, "Albedo Enhancement by Stratospheric Sulfur Injections," *Climate Change* 77 (2006): 211–19; Freeman Dyson, "The Question of Global Warming," *New York Review of Books* 55 (2008): 43–45.
19. Karl Marx, *Theories of Surplus Value*, vol. 3 (Moscow: Progress Publishers, 1971), 301–10; Mészáros, *Beyond Capital*, 174.
20. James Hansen, "The Threat to the Planet," *New York Review of Books* 53 (2006): 12–16.
21. Mészáros, "The Necessity of Planning," 28.
22. Marx, *Capital*, vol. 3, 911; Frederick Engels, *Anti-Dühring* (Moscow: Progress Publishers, 1969), 136–38.

CHAPTER THREE: CAPITALISM IN WONDERLAND

This chapter is adapted and revised for this book from the following article: Richard York, Brett Clark, and John Bellamy Foster, "Capitalism in Wonderland," *Monthly Review* 61/1 (May 2009): 1–18.

1. Jean-Philippe Bouchaud, "Economics Needs a New Scientific Revolution," *Nature* 455 (October 30, 2008): 1181.
2. See Naomi Klein, *The Shock Doctrine: The Rise of Disaster Capitalism* (New York: Henry Holt, 2007). "*Après moi le déluge!* is the watchword of every capitalist and every capitalist nation. Capital therefore takes no account of the health and length of life of the workers unless society forces it to do so." Karl Marx, *Capital*, vol. 1 (New York: Vintage, 1976), 381.
3. Paul R. Ehrlich, "An Economist in Wonderland," *Social Science Quarterly* 62 (1981): 44–49; Julian L. Simon, "Resources, Population, Environment: An Oversupply of False Bad News," *Science* 208 (June 27, 1980): 1431–37, "Bad News: Is It True?" *Science* 210 (December 19, 1980): 1305–8, "Environmental Disruption or Environmental Improvement?" *Social Science Quarterly* 62 (1981): 30–43, *The Ultimate Resource* (Princeton: Princeton University Press, 1981), "Paul Ehrlich Saying It Is So Doesn't Make It So," *Social Science Quarterly* 63 (1982): 381–5. For the rest of Ehrlich and colleagues' side of the exchanges, see Ehrlich, "Environmental Disruption: Implications for the Social Sciences," *Social Science Quarterly* 62 (1981): 7–22, "That's Right—You Should Check It for Yourself," *Social Science Quarterly* 63 (1982): 385–7, John P. Holdren, Paul R. Ehrlich, Anne H. Ehrlich, and John Harte, "Bad News: Is It True?" *Science* 210 (December 19, 1980): 1296–1301.
4. Bjørn Lomborg, *The Skeptical Environmentalist* (Cambridge: Cambridge University Press, 2001); Stuart Pimm and Jeff Harvey, review of *The*

Skeptical Environmentalist, *Nature* 414 (November 8, 2001): 149-50; Stephen Schneider, John P. Holdren, John Bogaars, and Thomas Lovejoy in *Scientific American* 286/1 (January 2002), 62–72; "Defending Science," *The Economist*, January 31, 2002, 15–16.

5. Bjørn Lomborg, *Global Crises, Global Solutions* (Cambridge: Cambridge University Press, 2004), 6.
6. Bjørn Lomborg, *Cool It: The Skeptical Environmentalist's Guide to Global Warming* (New York: Alfred A. Knopf, 2007), 37. See also Frank Ackerman, "Hot, It's Not: Reflections on *Cool It*, by Bjørn Lomborg," *Climatic Change* 89 (2008), 435–46.
7. Milton Friedman in Carla Ravaioli, *Economists and the Environment* (London; Zed Press, 1995), 32, 64–65.
8. Stephen H. Schneider, *Laboratory Earth* (New York: Basic Books, 1997), 129–35; William D. Nordhaus, "An Optimal Transition Path for Controlling Greenhouse Gases," *Science* 258 (November 20, 1992): 1318; Stephen Schneider, "Pondering Greenhouse Policy," *Science* 259 (March 5, 1993): 1381. The discussion here borrows from the introduction to John Bellamy Foster, *The Ecological Revolution* (New York: Monthly Review Press, 2009), 24–25.
9. Thomas C. Schelling, "The Greenhouse Effect," *The Concise Encyclopedia of Economics*, http://www.econlib.org/library/Enc1/GreenhouseEffect.html; Schelling in Lomborg, *Global Crises, Global Solutions*, 630. Schelling is often "credited" with having been the leading "strategist" of the Vietnam War.
10. Gordon Rattray Taylor, *The Doomsday Book* (Greenwich, Conn.: Fawcett Publications, 1970), 32–33.
11. After the memo was leaked Summers claimed that he was being "ironic," but his position conformed to both mainstream economic analysis and other statements that he had argued explicitly and publicly and thus belied that claim. See Summers's memo and its critique in John Bellamy Foster, *Ecology Against Capitalism* (New York: Monthly Review Press, 2002), 60–68.
12. William Nordhaus, *A Question of Balance: Weighing the Options on Global Warming Policies* (New Haven: Yale University Press, 2008).
13. Coastal Services Center, National Oceanic and Atmospheric Association, "Restoration Economics: Discounting and Time Preference," http://www.csc.noaa.gov/coastal/economics/discounting.htm.
14. William Nordhaus, "Critical Assumptions in the Stern Review on Climate Change," *Science* 317 (2007): 201–2; Coastal Services Center, "Restoration Economics."
15. Ackerman, "Hot, It's Not," 443.
16. Roy Harrod, *Towards a Dynamic Economy* (New York: St. Martin's Press, 1948), 40; Stern, *The Economics of Climate Change*, 35–36; William Cline, "Climate Change," in Lomborg, *Global Crises, Global Solutions*, 16.
17. Nordhaus, *A Question of Balance*, 13–14.

18. John Browne, "The Ethics of Climate Change: The Stern Review," *Scientific American* 298/6 (June 2008): 97–100.
19. Nicholas Stern, *The Economics of Climate Change: The Stern Review* (Cambridge: Cambridge University Press, 2007).
20. Bjørn Lomborg, "Stern Review: The Dodgy Numbers Behind the Latest Warming Scare," *Wall Street Journal*, November 2, 2006, and *Cool It!*, 31.
21. Nordhaus, *A Question of Balance*, 13–14; Simon Dietz and Nicholas Stern, "On the Timing of Greenhouse Gas Emissions Reductions: A Final Rejoinder to the Symposium on 'The Economics of Climate Change: The Stern Review and Its Critics,'" *Review of Environmental Economics and Policy* 3/1 (Winter 2009), 138–40; Intergovernmental Panel on Climate Change (IPCC), *Summary for Policymakers* in *Climate Change 2007: Mitigation* (Cambridge: Cambridge University Press, 2007), 15; Mark Lynas, *Six Degrees* (Washington, D.C.: National Geographic, 2008), 241.
22. James and Anniek Hansen, "Dear Barack and Michelle: An Open Letter to the President and the First Lady from the Nation's Top Climate Scientist," *Gristmill*, January 2, 2009, http://www.grist.org; IPCC, *Summary for Policymakers* in *Climate Change 2007*, 15; Stern, *The Economics of Climate Change*, 16.
23. IPCC, *Summary for Policymakers* in *Climate Change 2007*, 15; Dietz and Stern, "On the Timing," 139; Stern, *The Economics of Climate Change*, 16. Rather than using atmospheric carbon dioxide concentration, like Hansen and Nordhaus, the *Stern Review* focuses on carbon dioxide equivalent concentration, which includes the six Kyoto greenhouse gases (carbon dioxide, methane, nitrous oxide, hydrofluorocarbons, perfluorocarbons, and sulfur hexafluoride) all expressed in terms of the equivalent amount of carbon dioxide. For the sake of consistency, we present here the carbon dioxide concentration and then in parentheses the corresponding carbon dioxide equivalent concentration.
24. Stern, *The Economics of Climate Change*, 231. See also chapter 6 in this book.
25. Lewis Carroll, *The Annotated Alice: The Definitive Edition*, ed. Martin Gardner (New York: Norton, 2000), 204.
26. Joseph A. Schumpeter, *A History of Economic Analysis* (New York: Oxford University Press, 1951), 41–42.
27. Herman Daly, *Beyond Growth* (Boston: Beacon Press, 1996), 5–6. Summers himself, Daly explains, later denied that the economy should be seen as a subset of the biosphere.
28. In addition to Daly's book cited above see Paul Burkett, *Marxism and Ecological Economics* (Boston: Brill, 2006).
29. Herman E. Daly, "Economics in a Full World," *Scientific American* 293/3 (September 2005), 102.
30. Karl Marx, *Capital*, vol. 1 (New York: Vintage, 1976), 134, 637–638, and

Capital, vol. 3 (New York: Vintage, 1981), 754; Frederick Engels, *The Dialectics of Nature* (Moscow: Progress Publishers, 1966), 179–180. See also John Bellamy Foster, *Marx's Ecology* (New York: Monthly Review Press, 2000), 141–77.

31. Josep G. Canadell, Corinne Le Quéré, Michael R. Raupach, Christopher B. Field, Erik T. Buitenhuis, Philippe Ciais, Thomas J. Conway, Nathan P. Gillett, R. A. Houghton, and Gregg Marland, "Contributions to Accelerating Atmospheric CO_2 Growth from Economic Activity, Carbon Intensity, and Efficiency of Natural Sinks," *Proceedings of the National Academy of Sciences* 104/47 (2007): 18866–18870.
32. See John Bellamy Foster, "A Failed System," *Monthly Review* 60/10 (March 2009): 1–23.
33. Paul Ehrlich, "Environmental Disruption," 12–14.
34. Fred Krupp and Miriam Horn, *Earth: The Sequel* (New York: W. W. Norton, 2009), 261. For a treatment of the views of Friedman, Gingrich, and the Breakthrough Institute see the introduction to Foster, *The Ecological Revolution*.
35. See Brett Clark and John Bellamy Foster, "William Stanley Jevons and *The Coal Question*: An Introduction to Jevons's 'Of the Economy of Fuel,'" *Organization and Environment* 14/1 (March 2001): 93–98; see also chapter 7 below.
36. David G. Victor, M. Granger Morgan, Jay Apt, John Steinbruner, and Katharine Ricke, "The Geoengineering Option," *Foreign Affairs* 88/2 (March–April 2009): 64–76.
37. Rachel Carson, *Silent Spring* (Boston: Houghton Mifflin, 1994); Carson, *The Sense of Wonder* (New York: Harper Row, 1965).

CHAPTER FOUR: THE MIDAS EFFECT

This chapter is adapted and revised for this book from the following article: John Bellamy Foster, Brett Clark, and Richard York, "The Midas Effect: A Critique of Climate Change" *Development and Change* 40/6 (2009): 1085–97.

1. James Hansen, "Tipping Point," in *State of the Wild 2008–2009*, ed. Eva Fearn (Washington, D.C.: Island Press, 2008), 7–15.
2. Intergovernmental Panel on Climate Change, *Climate Change 2007: The Physical Science Basis* (Cambridge: Cambridge University Press, 2007).
3. Eli Kintisch, "Projections of Climate Change Go from Bad to Worse, Scientists Report," *Science* 323 (2009): 1546–47; Randolph E. Schmid, "Climate Warming Gases Rising Faster than Expected," *Guardian* (February 15, 2009), www.guardian.co.uk.
4. Josep G. Canadell, Corinne Le Quéré, Michael R. Raupach, Christopher

B. Field, Erik T. Buitenhuis, Philippe Ciais, Thomas J. Conway, Nathan P. Gillett, R. A. Houghton, and Gregg Marland, "Contributions to Accelerating Atmospheric CO_2 Growth from Economic Activity, Carbon Intensity, and Efficiency of Natural Sinks," *Proceedings of the National Academy of Sciences* 104/47 (2007): 18866–18870.

5. Kintisch, "Projections of Climate Change," 1546.
6. James Hansen, Makiko Sato, Pushker Kharecha, David Beerling, Valerie Masson-Delmotte, Mark Pagani, Maureen Raymo, Dana L. Royer, and James C. Zachos, "Target Atmospheric CO_2: Where Should Humanity Aim?" *Open Atmospheric Science Journal* 2 (2008): 217–31; Kintisch, "Projections of Climate Change."
7. Hansen et al., "Target Atmospheric," 217.
8. Hansen, "Tipping Point"; Kintisch, "Projections of Climate Change," 1547.
9. See chapter 5 in this book.
10. Joseph A. Schumpeter, *Essays* (Cambridge: Addison-Wesley, 1951), 293.
11. Nicholas Stern, *The Economics of Climate Change: The Stern Review* (Cambridge: Cambridge University Press, 2007).
12. Here we indicate a range in global temperature, given that the IPCC indicates that if the carbon dioxide equivalent concentration was stabilized at 535–590 ppm, the global mean temperature would increase 3.2°C above the preindustrial temperature. Stern, in contrast, estimates that a carbon dioxide equivalent concentration of 550 ppm would increase the global temperature 4.4°C. See Intergovernmental Panel on Climate Change, *Climate Change 2007: Mitigation of Climate Change* (2007), www.ipcc.ch; Stern, *The Economics of Climate Change*.
13. Stern, *The Economics of Climate Change*, 231.
14. William Nordhaus, *A Question of Balance* (New Haven: Yale University Press, 2008).
15. See James Lovelock, *The Revenge of Gaia* (New York: Basic Books, 2006).
16. Ovid, *Metamorphoses* (trans. Charles Martin) (New York: W.W. Norton, 2004), 373–75.
17. Elmar Altvater, *The Future of the Market* (New York: Verso, 1993), 184.
18. Julian Simon, "Resources, Population, Environment: An Oversupply of False Bad News," *Science* 208 (1980): 1431–37, quote on 1435; Julian Simon, *The Ultimate Resource* (Princeton: Princeton University Press, 1981), 47.
19. Julian Simon, "Answer to Malthus?: Julian Simon Interviewed by William Buckley," *Population and Development Review* 8/1 (1982): 205–18, quote on 207.
20. R. Kerry Turner, David Pearce, and Ian Bateman, *Environmental Economics* (Baltimore: Johns Hopkins University Press, 1993).
21. Robert M. Solow, "The Economics of Resources or the Resources of

Economics," *American Economic Review* 64/2 (1974): 1–14, quote on 11. Solow went on to consider the opposite case where substitutability was bounded. But the nature of his argument was to emphasize very high levels of substitutability: "Nordhaus's notion of a 'backstop technology,'" in which "at some finite cost, production can be freed from exhaustible resources altogether." Solow treated this not as absurd, but as somehow closer to the truth than its opposite. In another piece written at about the same time, Solow asked, "Why should we be concerned with the welfare of posterity, given the indubitable fact that posterity has never done a thing for us?" He added that productivity in natural resource use could increase "exponentially" indefinitely. Substitutability (for example, nuclear fission in place of oil) could compensate for loss of particular finite resources. He therefore asserted that there was no reason to think that there were environmental limits to growth. See Robert M. Solow, "Is the End of the World at Hand?," in *The Economic Growth Controversy*, ed. Andrew Weintraub, Eli Schwartz, and J. Richard Aronson (White Plains, N.Y.: International Arts and Sciences Press, 1973), 39–61. Nearly two decades after this, Solow was still arguing that the environmental impact problem was simply one of "substituting" rather than "reducing." See Solow interview in Carla Ravaioli, *Economists and the Environment* (Atlantic Highlands, N.J.: Zed Books, 1995), 72.

22. David Pearce, *Blueprint 3* (London: Earthscan, 1993), 8.
23. Charles Leadbeater, *The Weightless Society* (New York: Texere, 2000); Arthur P. J. Mol, "The Environmental Movements in an Era of Ecological Modernization," *Geoforum* 31 (2000): 45–56.
24. See chapter 7 below.
25. Paul Hawken, Amory Lovins, and L. Hunter Lovins, *Natural Capitalism* (Boston: Little, Brown, 1999).
26. Stephen Schneider, *Laboratory Earth* (New York: Basic Books, 1997), 134.
27. Epicurus, *The Epicurus Reader* (Indianapolis: Hackett, 1994), 39.
28. Karl Marx, *Capital*, vol. 3 (New York: Vintage, 1981), 754.
29. James Gustave Speth, *The Bridge at the Edge of the World* (New Haven: Yale University Press, 2008), 63. Speth takes a position that is explicitly "anti-capitalist" while being "non-socialist." He identifies socialism primarily with Soviet-type societies. In reality, many of his views are similar to the movement for a "socialism for the 21st century," which is aimed at the core values of social justice and environmental sustainability. See chapter 6 in this book for a discussion of Speth's views and their significance.
30. Paul Sweezy, "Capitalism and the Environment," *Monthly Review* 41/2 (1989): 1–10, quote on 6.
31. Evo Morales, "Climate Change: Save the Planet from Capitalism," (November 28, 2008), http://links.org.

32. John Bellamy Foster, *The Ecological Revolution* (New York: Monthly Review Press, 2009), 32–35. See also chapters 16 and 18 in this book.
33. Thomas L. Friedman, *Hot, Flat, and Crowded* (New York: Farrar, Straus and Giroux, 2008), 186–87.
34. Gregg Easterbrook, *A Moment on the Earth* (New York: Viking, 1995), 687–8.
35. James Hansen, "Carbon Tax & 100% Dividend vs. Tax & Trade," (2008), http://www.columbia.edu.
36. James Hansen, "Coal-Fired Power Stations Are Death Factories. Close Them," *The Observer* (February 15, 2009).
37. Anil Agarwal and Sunita Narain, *Global Warming in an Unequal World* (New Delhi: Centre for Science and Environment, 1991).
38. Tom Athanasiou and Paul Baer, *Dead Heat* (New York: Seven Stories Press, 2002).
39. Ibid., 84.
40. Lewis Mumford, *The Condition of Man* (New York: Harcourt Brace Jovanovich,1973), 419–23.
41. Herman E. Daly and John B. Cobb,Jr., *For the Common Good* (Boston: Beacon Press, 1989), 168–72.
42. Phillippe Buonarroti, *Babeuf's Conspiracy for Equality* (New York: Augustus M. Kelley, 1836), 364–74; István Mészáros, *The Challenge and Burden of Historical Time* (New York: Monthly Review Press, 2008), 258–59.
43. Aldo Leopold, *A Sand County Almanac* (New York: Oxford University Press, 1949); John Bellamy Foster, *Ecology Against Capitalism* (New York: Monthly Review Press, 2002), 31–32.

CHAPTER FIVE: CARBON METABOLISM AND GLOBAL CAPITAL ACCUMULATION

This chapter is extensively revised for this book from the following article: Brett Clark and Richard York, "Carbon Metabolism: Global Capitalism, Climate Change, and the Biospheric Rift," *Theory and Society* 34/4 (2005): 391–428.

1. George Perkins Marsh, *Man and Nature* (New York: Charles Scribner, 1864); Brett Clark and John Bellamy Foster, "George Perkins Marsh and the Transformation of Earth," *Organization & Environment* 15/2 (2002): 164–169; B. L. Turner II, William C. Clark, Robert W. Kates, John F. Richards, Jessica T. Mathews, and William B. Meyer, *The Earth as Transformed by Human Action* (New York: Cambridge University Press, 1991); Peter M.Vitousek, Harold A. Mooney, Jane Lubchenko, and Jerry A. Melilo, "Human Domination of Earth's Ecosystems," *Science* 277 (1997): 494–99; John Bellamy Foster, *Ecology Against Capitalism*; Frederick Buell,

From Apocalypse to Way of Life (New York: Routledge, 2003); Barry Commoner, *The Closing Circle* (New York: Alfred A. Knopf, 1971); Paul Ehrlich and John Holdren, "Impact of Population Growth," *Science* 171 (1971): 1212–17; Johan Rockström, Will Steffen, Kevin Noone, Åsa Persson, F. Stuart Chapin, III, Eric F. Lambin, Timothy M. Lenton, Marten Scheffer, Carl Folke, Hans Joachim Schellnhuber, Björn Nykvist, Cynthia A. de Wit, Terry Hughes, Sander van der Leeuw, Henning Rodhe, Sverker Sörlin, Peter K. Snyder, Robert Constanza, Uno Svedin, Malin Falkenmark, Louise Karlberg, Robert W. Corell, Victoria J. Fabry, James Hansen, Brian Walker, Diana Liverman, Katherine Richardson, Paul Crutzen, and Jonathan A. Foley, "A Safe Operating Space for Humanity," *Nature* 461/24 (September 2009), 472–75, and "Planetary Boundaries," *Ecology and Society* 14/2 (2009), http://ecologyand society.org.

2. Intergovernmental Panel on Climate Change, *Climate Change 2007: The Physical Science Basis* (Cambridge: Cambridge University Press, 2007).
3. John Bellamy Foster, "Marx's Theory of Metabolic Rift," *American Journal of Sociology* 105/2 (1999): 366–405.
4. John Bellamy Foster and Paul Burkett, "The Dialectic of Organic/Inorganic Relations: Marx and the Hegelian Philosophy of Nature," *Organization & Environment* 13/4 (2000): 403–25.
5. John Bellamy Foster, *Marx's Ecology: Materialism and Nature* (New York: Monthly Review Press, 2000), 159.
6. Ibid., 159.
7. Justus von Liebig, *Letters on Modern Agriculture* (London: Walton and Maberly, 1859).
8. Foster, *Marx's Ecology*, 158.
9. Karl Marx, *Capital*, vol. 1 (New York: Vintage, 1976), 637–38.
10. Karl Marx, *Economic and Philosophic Manuscripts of 1844* (New York: International Publishers, 1964), 112.
11. Karl Marx, *Capital*, vol. 3 (New York: Penguin Books, 1991), 949–50.
12. Marx, *Capital*, vol. 1, 283–90.
13. Foster, *Marx's Ecology*; Foster, "Marx's Theory of Metabolic Rift."
14. Karl Marx, *The Poverty of Philosophy* (New York: International Publishers, 1971), 162–63, 223. See also Daniel Hillel, *Out of the Earth* (Berkeley: University of California Press, 1992).
15. For a useful discussion of the accumulation of waste within cities at this point in time, see Edwin Chadwick, *Report on the Sanitary Conditions of the Labouring Population of Great Britain* (Edinburgh: University Press, 1965 [1842]); Nicholas Goddard, "19th-Century Recycling: The Victorians and the Agricultural Utilisation of Sewage," *History Today* 31/6 (1981): 32–36; Patrick Joyce, *The Rule of Freedom: Liberalism and the Modern City* (London: Verso, 2003).

16. For a discussion of the metabolic rift and the guano trade in the nineteenth century, see chapter 15 in this book.
17. Marx, *Capital*, vol. 1, 637–38.
18. John Bellamy Foster and Fred Magdoff, "Liebig, Marx, and the Depletion of Soil Fertility: Relevance for Today's Agriculture," in *Hungry For Profit: The Agribusiness Threat to Farmers, Food, and the Environment*, ed. Fred Magdoff, John Bellamy Foster, and F. H. Buttel (New York: Monthly Review Press, 2000); Fred Magdoff, Les Lanyon, and Bill Liebhardt, "Nutrient Cycling, Transformation and Flows," *Advances in Agronomy* 60 (1997): 1–73.
19. Marx, *Capital*, vol. 1, 637–38; Marx, *Capital*, vol. 3, 959.
20. Jason W. Moore, "The Crisis of Feudalism: An Environmental History," *Organization & Environment* 15/3 (2002): 301–22; Jason W. Moore, "The Modern World-System as Environmental History," *Theory and Society* 32/3 (2003): 307–77; Richard York, Eugene A. Rosa, and Thomas Dietz, "A Rift in Modernity?: Assessing the Anthropogenic Sources of Global Climate Change with the STIRPAT Model," *International Journal of Sociology and Social Policy* 23/10: 31–51; Paul Burkett, *Marx and Nature* (New York: St. Martin's Press, 1999); Foster and Magdoff, "Liebig, Marx, and the Depletion of Soil Fertility," 23–41; Philip Mancus, "Nitrogen Fertilizer Dependency and Its Contradictions," *Rural Sociology* 72/2 (2007): 269–88.
21. Foster, "Marx's Theory of Metabolic Rift," 375–83.
22. Eugene A. Rosa, "Global Climate Change: Background and Sociological Contributions," *Society & Natural Resources* 14 (2001): 491–99, quote on 494.
23. Foster, *The Vulnerable Planet* (New York: Monthly Review Press, 1994), 90–96.
24. Here the first law of thermodynamics still holds, but in an entropic kind of way. See Nicholas Georgescu-Roegen, *The Entropy Law and the Economic Process* (Cambridge: Harvard University Press, 1971); Georgescu-Roegen, *Energy and Economic Myths* (New York: Pergamon, 1976); Herman E. Daly, "The Economics of the Steady State," *American Economic Review* 64/2 (1974): 15–21; Daly, *Steady-State Economics* (San Francisco: W. H. Freeman and Company, 1977).
25. Vladimir I. Vernadsky, *The Biosphere* (New York: Copernicus, 1998); J. D. Bernal, *The Origins of Life* (New York: World Publishing Company, 1967); A. I. Oparin, *Origin of Life* (New York: Dover, 1965); A. I. Oparin, *Life: Its Nature, Origin, and Development* (New York: Academic Press, 1966).
26. Richard Levins and Richard Lewontin, *The Dialectical Biologist* (Cambridge: Harvard University Press, 1985), 46–49.
27. Vernadsky, *The Biosphere.*

28. Fritjof Capra, *The Web of Life* (New York: Anchor Books, 1996), 236–42; Lynn Margulis and Dorion Sagan, *Microcosmos* (New York: Summit, 1986).
29. Margulis and Sagan, *Microcosmos*.
30. Barbara Freese, *Coal: A Human History* (Cambridge: Perseus Publishing, 2003).
31. Eugene A. Rosa and Thomas Dietz, "Climate Change and Society," *International Sociology* 13/4 (1998): 421–55.
32. James Rodger Fleming, *Historical Perspectives on Climate Change* (New York: Oxford University Press, 1998), 58.
33. Ibid., 60–61.
34. Rosa and Dietz, "Climate Change and Society"; Andrew Simms, *Ecological Debt* (London: Pluto Press, 2005), 16; Spencer R. Weart, "From the Nuclear Frying Pan into the Global Fire," *Bulletin of the Atomic Scientists* 48 (1992): 19–27; Spencer R. Weart, *The Discovery of Global Warming* (Cambridge: Harvard University Press, 2003).
35. Weart, *The Discovery of Global Warming*, 2–3.
36. Ibid., 3.
37. Fleming, *Historical Perspectives on Climate Change*, 67.
38. Weart, *The Discovery of Global Warming*, 3.
39. Gale E. Christianson, *Greenhouse: The 200-Year Story of Global Warming* (New York: Walker and Company, 1999).
40. John Tyndall, "On Radiation through the Earth's Atmosphere," *Philosophical Magazine* 4 (1863): 200–207, esp. 204–5.
41. Pradip Baksi, "Karl Marx's Study of Science and Technology," *Nature, Society, and Thought* 9/3 (1996): 261–96; Pradip Baksi, "MEGA IV//31: Natural Science Notes of Marx and Engels, 1877–1883," *Nature, Society, and Thought* 17/1 (2001): 377–90; Paul Burkett and John Bellamy Foster, "Metabolism, Energy, and Entropy in Marx's Critique of Political Economy: Beyond the Podolinsky Myth," *Theory and Society* 35 (2006): 109–56.
42. August Arrenhius, "On the Influence of Carbonic Acid in the Air upon the Temperature of the Ground," *Philosophical Magazine* 41 (1896): 237–76; Christianson, *Greenhouse*, 108–15; Fleming, *Historical Perspectives on Climate Change*; John Imbrie and Katherine Palmer Imbrie, *Ice Ages: Solving the Mystery* (Short Hills, N.J.: Enslow Publishers, 1979); Simms, *Ecological Debt*, 14–18; Weart, *The Discovery of Global Warming*, 5–7.
43. Fleming, *Historical Perspectives on Climate Change*.
44. R. Revelle, "Introduction: The Scientific History of Carbon Dioxide," in *The Carbon Cycle and Atmospheric* CO_2 *: Natural Variations Archean to Present*, ed. E.T. Sundquist and W. S. Broecker (Washington D.C.: American Geophysical Union, 1985), 1–4.

45. In 1840, Liebig firmly established the position that the carbon assimilated by plants was carbon dioxide from the atmosphere. See Revelle, "Introduction."
46. P. Falkowski, R. J. Scholes, E. Boyle, J. Canadell, D. Canfield, J. Elser, N. Gruber, K. Hibbard, P. Högberg, S. Linder, F. T. Mackenzie, B. Moore III, T. Pedersen, Y. Rosenthal, S. Seitzinger, V. Smetacek, and W. Steffen, "The Global Carbon Cycle," *Science* 290 (2000): 291–96. It should be noted that there have been natural climate variations in global temperatures through the centuries, which contributed to a "Little Ice Age," but there is widespread agreement among scientists that the accumulation of carbon in the atmosphere driving global climate change is the result of human activities. For a discussion of the "Little Ice Age," see Richard H. Grove, *Ecology, Climate and Empire* (Knapwell, Cambridge, UK: White Horse Press, 1997); Brian Fagan, *The Little Ice Age: How Climate Made History, 1300–1850* (New York: Basic Books, 2000); Jean Grove, *The Little Ice Age* (London: Methuen, 1988); John F. Richards, *The Unending Frontier* (Berkeley: University of California Press, 2003). Scientists make great efforts to distinguish between natural variations and human induced variations within systems.
47. Thomas R. Karl and Kevin Trenberth, "Modern Global Climate Change," *Science* 302 (2003): 1719–1723; also see Peter M. Vitousek, "Beyond Global Warming: Ecology and Global Change," *Ecology* 75 (1994): 1861–1876.
48. Falkowski et al., "The Global Carbon Cycle."
49. Rosa and Dietz, "Climate Change and Society"; Elizabeth Crawford, *From Ionic Theory to the Greenhouse Effect* (Nantucket, Mass.: Science History Publications, 1996).
50. Furthermore, the concentration of other greenhouse gases in the atmosphere, since preindustrial times, has increased: methane by 145 percent and nitrous oxide by 15 percent. See Intergovernmental Panel on Climate Change, *IPCC Second Assessment: Climate Change 1995*, 21; also see Karl and Trenberth, "Modern Global Climate Change."
51. Foster, *The Vulnerable Planet*; Jared Diamond, *Guns, Germs, and Steel* (New York: W. W. Norton, 1999); Clive Ponting, *The Green History of the World* (New York: St. Martin's Press, 1991); Frederick Buell, *From Apocalypse to Way of Life* (New York: Routledge, 2003); Mike Davis, *Late Victorian Holocausts* (London: Verso, 2001); Franz J. Broswimmer, *Ecocide* (London: Pluto Press, 2002).
52. Moore, "*The Modern World-System* as Environmental History?" For a detailed and useful study of the development of global capitalism, see Immanuel Wallerstein, *The Modern World-System I: Capitalist Agriculture and the Origins of the European World-Economy in the Sixteenth Century* (New York: Academic Press, 1974).

53. Foster, *Marx's Ecology*.
54. See chapter 15 in this book.
55. Capitalism is a dynamic system, but the drive to accumulate capital dictates throughout its history. In this discussion we draw upon Marx to situate the logic of capital and understand how this fueled technological development. As capitalism developed and as capital was concentrated and centralized, the characteristics of monopoly capital came to dominate in the twentieth century, but still the drive to accumulate capital was central.
56. In addition to its place in the writings of Marx, Engels, and Liebig, the division between town and country remained a dominant concern in the work of social critics and activists. See Ebenezer Howard, *Garden Cities of To-morrow* (London: Swan Sonnenschein & Co., 1902); Peter Kropotkin, *Fields, Factories and Workshops: Or Industry Combined with Agriculture and Brain Work with Manual Work* (New York: G. P. Putnam's Sons, 1913); William Morris, *News from Nowhere and Selected Writings and Designs* (Harmondsworth, Middlesex, UK: Penguin, 1962); William Morris, *William Morris on Art and Socialism* (Mineola, N.Y.: Dover Publications, 1999); Raymond Williams, *The Country and the City* (New York: Oxford University Press, 1975).
57. Boris Hessen, "The Social and Economic Roots of Newton's 'Principia'," in *Science at the Cross Roads*, ed. N. I. Bukharin (London: Frank Cass and Company, 1971); Robert K. Merton, *Science, Technology and Society in Seventeenth Century England* (New York: Harper & Row Publishers, 1970); Eric Hobsbawm, *The Age of Capital, 1848–1875* (New York: Vintage Books, 1996); Daniel R. Headrick, *The Tentacles of Progress* (New York: Oxford University Press, 1988), Headrick; *The Tools of Empire* (New York: Oxford University Press, 1981).
58. Marx, *Capital*, vol. 1, 494–95.
59. Ibid., 497.
60. Ibid., 494–501.
61. Karl Marx, "Wages," in *Collected Works* (New York: International Publishers, 1976), vol. 6, 431; also see Burkett, *Marx and Nature*, 108–12.
62. Karl Marx, *Capital*, volume 2 (New York: Penguin, 1992), 218–19.
63. Marx, "Wages," 431.
64. Foster, *Ecology Against Capitalism*, 19.
65. Daly, *Steady-State Economics*, 23.
66. Ibid.
67. Paul Burkett, "Natural Capital, Ecological Economics, and Marxism," *International Papers in Political Economy* 10 (2003): 1–61.
68. Foster, *Ecology Against Capitalism*, 36; also see Burkett, *Marx and Nature*; Elmar Altvater, *The Future of the Market* (London: Verso, 1993); Allan Schnaiberg and Kenneth A. Gould, *Environment and Society* (New York:

St. Martin's Press, 1994); Paul Sweezy, *The Theory of Capitalist Development* (New York: Monthly Review Press, 1970); Paul A. Baran, *The Political Economy of Growth* (New York: Monthly Review Press, 1968); Paul A. Baran and Paul Sweezy, *Monopoly Capital* (New York: Monthly Review Press, 1966) for discussions of capitalism as a system of constant expansion.

69. Paul Sweezy, "Capitalism and the Environment," *Monthly Review* 56/5 (2004): 86–93.
70. Burkett, *Marx and Nature*; Paul Burkett, "Nature's 'Free Gifts' and Ecological Significance of Value," *Capital and Class* 68 (1999): 89–110.
71. Burkett, "Natural Capital, Ecological Economics, and Marxism"; Moore, "The Crisis of Feudalism."
72. Marx, *Capital*, volume 3, 955.
73. Marx, *Capital*, vol. 1. Marx's adherence to the labor theory of value was not a shortcoming but key to understanding the operations of capitalism both socially and ecologically. Marx illuminates how capital mystifies nature's contribution to wealth. Burkett argues that simply ascribing a social value to nature will not solve this contradiction, because "any attempt to directly attribute value to nature, without taking account of the historical specificity of wealth's social forms, results in an inability to specify the precise *value-form* taken by nature (value in terms of what, and for whom?) without running into serious theoretical difficulties." See Burkett, "Nature's 'Free Gifts' and Ecological Significance of Value"; Foster, *Ecology Against Capitalism.*
74. Karl Marx, *A Contribution to the Critique of Political Economy* (New York: International Publishers, 1972), 47; Marx, *Grundrisse*, 145.
75. Burkett, *Marx and Nature*, 84.
76. Marx noted that capital only made efforts to reduce waste (or refuse) when it was profitable. See Marx, *Capital*, vol. 3, 195–98. Yet these efforts do not resolve the ecological crisis, because they only offer new ways to pursue the accumulation of wealth (in the short run) while increasing the demands placed upon the environment for further production, generating more waste. Thus the pursuit of profit threatens the ability of nature to reproduce itself and to absorb waste and energy. See Foster, *Ecology Against Capitalism*; Barry Commoner, *The Closing Circle* (New York: Alfred A. Knopf, 1971); Doug Dowd, *The Waste of Nations* (Boulder, Colo.: Westview Press, 1989).
77. Foster, *The Vulnerable Planet.*
78. Burkett, "Natural Capital, Ecological Economics, and Marxism," 47; István Mészáros, *Socialism or Barbarism* (New York: Monthly Review Press, 2001).
79. Foster, *Ecology Against Capitalism*, 66.

80. Rosa and Dietz, "Climate Change and Society," 423.
81. Michael Perelman, "Myths of the Market: Economics and the Environment," *Organization & Environment* 16/2 (2003): 168–226; Harry Magdoff, *The Age of Imperialism* (New York: Monthly Review Press, 1969), *Imperialism* (New York: Monthly Review Press, 1978).
82. Andrew Simms, Aubrey Meyer, and Nick Robins, *Who Owes Who? Climate Change, Debt, Equity and Survival* (1999), www.jubilee2000uk.org.
83. For a review of this literature, see Soumyananda Dinda, "Environmental Kuznets Curve Hypothesis: A Survey," *Ecological Economics* 49 (2004): 431–55; Theresa A. Cavlovic, Kenneth H. Baker, Robert P. Berrens, and Kishore Gawande, "A Meta-Analysis of Environmental Kuznets Curve Studies," *Agricultural and Resource Economics Review* 29 (2000): 32–42; and David I. Stern, "Progress on the Environmental Kuznets Curve?" *Environment and Development Economics* 3 (1998): 173–96.
84. Paul Hawken, "Foreword," in *Natural Capital and Human Economic Survival*, Thomas Prugh, editor (Solomons, Md.: International Society for Ecological Economics, 1995); Paul Hawken, Amory Lovins, and L. Hunter Lovins, *Natural Capitalism: Creating the Next Industrial Revolution* (Boston: Little, Brown, 1999).
85. Paul Hawken, "Natural Capitalism," *Mother Jones Magazine* (April 1997): 40–53, 59–62.
86. Arthur P. J. Mol, *The Refinement of Production* (Utrecht, Netherlands: Van Arkel, 1995); Gert Spaargaren, "The Ecological Modernization of Production and Consumption" (Ph.D. diss., Wageningen University, Netherlands, 1997).
87. Arthur P. J. Mol, "Ecological Modernisation and Institutional Reflexivity," *Environmental Politics* 5 (1996): 302–23, *Globalization and Environmental Reform* (Cambridge: MIT Press, 2001).
88. This account depends upon Sebastian Budgen's review of Luc Boltanski and Ève Chiapello's book, *Le Nouvel Esprit de Capitalisme* (Gallimard: Paris, 1999). It should also be noted that, in concluding the review, Budgen states that Boltanski and Chiapello's belief that enlightened capitalists will pursue the development of a society premised on social needs and the common good is "the point at which it [their argument] deserts any sense of realism." See Sebastian Budgen, "A New 'Spirit of Capitalism,'" *New Left Review* (January–February 2000): 149–56.
89. Anqing Shi, "The Impact of Population Pressure on Global Carbon Dioxide Emissions, 1975–1996: Evidence from Pooled Cross-country Data," *Ecological Economics* 44 (2003): 29–42; York, Rosa, and Dietz, "A Rift in Modernity?"; Richard York, Eugene A. Rosa, and Thomas Dietz, "STIRPAT, IPAT, and ImPACT: Analytic Tools for Unpacking the Driving Forces of Environmental Impacts," *Ecological Economics* 46 (2003):

351-65.

90. Richard York, Eugene A. Rosa, and Thomas Dietz, "The Ecological Footprint Intensity of National Economies," *Journal of Industrial Ecology* 8 (2004): 139-54.
91. William Stanley Jevons, *The Coal Question: An Enquiry Concerning the Progress of the Nation, and the Probable Exhaustion of Our Coal-mines* (London: Macmillan, 1865).
92. Mario Giampietro and Kozo Mayumi, "Another View of Development, Ecological Degradation, and North-South Trade," *Review of Social Economy* 56 (1998): 20-36. See chapter 7 below.
93. Dowd, *The Waste of Nations*; also see Baran, *The Political Economy of Growth*; and Baran and Sweezy, *Monopoly Capital*.
94. Richard York and Eugene Rosa, "Key Challenges to Ecological Modernization Theory," *Organization & Environment* 16 (2003): 273-88; York, Rosa, and Dietz, "A Rift in Modernity?"; York, Rosa, and Dietz, "STIRPAT, IPAT, and ImPACT"; Richard York, Eugene A. Rosa, and Thomas Dietz, "Footprints on the Earth," *American Sociological Review* 68 (2003): 279-300; York, Rosa, and Dietz, "The Ecological Footprint Intensity of National Economies"; Foster, *Ecology Against Capitalism*; Schnaiberg and Gould, *Environment and Society*; Stephen Bunker, "Raw Material and the Global Economy," *Society and Natural Resources* 9 (1996): 419-29.
95. Data used to make these calculations are from Emily Matthews, Christof Amann, Stefan Bringezu, Marina Fischer-Kowalski, Walter Hüttler, René Kleijn, Yuichi Moriguchi, Christian Ottke, Eric Rodenburg, Don Rogich, Heinz Schandl, Helmut Schütz, Ester van der Voet, and Helga Weis, *The Weight of Nations* (Washington D.C: World Resources Institute, 2000). GDP is measured in constant units of each nation's own currency. Carbon dioxide emissions are those from combustion of fossil fuels, including bunker fuels. We present data for this selection of nations because Matthews et al. provide data for only these four nations and Germany. The data for Germany are problematic for this presentation since, for years prior to reunification (1990-1991), data are only presented for the former West Germany.
96. Anqing Shi, "The Impact of Population Pressure on Global Carbon Dioxide Emissions, 1975-1996"; York, Rosa, and Dietz, "A Rift in Modernity?"; York, Rosa, and Dietz, "STIRPAT, IPAT, and ImPACT."
97. Marx, *Capital*, vol. 1; also see Edward Crenshaw and J. Craig Jenkins, "Social Structure and Global Climate Change," *Sociological Focus* 29 (1996): 341-58.
98. The calculations are based on data presented by Matthews et al., *The Weight of Nations*.

99. Foster, *Ecology Against Capitalism*; Loren Lutzenhiser, "The Contours of U.S. Climate Non-Policy," *Society and Natural Resources* 14 (2001): 511–23.
100. Crenshaw and Jenkins, "Social Structure and Global Climate Change."
101. Burkett, "Natural Capital, Ecological Economics, and Marxism," 27.
102. J. Timmons Roberts, "Global Inequality and Climate Change," *Society & Natural Resources* 14 (2001): 501–9; Karl and Trenberth, "Modern Global Climate Change."
103. See Nicholas Georgescu-Roegen, *The Entropy Law and the Economic Process* (Cambridge: Harvard University Press, 1971); Nicholas Georgescu-Roegen, *Energy and Economic Myths* (New York: Pergamon, 1976); Nicholas Georgescu-Roegen, "Energy Analysis and Economic Valuation," *Southern Economic Journal* 45 (1979): 1023–58.
104. Rachel Carson, *The Sea Around Us* (New York: Oxford University Press, 1991), 170.
105. Falkowski et al., "The Global Carbon Cycle."
106. For analyses and discussion of the carbon sequestration processes in the oceans, see Taro Takahashi, "The Fate of Industrial Carbon Dioxide," *Science* 305 (2004): 352–53; Richard A. Freely et al., "Impact of Anthropogenic CO_2 on the C_aCO_3 System in the Oceans," *Science* 305 (2004): 362–66; Christopher L. Sabine et al., "The Oceanic Sink for Anthropogenic CO_2," *Science* 305(2004): 367–71.
107. Stephen H. Schneider and Randi Londer, *The Coevolution of Climate and Life* (San Francisco: Sierra Club Books, 1984), 309.
108. World Bank, *World Development Indicators 2002* (Washington, D.C.: World Bank, 2002), 148; also see R. A. Houghton and J. L. Hackler, "Emissions of Carbon from Forestry and Land-use Change in Tropical Asia," *Global Change Biology* 5 (1995): 481–92; R. A. Houghton, "Land-use Change and the Carbon Cycle," *Global Change Biology* 1 (1995): 275–87.
109. Thomas J. Burns, Edward Kick, David A. Murray, and Dixie Murray, "Demography, Development and Deforestation in a World-System Perspective," *International Journal of Comparative Sociology* 35/3-4 (1994): 221–39; Edward Kick, Thomas Burns, Byron Davis, David Murray, and Dixie Murray, "Impacts of Domestic Population Dynamics and Foreign Wood Trade on Deforestation," *Journal of Developing Societies* 12 (1996): 68–87; Thomas K. Rudel, "Population, Development, and Tropical Deforestation: A Cross-National Study," *Rural Sociology* 54 (1989): 327–38; Thomas K. Rudel and Jill Roper, "The Paths to Rain Forest Destruction," *World Development* 25 (1997): 53–65.
110. Falkowski et al., "The Global Carbon Cycle," 293–94.
111. Roberts, "Global Inequality and Climate Change," 502; see also Bruce Podobnik, "Energy Demands and Consumption," in *Cities in a*

Globalizing World: Global Report on Human Settlements 2001, ed. United Nations (Sterling, Va.: Earthscan, 2001); Bruce Podobnik, "Global Energy Inequalities," *Journal of World-Systems Research* 8 (2002): 252–74. For a useful discussion of the historical aspects of uneven development and ecological crisis, see James O'Connor, *Natural Causes* (New York: Guilford Press, 1998), 187–99.

112. U.S. Department of Energy, Energy Information Administration, International Energy Annual, "World Carbon Dioxide Emissions from the Consumption and Flaring of Fossil Fuels" (Washington, D.C.: U.S. Department of Energy, 2002). Note that this corresponds with approximately 20 metric tons per capita per year of carbon dioxide.
113. Foster, *Ecology Against Capitalism*, 18.
114. Peter Grimes and Jeffrey Kentor, "Exporting the Greenhouse: Foreign Capital Penetration and CO_2 Emissions 1980–1996," *Journal of World-Systems Research* 9 (2003): 261–75.
115. Marx, *Capital*, vol. 1, 475.
116. Richard B. Alley, *The Two-mile Time Machine* (Princeton: Princeton University Press, 2000); Al Gore, *An Inconvenient Truth* (Emmaus, Penn.: Rodale, 2006).
117. Vernadsky, *The Biosphere*.
118. In 1987, Paul Sweezy stated that the expanding scale of capitalist production created ever-greater threats to the natural world, especially to nature's resiliency. See Sweezy, "Capitalism and the Environment"; Commoner, *The Closing Circle*.
119. Marten Scheffer, Steve Carpenter, Jonathan A. Foley, Carl Folke, and BrianWalker, "Catastrophic Shifts in Ecosystems," *Nature* 413 (2001): 591–96; Roldan Muradian, "Ecological Thresholds: A Survey," *Ecological Economics* 38 (2001): 7–24.
120. James Hansen, "The Threat to the Planet," *New York Review of Books* 53 (2006): 12–16.
121. Gore, *An Inconvenient Truth*, 176–95; Orin H. Pilkey and Rob Young, *The Rising Sea* (Washington, D.C.: Island Press, 2009), 70–78.
122. Hansen, "The Threat to the Planet"; Simms, *Ecological Debt*.
123. Vitousek, "Beyond Global Warming: Ecology and Global Change," 1864.
124. Hansen, "The Threat to the Planet," 12.
125. Krajick, "All Downhill from Here?"
126. Broswimmer, *Ecocide*; Niles Eldredge, *The Miner's Canary: Unraveling the Mysteries of Extinction* (Princeton: Princeton University Press, 1991); Niles Eldredge, *Dominion* (New York: Henry Holt, 1995); Tim Flannery, *The Weather Makers* (New York: Atlantic Monthly Press, 2005); Hansen, "The Threat to the Planet"; Richard Leakey and Roger Lewin, *The Sixth Extinction* (New York: Anchor Books, 1996).

127. Carl Folke, Steve Carpenter, Brian Walker, Marten Scheffer, Thomas Elmqvist, Lance Gunderson, C. S. Holling, "Regime Shifts, Resilience, and Biodiversity in Ecosystem Management," *Annual Review of Ecology, Evolution, & Systematics* 35 (2004): 557–81, see esp. 557.
128. Vitousek, "Beyond Global Warming," 1862.
129. Richard Harris, "Dust Storms Threaten Snow Packs," National Public Radio, *Morning Edition* (May 30, 2006).
130. Gore, *An Inconvenient Truth*, 168; see also Tim P. Barnett, David W. Pierce, and Reiner Schnur, "Detection of Anthropogenic Climate Change in the World's Oceans," *Science* 292 (2001): 270–74.
131. Folke et al., "Regime Shifts, Resilience, and Biodiversity in Ecosystem Management," 568.
132. Buell, *From Apocalypse to Way of Life.*

CHAPTER SIX: THE PLANETARY MOMENT OF TRUTH

This chapter is an extensively rewritten and updated version of an argument introduced in John Bellamy Foster, Brett Clark, and Richard York, "Ecology: The Moment of Truth," *Monthly Review* 60/3 (July–August 2008): 1–11.

1. James Hansen, "Tipping Point," in *State of the Wild 2008-2009*, ed. E. Fearn (Washington, D.C.: Island Press, 2008), 7–15. See also James Hansen, "The Threat to the Planet," *New York Review of Books*, July 13, 2006.
2. Percentages of bird, mammal, and fish species "vulnerable or in immediate danger of extinction" are "now measured in double digits." Lester R. Brown, *Plan B 3.0* (New York: W. W. Norton, 2008), 102. The share of threatened species in 2007 was 12 percent of the world's bird species, 20 percent of the world's mammal species, and 39 percent of the world's fish species evaluated. See International Union for the Conservation of Nature (IUCN), *IUCN Red List of Threatened Species*, Table 1, "Numbers of Threatened Species by Major Groups of Organisms," http://www.iucnredlist.org/info/stats. Additionally, climate change is having significant effects on plant diversity. "Recent studies predict that climate change could result in the extinction of up to half the world's plant species by the end of the century." See Belinda Hawkins, Suzanne Sharrock, and Kay Havens, *Plants and Climate Change* (Richmond, UK: Botanic Gardens Conservation International, 2008), 9.
3. James Hansen, *Storms of Our Grandchildren* (New York: Bloomsbury, 2010), 164–68; David Spratt and Philip Sutton, *Climate Code Red* (Fitzroy, Australia: Friends of the Earth, 2008), 4, http://www.climatecodered.net; Brown, *Plan B 3.0*, 3; James Hansen, Makiko Sato, Pushker Kharecha, Gary Russell, David W Lea, and Mark Siddall, "Climate Change and Trace Gases," *Philosophical*

Transactions of the Royal Society 365 (2007): 1925–54; James Lovelock, *The Revenge of Gaia* (New York: Basic Books, 2006), 34; Minqi Li, "Climate Change, Limits to Growth, and the Imperative for Socialism," *Monthly Review* 60/3 (July–August 2008): 51–67; Orin H. Pilkey and Rob Young, *The Rising Sea* (Washington, D.C.: Island Press, 2009), 40.

4. Hansen, "Tipping Point," 7–8.
5. Brown, *Plan B 3.0*, 4–5. Brown correctly depicts the seriousness of the ecological problem, but as a mainstream environmentalist he insists that all can easily be made well without materially altering society by a clever combination of technological fixes and the magic of the market. See Li, "Climate Change, the Limits to Growth, and the Imperative for Socialism."
6. Nicholas Stern, *The Economics of Climate Change: The Stern Review* (Cambridge: Cambridge University Press, 2007).
7. The *Stern Review* has been criticized by more conservative mainstream economists, including William Nordhaus, for its ethical choices, which, it is claimed, place too much emphasis on the future as opposed to present-day values by adopting a much lower discount rate on future costs and benefits as compared to other, more standard economic treatments such as that of Nordhaus. This then gives greater urgency to today's environmental problem. Nordhaus discounts the future at 6 percent a year; Stern by less than a quarter of that at 1.4 percent. This means that for Stern having a trillion dollars a century from now is worth $247 billion today, while for Nordhaus it is only worth $2.5 billion. Nordhaus calls the *Stern Review* a "radical revision of the economics of climate change" and criticizes it for imposing "excessively large emissions reductions in the short run." John Browne, "The Ethics of Climate Change," *Scientific American* 298/6 (June 2008): 97–100; William Nordhaus, *A Question of Balance* (New Haven: Yale University Press, 2008), 18, 190.
8. James Hansen, Makiko Sato, Pushker Kharecha, David Beerling, Valerie Masson-Delmotte, Mark Pagani, Maureen Raymo, Dana L. Royer, and James C. Zachos, "Target Atmospheric CO_2: Where Should Humanity Aim?" *Open Atmospheric Science Journal* 2 (2008): 217–31; Nicholas Stern, *The Global Deal* (New York: Perseus, 2009), 150. Even before this, Hansen and his colleagues at NASA's Goddard Institute argued that due to positive feedbacks and climatic tipping points global average temperature increases had to be kept to less than 1°C below 2000 levels. This meant that atmospheric carbon dioxide needed to be kept to 450 ppm or below. See Pushker A. Kharecha and James E. Hansen, "Implications of 'Peak Oil' for Atmospheric CO_2 and Climate," *Global Biogeochemistry* 22 (2008): http://pubs.giss.nasa.gov/abstracts/inpress/Kharecha_Hansen.html.
9. Stern, *The Economics of Climate Change*, 4–5, 11–16, 95, 193, 220–34, 637, 649–51; "Evidence of Human-Caused Global Warming Is Now

'Unequivocal,'" *Science Daily*, http://www.sciencedaily.com; Browne, "The Ethics of Climate Change," 100; Spratt and Sutton, *Climate Code Red*, 30; Editors, "Climate Fatigue," *Scientific American* 298/6 (June 2008): 39; Ted Trainer, "A Short Critique of the *Stern Review*," *Real-World Economics Review*, 45 (2008): 51–67, http://www.paecon.net. Despite the *Stern Review*'s presentation of France's nuclear switch as a greenhouse gas success story there are strong environmental reasons for not proceeding along this path. See Robert Furber, James C. Warf, and Sheldon C. Plotkin, "The Future of Nuclear Power," *Monthly Review* 59/9 (February 2008): 38–48; See Working Group III, "Summary for Policy Makers," *Fourth Assessment Report of the International Group on Climate Change* (2007), 15. For a discussion of the reduction in carbon dioxide emissions in post-Soviet Union republics, see Richard York, "De-Carbonization in Former Soviet Republics, 1992–2000: The Ecological Consequences of De-Modernization," *Social Problems* 55/3 (2008): 370–90.

10. Stern, *The Global Deal*, 145–52.
11. Paul M. Sweezy, "Capitalism and the Environment," *Monthly Review* 41/2 (June 1989): 1–10.
12. Michael Shellenberger and Ted Nordhaus, "The Death of Environmentalism," Environmental Grantmakers Association, October 2004, http://thebreakthrough.org.
13. James Gustave Speth, *The Bridge at the Edge of the World: Capitalism, the Environment, and Crossing from Crisis to Sustainability* (New Haven: Yale University Press, 2008).
14. James Gustave Speth, "Can the World Be Saved?" in *Environment in Peril*, ed. Anthony B. Wolbarst (Washington: Smithsonian Institution Press, 1991), 64–65.
15. James Gustave Speth, *The Red Sky at Morning: America and the Crisis of the Global Environment* (New Haven: Yale University Press, 2004), 157.
16. Ibid., 161–62.
17. James Gustave Speth and Peter M. Haas, *Global Environmental Governance* (Washington, D.C.: Island Press, 2006). Speth and Haas refer very briefly to "natural capitalism" in the sense promoted by Paul Hawken and others. But since this is capitalism lifted out of any realistic context it is merely another way of avoiding the system. See Speth and Haas, *Global Environmental Governance*, 141–43.
18. For the effect of the failure of the Johannesburg conference on Speth's thinking see Speth and Haas, *Global Environmental Governance*, 76–78. For a similar critique see John Bellamy Foster, *The Ecological Revolution* (New York: Monthly Review Press, 2009), 129–40.
19. Speth, *The Bridge at the Edge of the World*, xi–xiii, 48–63, 107, 194–98; Samuel Bowles and Richard Edwards, *Understanding Capitalism* (New

York: Oxford University Press, 1985), 119, 148–52. On the Global Scenario Group, see John Bellamy Foster, *The Ecological Revolution* (New York: Monthly Review Press, 2009).

20. Speth, *The Bridge at the Edge of the World*, 190–94; Peter Barnes, *Capitalism 3.0* (San Francisco: Berrettt-Koehler, 2006).
21. Gus Speth, "Towards a New Economy and New Politics," *Solutions*, 5 (May 28, 2010), http://thesolutionsjournal.com. A major influence on Speth's argument both in *The Bridge at the Edge of the World* and even more perhaps in his recent *Solutions* article is Clive Hamilton, in *Growth Fetish* (London: Pluto Press, 2004). Hamilton attempts to address the question of growth while avoiding the issue of ownership and control of production, and thus the question of capitalism or socialism.
22. On the nature of this crisis see John Bellamy Foster and Fred Magdoff, *The Great Financial Crisis* (New York: Monthly Review Press, 2009).
23. Paul A. Baran, *The Political Economy of Growth* (New York: Monthly Review Press, 1957), 41–43.
24. It is noteworthy that classical economists generally saw the stationary state (or no-growth economy) as the negation of capitalism. It was in this context that John Stuart Mill—in what he called his "socialist" phase (influenced by such utopian socialists as Robert Owen and Claude Henri de Rouvroy, comte de Saint-Simon) as reflected in his *Principles of Political Economy*, and particularly in the section on "Distribution"—introduced the notion that such a steady-state economy was consistent with qualitative development and environmental sustainability. See John Stuart Mill, *Principles of Political Economy* (New York: Longhams, Green, and Co., 1904), 453–55, and *The Autobiography of John Stuart Mill* (New York: Columbia University Press, 1924), 161–64; John Bellamy Foster, *The Ecological Revolution* (New York: Monthly Review Press, 2009), 29–30; Herman E. Daly, *Beyond Growth* (Boston: Beacon Press, 1996), 3–4.
25. Speth does point out that "to destroy the planet's climate and biota and leave a ruined world to future generations" all that society has to do "is to keep doing exactly what is being done today, with no growth in the human population or the world economy." Current levels of consumption and exploitation of the earth, he rightly says, are unsustainable. But he does not go beyond this in his 2010 article to question the system itself, which obviously would have to be radically transformed if society were to shift to a degrowth system. See Speth, "Towards a New Economy and a New Politics."
26. "Epicureans made statements . . . that the world must be *disillusioned*, and especially freed from fear of gods, for the world is my *friend*." Karl Marx and Frederick Engels, *Collected Works* (New York: International Publishers, 1975), vol. 5, 141–42.

CHAPTER SEVEN: THE RETURN OF THE JEVONS PARADOX

1. Sir William George Armstrong, Presidential Address, Report of the 33rd Meeting of the British Association for the Advancement of Science, Held at Newcastle-upon-Tyne (London: John Murray, 1864), li–lxiv. See also William Stanley Jevons, *The Coal Question: An Inquiry concerning the Progress of the Nation, and the Probable Exhaustion of Our Coal-Mines*, ed. A. W. Flux (London: Macmillan, 1906 [1865]), 32–36.
2. Jevons, *The Coal Question*, xxxi, 274.
3. Herschel quoted in Juan Martinez-Alier, *Ecological Economics* (Oxford: Basil Blackwell, 1987), 161–62.
4. Michael V. White, "Frightening the 'Landed Fogies' Parliamentary Politics and the *Coal Question*," *Utilitas* 3/2 (November 1991): 289–302; Leonard H. Courtney, "Jevons's Coal Question: Thirty Years After," *Journal of the Royal Statistical Society* 60/4 (December 1897): 789; John Maynard Keynes, *Essays and Sketches in Biography* (New York: Meridian Books, 1956), 132. Gladstone's approach to Jevons's work was primarily a tactical ploy, used politically to justify a debt reduction argument that was never actually implemented in the budget.
5. Courtney, "Jevons's Coal Question," 797.
6. Jevons was not alone in making such an error. John Tyndall, one of the premier physicists of the day, observed in 1865: "I see no prospect of any substitute being found for coal, as a source of motive power." Quoted in Jevons, *The Coal Question*, x1. It is worth noting that the drilling of Edwin Drake's historic oil well in northwestern Pennsylvania had occurred only six years before, in 1859, and its full significance was not yet understood.
7. Keynes, *Essays and Sketches in Biography*, 128.
8. Mario Giampietro and Kozo Mayumi, "Another View of Development, Ecological Degradation, and North–South Trade," *Review of Social Economy*, 56/1 (1998): 24–26; John M. Polimeni, Kozo Mayumi, Mario Giampetro, and Blake Alcott, eds., *The Jevons Paradox and the Myth of Resource Efficiency Improvements* (London: Earthscan, 2008).
9. Jevons, *The Coal Question*, 137–41.
10. Ibid., 141–43.
11. Ibid., 152–53.
12. As late as 1842, British fireplaces still consumed two-thirds of the country's coal, but by the time Jevons wrote his book, more than two decades later, this had diminished to about a fifth of national consumption and barely entered his argument, which focused on the industrial demand for coal as the major and indispensable source of demand. As Jevons said, "I speak not here of the *domestic consumption of coal*. This is undoubtedly capable of being cut down without other harm than curtailing our home comforts, and somewhat alter-

ing our confirmed national habits." See Jevons, *The Coal Question*, 138–39; Eric J. Hobsbawm, *Industry and Empire* (London: Penguin, 1969), 69.

13. Eric J. Hobsbawm, *The Age of Capital, 1848–1873* (New York: Vintage, 1996), 39–40.
14. Jevons, *The Coal Question*, 140–42.
15. The data was for 1869 was provided in A. W. Flux's annotated edition of Jevons's work. By 1903 the relationships had changed, with the iron and steel industries accounting for 28 million tons of coal consumption (less than in Jevons's day), while the consumption of general manufactures had grown to 53 million tons and the railways to 13 million tons. See Jevons, *The Coal Question*, 138–39.
16. Hobsbawm, *Industry and Empire*, 70–71.
17. Jevons, *The Coal Question*, 245.
18. Ibid., 156.
19. Ibid., 195, 234–41; Thomas Jevons, *The Prosperity of the Landholders Not Dependent on the Corn Laws* (London: Longmans, 1840).
20. Malthus himself denied the possibility of scarcity in minerals, arguing that raw materials in contrast to food "are in great plenty" and "a demand . . . will not fail to create them in as great a quantity as they are wanted." See Thomas Robert Malthus, *An Essay on the Principle of Population and a Summary View of the Principle of Population* (London: Penguin, 1970), 100.
21. Keynes, *Essays and Sketches in Biography*, 128–29.
22. Jevons, *The Coal Question*, 195–96. Jevons's discussion of industrial development in terms of various staple products anticipated the work of Harold Innis and the staple theory of economic growth. See Mel Watkins, *Staples and Beyond* (Montreal: McGill-Queens University Press, 2006).
23. Jevons, *The Coal Question*, 142.
24. Charles Dickens, *The Old Curiosity Shop* (New York: E. P. Dutton and Co., 1908), 327.
25. Frederick Engels, *The Condition of the Working Class in England* (Chicago: Academy Publishers, 1984). See also John Bellamy Foster, *The Vulnerable Planet* (New York: Monthly Review Press, 1994), 50–59; Brett Clark and John Bellamy Foster, "The Environmental Conditions of the Working Class: An Introduction to Selections from Frederick Engels's *The Condition of the Working Class in England in 1844*," *Organization & Environment* 19/3 (2006): 375–388.
26. Jevons, *The Coal Question*, 164–71.
27. Ibid., 459–60.
28. Keynes, *Essays and Sketches in Biography*, 132.
29. Karl Marx and Frederick Engels, *Collected Works* (New York: International Publishers, 1975), vol. 46, 411.
30. Blake Alcott, "Historical Overview of the Jevons Paradox in the

Literature," in *The Jevons Paradox*, 8, 63. For the Club of Rome study, see Donella H. Meadows, Dennis L. Randers, Jørgen Randers, William W. Behrens III, *The Limits to Growth* (New York: Universe Books, 1972).

31. Juliet B. Schor, *Plenitude* (New York: Penguin Press, 2010), 88–90. For a detailed discussion of the empirical data on the Jevons Paradox, see John M. Polimeni, "Empirical Evidence for the Jevons Paradox," in *The Jevons Paradox*, 141–71.
32. Mario Giampietro and Kozo Mayumi, "The Jevons Paradox," in *The Jevons Paradox*, 80–81.
33. For a discussion of epoch-making innovations, see Paul A. Baran and Paul M. Sweezy, *Monopoly Capital* (New York: Monthly Review Press, 1966), 219–22.
34. Alfred J. Lotka, "Contributions to the Energetics of Evolution," *Proceedings of National Academy of Sciences* 8 (1922): 147–51; Giampietro and Mayumi, "The Jevons Paradox," 111–15.
35. Karl Marx, *Capital*, vol. 1 (New York: Vintage, 1976), 742.
36. John Bellamy Foster, *Ecology Against Capitalism* (New York: Monthly Review Press, 2002), 22–24.
37. Stephen G. Bunker, "Raw Materials and the Global Economy," *Society and Natural Resources* 9/4 (July–August 1996): 421.
38. Robert L. Heilbroner, *The Nature and Logic of Capitalism* (New York: W. W. Norton, 1985).
39. Emily Matthews, Christof Amann, Stefan Bringezu, Marina Fischer-Kowalski, Walter Hüttler, René Kleijn, Yuichi Moriguchi, Christian Ottke, Eric Rodenburg, Don Rogich, Heinz Schandl, Helmut Schütz, Ester van der Voet, and Helga Weis, *The Weight of Nations* (Washington, D.C: World Resources Institute, 2000), 35.

CHAPTER EIGHT: THE PAPERLESS OFFICE

This chapter is adapted and revised for this book from the following article: Richard York, "Ecological Paradoxes: William Stanley Jevons and the Paperless Office," *Human Ecology Review* 13/2 (2006): 143–47.

1. Here we are referring to absolute dematerialization. Some scholars focus on relative dematerialization—the reduction in resource consumption per unit of production. It is possible to have relative dematerialization while total material consumption increases because the scale of production increases faster than efficiency of material use improves.
2. Typically this entails substituting one type of natural resource for another, such as plastic for steel or wood; or in the specific example used here, the resources embedded in computer hardware and electronic storage mediums for those used in hard-copy storage mediums such as paper.

3. William Stanley Jevons, *The Coal Question: An Inquiry Concerning the Progress of the Nation, and the Probable Exhaustion of Our Coal-Mines* (London: Macmillan, 1906 [1865]).
4. John Bellamy Foster, *Ecology Against Capitalism* (New York: Monthly Review Press, 2002), 92–103; Allan Schnaiberg and Kenneth A. Gould. *Environment and Society: The Enduring Conflict* (New York: St. Martin's Press, 1994), 45–67.
5. For an example of ecological modernization, see Arthur P. J. Mol and Gert Spaargaren, "Ecological Modernization Theory in Debate," *Environmental Politics* 9/1 (2000): 17–49.
6. Blake Alcott, "Jevons' Paradox," *Ecological Economics* 54 (2005): 9–21. For an interesting discussion on similar issues, see also J. Daniel Khazzoom, "Economic Implications of Mandated Efficiency in Standards for Household Appliances," *Energy Journal* 1/4 (1980): 21–39.
7. Stephen G. Bunker, "Raw Material and the Global Economy," *Society and Natural Resources* 9 (1996): 419–29.
8. Richard York, Eugene A. Rosa, and Thomas Dietz, "The Ecological Footprint Intensity of National Economies," *Journal of Industrial Ecology* 8/4 (2004): 139–54.
9. Richard York, Eugene A. Rosa, and Thomas Dietz, "STIRPAT, IPAT, and ImPACT: Analytic Tools for Unpacking the Driving Forces of Environmental Impacts," *Ecological Economics* 46/3 (2003): 351–65.
10. It is possible that in some instances total resource consumption expands in spite of rather than because of improvements in efficiency. Establishing the nature of causal processes is, of course, a difficult task in non-experimental sciences. Determining the extent to which the link between efficiency and total resource consumption is causal, and which direction the causality flows, will require both further empirical work and nuanced theoretical development.
11. Richard York and Eugene A. Rosa, "Key Challenges to Ecological Modernization Theory," *Organization & Environment* 16/3 (2003): 273–88. Of course, factors like population growth also contribute to driving the expansion of resource consumption. The focus here is on the extent to which a connection between efficiency and the dynamics of economies can lead to escalation of resource consumption independently of other forces.
12. All data for these calculations are from NTHSA (National Traffic Highway Safety Administration) 2005. See http://www.nhtsa.dot.gov/cars/rules/cafe/index.htm. Last accessed January 10, 2006.
13. This increase in light trucks did not happen by chance. It was strongly pushed by the auto industry to circumvent CAFE standards. See Keith Bradsher, *High and Mighty: SUVs—The World's Most Dangerous Vehicles and How They Got That Way* (New York: Public Affairs, 2002).

14. Vaclav Smil, *Energy at the Crossroads: Global Perspectives and Uncertainties* (Cambridge: MIT Press, 2003), 326.
15. World Bank, *World Development Indicators*, 2005, CD-ROM.
16. Abigail J. Sellen and Richard H. R. Harper, *The Myth of the Paperless Office* (Cambridge: MIT Press, 2002).
17. Ibid., 11.
18. Ibid., 13. For a counterexample, see M. J. Hoogeveen and Lucas Reijnders, "E-Commerce, Paper and Energy Use: A Case Study Concerning a Dutch Electronic Computer Retailer," *International Journal of Global Energy Issues* 18/2 (2002): 294–301. They provide an analysis of the effects of a Dutch electronic retailer's application of e-commerce on paper and energy consumption.
19. Sellen and Harper, *The Myth of the Paperless Office.*
20. Vaclav Smil, *Energy in World History* (Boulder, Colo.: Westview Press, 1994).

CHAPTER NINE: THE TREADMILL OF ACCUMULATION

This chapter is an extensively rewritten, updated version of John Bellamy Foster, "The Treadmill of Accumulation: Schnaiberg's Environment and Marxian Political Economy," *Organization & Environment* 18/1 (2005): 7–18.

1. Allan Schnaiberg and Kenneth A. Gould, *Environment and Society* (New York: St. Martin's, 1994).
2. Allan Schnaiberg, *The Environment: From Surplus to Scarcity* (New York: Oxford University Press, 1980).
3. John Bellamy Foster, "Global Ecology and the Common Good," *Monthly Review* 46/9 (1995): 1–10, quote on 2.
4. Schnaiberg and Gould, *Environment and Society*, 69.
5. Foster, "Global Ecology and the Common Good," 9.
6. Eugene S. Ferguson, "The Measurement of the 'Man-Day,'" *Scientific American* 225/4 (October 1971): 96–103. Use of treadmills in houses of correction in England preceded the year 1818 referred to by Ferguson. They were already employed in Bridewell and other early houses of correction/workhouses (known as Bridewells) probably as early as the seventeenth century. See Karen Farrington, *Dark Justice: A History of Punishment and Torture* (New York: Smithmark Books, 1996), 6, 95–99.
7. Karl Marx, *Early Writings* (New York: Vintage, 1974), 360.
8. Karl Marx and Frederick Engels, *Collected Works* (New York: International Publishers, 1975), vol. 6, 434.
9. Marx and Engels, *Collected Works*, vol. 8, 218. For a more detailed treatment of Marx on the treadmill and its relation to his concept of modern

barbarism, see John Bellamy Foster and Brett Clark, "Empire of Barbarism," *Monthly Review* 56/7 (December 2004): 1–15.

10. Schnaiberg, *The Environment*, 145.
11. Gabriel Kolko, *The Triumph of Conservatism* (New York: Free Press, 1963); Gabriel Kolko, *Maincurrents in Modern American History* (New York: Harper and Row, 1976). These calculations are made by adding the separate references to these thinkers in the texts to the number of times they are cited in the footnotes.
12. Schnaiberg, *The Environment*, 205–73.
13. John Kenneth Galbraith, *The Affluent Society* (New York: New American Library, 1958), 125.
14. Schnaiberg, *The Environment*, 227.
15. Ibid., 230.
16. See Harry Magdoff, "Capital, Technology, and Development," *Monthly Review* 27/8 (1976): 1–11, specifically 3.
17. Schnaiberg, *The Environment*, 230.
18. Paul A. Baran and Paul M. Sweezy, *Monopoly Capital* (New York: Monthly Review Press, 1966); Schnaiberg, *The Environment*, 228, 245.
19. Schnaiberg, *The Environment*, 247.
20. Baran and Sweezy, *Monopoly Capital*; James O'Connor, *The Fiscal Crisis of the State* (New York: St. Martin's, 1973).
21. Schnaiberg's use of Magdoff and Sweezy's analysis of stagnation and the financial explosion was extensive. In all, he cited and quoted from seven works coauthored by these two authors, four additional works by Magdoff, five additional works written by Sweezy, along with *Monopoly Capital.*
22. Schnaiberg, *The Environment*, 223.
23. Ibid., 230.
24. Ibid., 249; Paul M. Sweezy, "Some Problems in the Theory of Capital Accumulation," *Monthly Review* 26/1 (1974): 38–55, specifically 54.
25. Schnaiberg, *The Environment*, 250.
26. Ibid., 250.
27. See Riley Dunlap, "The Evolution of Environmental Sociology," in *The International Handbook of Environmental Sociology*, ed. Michael Redclift and Graham Woodgate (Northampton, Mass.: Edward Elgar, 1997), 21–39.
28. Schnaiberg and Gould, *Environment and Society*, 107.
29. Kenneth A. Gould, David N. Pellow, and Allan Schnaiberg, *The Treadmill of Production* (Boulder, Colo.: Paradigm Publishers, 2008); Frederick Buttel, "The Treadmill of Production: An Appreciation, Assessment, and Agenda for Research," *Organization & Environment* 17/3 (September 2004): 329.
30. Sweezy, "Capitalism and the Environment," 7–8.

31. Karl Marx and Frederick Engels, *Selected Works in One Volume* (New York: International Publishers, 1968), 90.
32. Foster, *The Vulnerable Planet*, 111–12.
33. Joan Robinson, *Contributions to Modern Economics* (Oxford: Basil Blackwell, 1978), 1–19; Baran and Sweezy, *Monopoly Capital*; Henryk Szlajfer, "Waste, Marxian Theory, and Monopoly Capital," in *The Faltering Economy*, ed. John Bellamy Foster and Henryk Szlajfer (New York: Monthly Review Press, 1984), 297–321.
34. Schnaiberg's classical contribution, like Veblen's work, remained intact. He adopted a consistent, principled resistance to the demands by the publisher of *The Environment* to write a new, altered edition, preferring that the book retain its original form even at the cost of remaining out of print. He eventually made it freely available on the Internet.
35. The empirical critique is developed further by the well-known authors of the STIRPAT statistical model for assessing environmental impacts, who have demonstrated that an empirical approach relying heavily on the treadmill model (along with other key concepts such as the ecological footprint and the metabolic rift) constituted the most effective refutation of those who argued that the system would (or even could) simply modernize so as to be compatible with the earth. See Richard York, Eugene A. Rosa, and Thomas Dietz, "Footprints on the Earth," *American Sociological Review* 68/2 (2003): 279–300.
36. Buttel, "The Treadmill of Production," 335.

CHAPTER TEN: THE ABSOLUTE GENERAL LAW OF ENVIRONMENTAL DEGRADATION UNDER CAPITALISM

This chapter is adapted and revised for this book from the following article: John Bellamy Foster, "The Absolute General Law of Environmental Degradation Under Capitalism," *Capitalism, Nature, Socialism* 3/3 (1992): 77–82.

1. "Like all other laws," Marx wrote of this absolute general law, "it is modified in its working by many circumstances, the analysis of which does not concern us here." See Karl Marx, *Capital*, vol. 1 (New York: International Publishers, 1967), 644. The term "absolute" is used, as Paul Sweezy notes, "in the Hegelian sense of 'abstract.'" See Paul M. Sweezy, *The Theory of Capitalist Development* (New York: Monthly Review Press, 1942), 19.
2. James O'Connor, "On the Two Contradictions of Capitalism," *Capitalism, Nature, Socialism* 2/3 (1991): 107–9.
3. Marx, *Capital*, vol. 1, 645, 769; Harry Magdoff and Paul M. Sweezy, *Stagnation and the Financial Explosion* (New York: Monthly Review Press, 1987), 204.

4. James O'Connor, "Capitalism, Nature, Socialism: A Theoretical Introduction," *Capitalism, Nature, Socialism* 1/1 (1988): 11–38, quote on 16–17.
5. See Nicholas Georgescu-Roegen, "Afterword," in Jeremy Rifkin, *Entropy* (New York: Bantam, 1989), 305; Narindar Singh, *Economics and the Crisis of Ecology* (New Delhi: Oxford, 1976), 20–24, 30–35.
6. Frederick Engels, *The Dialectics of Nature* (New York: International Publishers, 1940), 291–92; Marx, *Capital*, vol. 1, 43, 505–6.
7. Paul M. Sweezy, "Capitalism and the Environment," *Monthly Review* 41/2 (1989): 1–10; Michael Lebowitz, "The General and Specific in Marx's Theory of Crisis," *Studies in Political Economy* 7 (1982): 5–25.
8. Karl Marx, *Grundrisse* (New York: Vintage, 1973), 408.
9. See Paul Baran and Paul Sweezy, *Monopoly Capital* (New York: Monthly Review Press, 1966), 131–39; Thorstein Veblen, *Absentee Ownership and Business Enterprise in Modern Times* (New York: Augustus M. Kelley, 1923), 284–319; Michael Dawson, *The Consumer Trap* (Urbana: University of Illinois Press, 2003), 11–14.
10. Rifkin, *Entropy*, 148–49.
11. On "commodity chains" see Immanuel Wallerstein, *Historical Capitalism* (London: Verso, 1983), 15–16.
12. See Barry Commoner, *The Closing Circle* (New York: Bantam, 1971), 138–75.
13. Joan Robinson, *Contributions to Modern Economics* (Oxford: Basil Blackwell, 1978), 1–13.
14. Karl Polanyi, *The Great Transformation* (Boston: Beacon, 1944).
15. See John Bellamy Foster and Fred Magdoff, *The Great Financial Crisis* (New York: Monthly Review Press), 2009.
16. James O'Connor, "The Second Contradiction of Capitalism: Causes and Consequences," *Conference Papers* (Santa Cruz: CES/CNS Pamphlet 1), 10.

CHAPTER ELEVEN: THE DIALECTICS OF NATURE AND MARXIST ECOLOGY

This chapter is adapted and revised for this book from the following: John Bellamy Foster, "The Dialectics of Nature and Marxist Ecology," in *Dialectics for the New Century*, ed. Bertell Ollman and Tony Smith (New York: Palgrave, 2008).

1. In conformity with what has become a common practice, "Western Marxism" is used here to refer to a specific philosophical tradition, dating back to the work of Georg Lukács, Karl Korsch, and Antonio Gramsci in the 1920s. The term was first introduced in Maurice Merleau-Ponty, *Adventures of the Dialectic* (Evanston, Ill.: Northwestern University Press,

1973). Western Marxism drew its principal inspiration from the rejection of positivistic influences in Marxism including the concept of the dialectics of nature, which was seen as inherently positivistic in content.

2. Georg Lukács, *History and Class Consciousness* (London: Merlin Press, 1971; originally published in 1923).
3. Ibid., 24; see also Russell Jacoby, "Western Marxism," in *A Dictionary of Marxist Thought*, ed. Tom Bottomore (Oxford: Blackwell, 1983), 525; Steven Vogel, *Against Nature* (Albany: State University of New York Press, 1996), 15.
4. Georg Lukács, *Tactics and Ethics* (New York: Harper and Row, 1972), 139. Engels's statement, which Lukács must have quoted from memory, was considerably more complex: "Dialectics . . . is nothing more than the science of the general laws of motion and development of nature, human society and thought." See Karl Marx and Frederick Engels, *Collected Works* (New York: International Publishers, 1975), vol. 25, 131.
5. Marx and Engels, *Collected Works*, vol. 25, 23.
6. Lukács, *Tactics and Ethics*, 142.
7. Antonio Gramsci, *Selections from the Prison Notebooks* (New York: International Publishers, 1971), 448.
8. Antonio Gramsci, *Further Selections from the Prison Notebooks* (Minneapolis: University of Minnesota Press, 1995), 293.
9. Lucio Colletti, *Marxism and Hegel* (London: Verso, 1973), 191–92.
10. Lukács, *History and Class Consciousness*, 207.
11. Lukács, *Tactics and Ethics*, 144.
12. Georg Lukács, *A Defence of History and Class Consciousness: Tailism and the Dialectic* (London: Verso, 2000), 102–7.
13. See Paul Burkett, "Lukács on Science: A New Act in the Tragedy," *Economic and Political Weekly* 36/48 (December 1, 2001): 4485–89.
14. Lukács, *A Defence of History*, 129–31; Karl Marx, *The Letters of Karl Marx*, ed. Saul Padover (Englewood Cliffs, N.J.: Prentice-Hall, 1979), 418, 422.
15. Lukács, *History and Class Consciousness*, ix–xviii.
16. Georg Lukács, *Conversations with Lukács*, ed. Theo Pinkus (Cambridge: MIT Press, 1974), 21.
17. Ibid., 43.
18. An example of the latter is Vogel, *Against Nature*. He argues in a cross between critical theory and post-structuralism that in rejecting the positivistic "misapplication" of science to society Lukács left natural science and nature itself (that is, the domain of natural-physical science) unquestioned. For Vogel the object should be to extend the argument by deconstructing natural scientific conceptions of "nature," which should be viewed as mere discursive constructions.
19. Herbert Marcuse, *Reason and Revolution* (Boston: Beacon Press, 1960), 314.

20. Sidney Hook, *From Hegel to Marx* (Ann Arbor: University of Michigan Press, 1950), 75–76.
21. Jean-Paul Sartre, *Critique of Dialectical Reason*, vol. 1 (London: Verso, 2004), 32.
22. Sebastiano Timpanaro, *On Materialism* (London: Verso, 1975), 29.
23. Marx and Engels, *Collected Works*, vol. 1, 49.
24. Jean-Paul Sartre, *Literary and Philosophical Essays* (New York: Criterion Books, 1955), 207.
25. Marx and Engels, *Collected Works*, vol. 1, 64–65.
26. Karl Marx, *Early Writings* (New York: McGraw Hill, 1964), 217.
27. See John Bellamy Foster and Paul Burkett, "The Dialectic of Organic/Inorganic Relations," *Organization & Environment* 13/4 (2000): 403–25.
28. Marx, *Early Writings*, 127.
29. Marx and Engels, *Collected Works*, vol. 5, 31–32.
30. Marx, *Early Writings*, 207.
31. Alfred Schmidt, *The Concept of Nature in Marx* (London: Verso, 1971), 79.
32. Andrew Feenberg, *Lukács, Marx and the Sources of Critical Theory* (Totowa, N.J.: Rowman and Littlefield, 1981), 213.
33. Feenberg, *Lukács, Marx and the Sources of Critical Theory*, 214.
34. Ibid., 217–18; Marx, *Early Writings*, 160.
35. Alexei Mikhailovich Voden, in Institute of Marxism-Leninism, *Reminiscences of Marx and Engels* (Moscow: Foreign Language Publishing House, n.d.), 332–33.
36. Ludwig Feuerbach, *The Fiery Brook* (Garden City, N.Y.: Doubleday, 1972), 213, 224.
37. G. W. F. Hegel, *The Philosophy of Mind* (Oxford: Oxford University Press, 1971), 34.
38. Feenberg, *Lukács, Marx and the Sources of Critical Theory*, 218.
39. Marx, *Early Writings*, 160–61.
40. Marx and Engels, *Collected Works*, vol. 4,129.
41. Marx, *Early Writings*, 164.
42. Feenberg, *Lukács, Marx and the Sources of Critical Theory*, 219–20.
43. Diogenes Laertius, *Lives of Eminent Philosophers*, vol. 2 (Cambridge, Mass.: Harvard University Press, 1925), 563; Marx and Engels, *Collected Works*, vol. 1, 405–06. Marx was not only aware of the epistemological bases of Epicurus's thought but also of the way in which this was incorporated (often in less sophisticated forms) in early modern philosophy and science. Although in the history of philosophy Epicurus's epistemology and system of scientific inference was typically underestimated, in recent decades there has been a dramatic turnaround as new sources of Epicurean

philosophy drawn from the library of papyri in Herculaneum have increased the knowledge of this philosophy. Much of this has verified what Marx deduced in the analysis in his dissertation. See Elizabeth Asmis, *Epicurus' Scientific Method* (Ithaca, N.Y.: Cornell University Press, 1984); John T. Fitzgerald, "Introduction: Philodemus and the Papyri from Herculaneum," in *Philodemus and the New Testament World*, ed. John T. Fitzgerald, Dirk Obrinsk, and Glenn S. Holland (Boston: Brill, 2004), 1–12; John Bellamy Foster, *Marx's Ecology* (New York: Monthly Review Press, 2000), 21–65.

44. Feenberg, *Lukács, Marx and the Sources of Critical Theory*, 220–21. The "phenomenological" element in Marx's thought derives from his brand of materialism, which sees human existence and consciousness as corporeal, based on the body, the sense organs, and sense perception. Merleau-Ponty is in accord with Marx when he writes: "The body...is wholly animated, and all its functions contribute to the perception of objects—an activity long considered by philosophy to be pure knowledge." See Maurice Merleau-Ponty, *The Primacy of Perception* (Evanston, Ill.: Northwestern University Press, 1964), 5. On the corporeal aspects of Marx's materialism, see Joseph Fracchia, "Beyond the Human-Nature Debate: Human Corporeal Organisation as the 'First Fact' of Historical Materialism," *Historical Materialism* 13/1 (2005): 33–61.
45. Karl Marx, *Texts on Method* (Oxford: Basil Blackwell, 1975), 190–91.
46. Lukács, *Conversations*, 42.
47. Karl Marx, *Early Writings* (London: Penguin, 1974), 421–23.
48. Karl Marx, *Letters to Kugelmann* (New York: International Publishers, 1934), 112.
49. G. W. F. Hegel, *The Phenomenology of Spirit* (Oxford: Oxford University Press, 1977), 11.
50. Lucretius, *On the Nature of the Universe* (Oxford: Oxford University Press, 1997), 93; Karl Marx, *The Poverty of Philosophy* (New York: International Publishers, 1963), 110; Marx and Engels, *Collected Works*, vol. 1, 65, 473, 478.
51. Anne Fairchild Pomeroy, *Marx and Whitehead: Process, Dialectics, and the Critique of Capitalism* (Albany: State University of New York Press, 2004), 9, 50.
52. Marx and Engels, *Collected Works*, vol. 1, 65.
53. Marx and Engels, *Collected Works*, vol. 25, 21–24.
54. Bertell Ollman, *Dialectical Investigations* (New York: Routledge, 1993), 35.
55. Ollman, *Dialectical Investigations*, 11; also Bertell Ollman, *Dance of the Dialectic* (Urbana: University of Illinois Press, 2003).
56. Foster, *Marx's Ecology*, 141–77. See also chapter 15 in this book.

57. Marx, *Texts on Method*, 209.
58. Karl Marx, *Capital*, vol. 1 (New York: Vintage, 1976), 323.
59. Karl Marx, *Capital*, vol. 3 (New York: Vintage, 1981), 959.
60. Marx and Engels, *Collected Works*, vol. 25, 23.
61. Ibid., 25:356.
62. Ibid., 25:493.
63. Z. A. Jordan, *The Evolution of Dialectical Materialism* (New York: St. Martin's Press, 1967), 165–66.
64. Marx and Engels, *Collected Works*, vol. 25, 460–61.
65. Karl Marx, *Grundrisse* (London: Penguin, 1973), 489.
66. Marx, *Capital*, vol. 3, 949.
67. Foster, *Marx's Ecology*, 236–54.
68. Max Horkheimer and Theodor Adorno, *The Dialectic of Enlightenment* (New York: Continuum, 1972).
69. See John Bellamy Foster, *The Ecological Revolution* (New York: Monthly Review Press, 2009), chap. 8.
70. Hyman Levy, *A Philosophy for a Modern Man* (New York: Alfred A. Knopf, 1938), 126.
71. Levy, *A Philosophy for a Modern Man*, 91, 125–30, 199, 227.
72. See Paul Burkett, "Nature in Marx Reconsidered: A Silver Anniversary Assessment of Alfred Schmidt's *Concept of Nature in Marx*," *Organization & Environment* 10/3 (1997): 164–83.
73. Levy, *A Philosophy for a Modern Man*, 19, 24, 100, 130.
74. Richard Levins and Richard Lewontin, *The Dialectical Biologist* (Cambridge, Mass: Harvard University Press, 1985), 277.
75. Rachel Carson, *Lost Woods* (Boston: Beacon Press, 1998), 230.
76. J. D. Bernal, *The Origins of Life* (New York: World Publishing Co., 1967), 182.
77. David R. Keller and Frank B. Golley, "Introduction," in *The Philosophy of Ecology*, ed. Keller and Golley (Athens: University of Georgia Press, 2000), 1–2.
78. Alfred North Whitehead, *Science and the Modern World* (New York: Free Press, 1967), 54.
79. Levins and Lewontin, *The Dialectical Biologist*; Stephen Jay Gould, *The Structure of Evolutionary Theory* (Cambridge, Mass.: Harvard University Press, 2002).
80. See Paul Burkett, *Marx and Nature* (New York: St. Martin's Press, 1999); Peter Dickens, *Society and Nature* (Cambridge: Polity, 2004); Foster, *Marx's Ecology*; Levins and Lewontin, *The Dialectical Biologist*. Not all Marxist ecologists are willing to embrace the notion of the dialectics of nature, even when qualified by a critical realism/critical materialism. For example, Joel Kovel, editor of the ecological socialist journal *Capitalism,*

Nature, Socialism, while making some valid points, clearly leans toward the early Lukácsian rejection of the concept: "Efforts to read dialectic directly into non-human nature are undialectical. Since a 'dialectic of anything' must include praxis, a dialectic of nature posits that nature directly manifests the relation between theory and practice characteristic of human beings. This would make the human and non-human worlds identical in their most fundamental feature. . . . The disjunction between humanity and nature extends to the knowing of nature, which can never be fully realized." See Joel Kovel, "Dialectic as Praxis," *Science & Society* 62/3 (1998): 474–82, quote on 479.

81. Sartre, *Literary and Philosophical Essays*, 218.

CHAPTER TWELVE: DIALECTICAL MATERIALISM AND NATURE

This chapter is adapted and revised for this book from the following article: Brett Clark and Richard York, "Dialectical Materialism and Nature: An Alternative to Economism and Deep Ecology," *Organization & Environment* 18/3 (2005): 318–37.

1. William R. Catton, Jr. and Riley E. Dunlap, "Environmental Sociology: A New Paradigm," *The American Sociologist* 13 (1978): 41–49; Allan Schnaiberg, *The Environment: From Surplus to Scarcity* (New York: Oxford University Press, 1980).
2. John Bellamy Foster, *Marx's Ecology* (New York: Monthly Review Press, 2000).
3. Donella H. Meadows, *The Limits to Growth* (New York: Universe Books, 1972).
4. Barry Commoner, *The Closing Circle* (New York: Knopf, 1971); Paul R. Ehrlich, Anne H. Ehrlich, John P. Holdren, *Human Ecology* (San Francisco: Freeman, 1973).
5. Julian Simon, *The Ultimate Resource* (Princeton: Princeton University Press, 1981); Robert Solow, "The Economics of Resources or the Resources of Economics," *American Economic Review* 64/2 (1974): 1–14.
6. Robert Gottlieb, *Forcing the Spring* (Washington, D.C.: Island Press, 1993).
7. Frederick Buell, *From Apocalypse to Way of Life* (New York: Routledge, 2003).
8. Gene M. Grossman and Alan B. Krueger, "Economic Growth and the Environment," *Quarterly Journal of Economics* 110 (1995): 353–77.
9. Simon Kuznets, "Economic Growth and Income Inequality," *American Economic Review* 45 (1955): 1–28.

10. Arthur P. J. Mol, *The Refinement of Production* (Utrecht, Netherlands: Van Arkel, 1995), 42.
11. Maurie J. Cohen, "Sustainable Development and Ecological Modernization: National Capacity for Rigorous Environmental Reform," in *Environmental Policy and Societal Aims*, ed. D. Requier-Desjardins, C. Spash, and J. van der Staaten (Dordrecht, Netherlands: Kluwer, 1999), 103–28; Maarten A. Hajer, *The Politics of Environmental Discourse: Ecological Modernization and the Policy Process* (Oxford, UK: Clarendon, 1995); Mol, *The Refinement of Production*; Gert Spaargaren and Arthur P. J. Mol, "Sociology, Environment and Modernity: Ecological Modernization as a Theory of Social Change," *Society and Natural Resources* 5 (1992): 323–44.
12. Mol, *The Refinement of Production*; Gert Spaargaren, "The Ecological Modernization of Production and Consumption: Essays in Environmental Sociology," Ph.D. diss. (Wageningen University, Netherlands, 1997).
13. Charles Leadbeater, *The Weightless Society* (New York: Texere, 2000).
14. Arthur P. J. Mol, "Ecological Modernisation and Institutional Reflexivity," *Environmental Politics* 5/2 (1996): 302–23; Arthur P. J. Mol, *Globalization and Environmental Reform* (Cambridge, Mass.: MIT Press, 2001).
15. Jack Speer, "General Electric's Plans to Increase Spending on Cutting-Edge Environmental Technologies," National Public Radio, *Morning Edition* (May 10, 2005).
16. Mol, *The Refinement of Production.*
17. Ibid.
18. Paul Hawken, "Foreword," in *Natural Capital and Human Economic Survival*, ed. Thomas Prugh (Solomons, Md.: International Society for Ecological Economics, 1995), xi–xv; Paul Hawken, Amory Lovins, and L. Hunter Lovins, *Natural Capitalism* (Boston: Little, Brown, 1999).
19. Paul Hawken, *The Ecology of Commerce* (New York: Harper Business, 1993).
20. Paul Hawken, "Natural Capitalism," *Mother Jones Magazine* (April 1997): 40–53, 59–62.
21. John Bellamy Foster, *Ecology Against Capitalism* (New York: Monthly Review Press, 2002).
22. Paul Burkett, *Marx and Nature* (New York: St. Martin's, 1999), 64–65.
23. Ibid., 70–79.
24. Paul Burkett, "Natural Capital, Ecological Economics, and Marxism," *International Papers in Political Economy* 10/3 (2003): 1–61; Foster, *Ecology Against Capitalism.*
25. Richard York and Eugene A. Rosa, "Key Challenges to Ecological Modernization Theory," *Organization & Environment* 16/3 (2003): 273–88.

26. Theresa A. Cavlovic, Kenneth H. Baker, Robert P. Berrens, and Kishore Gawande, "A Meta-Analysis of Environmental Kuznets Curve Studies," *Agricultural and Resource Economics Review* 29/1 (2000): 32–42; Andrew K. Jorgenson and Brett Clark, "The Economy, Military, and Ecologically Unequal Exchange Relationships in Comparative Perspective: A Panel Study of the Ecological Footprints of Nations, 1975–2000," *Social Problems* 56/4 (2009): 621–46; Eugene A. Rosa, Richard York, and Thomas Dietz, "Tracking the Anthropogenic Drivers of Ecological Impacts," *Ambio* 33/8 (2004): 509–12; Anqing Shi, "The Impact of Population Pressure on Global Carbon Dioxide Emissions, 1975–1996," *Ecological Economics* 44 (2003): 29–42; Richard York, Eugene A. Rosa, and Thomas Dietz, "Footprints on the Earth: The Environmental Consequences of Modernity," *American Sociological Review* 68 (2003): 279–300; Richard York, Eugene A. Rosa, and Thomas Dietz, "The Ecological Footprint Intensity of National Economies," *Journal of Industrial Ecology* 8/4 (2004): 139–54.
27. Arne Naess, "The Shallow and the Deep, Long-Range Ecology Movements," *Inquiry* 1 (1973): 16.
28. Bill Devall, "The Deep, Long-Range Ecology Movement: 1960–2000—A Review," *Ethics & the Environment* 6/1 (2001): 18–41; Bill Devall and George Sessions, *Deep Ecology* (Salt Lake City, Utah: Peregrine Smith Books, 1985).
29. James Lovelock, *Gaia* (New York: Oxford University Press, 1979).
30. Talcott Parsons, *The Structure of Social Action* (New York: Free Press, 1937); Talcott Parsons, *The Social System* (New York: Free Press, 1951).
31. Lovelock, *Gaia*, 12.
32. The term *dialectical materialism* is often used derisively within contemporary Western Marxism and has come to symbolize Stalinist dogma. Our intention here is to rehabilitate the term by using it to refer exclusively to those inquiries that can be seen as genuine attempts to employ both dialectical and materialist methodologies in both the natural and the social realms. Materialism without dialectics tends toward mechanism and reductionism. Dialectics without materialism tends toward idealism and vitalism. Genuine dialectical materialism seeks to transcend these antinomies. It thus stands for a critical realism sorely lacking in conventional thought. See Brett Clark and Richard York, "Dialectical Nature," *Monthly Review* 57/1 (2005): 13–22; Foster, *Marx's Ecology*.
33. See V. I. Vernadsky, *The Biosphere* (New York: Copernicus, 1998); A. I. Oparin, *Origin of Life* (New York: Dover, 1965).
34. Lucio Colletti, *From Rousseau to Lenin* (London: New Left Books, 1972), 13–14.
35. Burkett, *Marx and Nature*; Foster, *Marx's Ecology*; John Bellamy Foster and Paul Burkett, "The Dialectic of Organic/Inorganic Relations: Marx

and the Hegelian Philosophy of Nature," *Organization & Environment* 13/4 (2000): 403–25; Yrjo Haila and Richard Levins, *Humanity and Nature* (London: Pluto Press, 1992); Karl Marx, *Early Writings* (New York: Vintage, 1974), 400.

36. For discussions of the intellectual perspectives of Gould, Lewontin, and Levins, particularly as they relate to nature, science, and society, see Richard York and Brett Clark, *The Science and Humanism of Stephen Jay Gould* (New York: Monthly Review Press, 2010).
37. Richard Dawkins, *The Selfish Gene* (New York: Oxford University Press, 1976); Daniel C. Dennett, *Darwin's Dangerous Idea* (New York: Simon & Schuster, 1995).
38. Richard Levins and Richard Lewontin, *The Dialectical Biologist* (Cambridge, Mass.: Harvard University Press, 1985).
39. Richard Lewontin, *The Triple Helix: Gene, Organism, and Environment* (Cambridge, Mass.: Harvard University Press, 2000), 17–18.
40. Ibid., 42.
41. Ibid., 47.
42. Foster, *Marx's Ecology*.
43. Lewontin, *The Triple Helix*, 41–42.
44. Ibid., 43–44.
45. Ibid., 48.
46. Ibid., 51.
47. Levins and Lewontin, *The Dialectical Biologist*.
48. Ibid., 89–106; Lewontin, *The Triple Helix*, 48–55.
49. Lewontin, *The Triple Helix*, 55.
50. Levins and Lewontin, *The Dialectical Biologist*; Richard Lewontin, *It Ain't Necessarily So: The Dream of the Human Genome and Other Illusions*, 2nd ed. (New York: New York Review Books, 2001).
51. Lewontin, *The Triple Helix*, 62–64, 66.
52. Ibid., 88.
53. Barry Commoner, "Unraveling the DNA Myth," *Harper's Magazine* 304/1821 (2002): 39–47, quote on 47.
54. Stephen Jay Gould, *The Mismeasure of Man* (New York: W. W. Norton, 1981); Richard Levins, "A Science of Our Own: Marxism and Nature," *History as It Happened: Selected Articles from* Monthly Review, *1949–1989*, ed. B. S. Ortiz (New York: Monthly Review Press, 1990), 235–42; Richard Lewontin and Richard Levins, *Biology Under the Influence* (New York: Monthly Review Press, 2007), chap. 17; Richard C. Lewontin, Stephen Rose, and Leon J. Kamin, *Not in Our Genes* (New York: Pantheon, 1984).
55. Niles Eldredge and Stephen Jay Gould, "Punctuated Equilibria: An Alternative to Phyletic Gradualism," in *Models of Paleobiology*, ed. Thomas

J. M. Schopf (San Francisco: Freeman, Cooper & Co., 1972), 82–115; Stephen Jay Gould and Niles Eldredge, "Punctuated Equilibria: The Tempo and Mode of Evolution Reconsidered," *Paleobiology* 3 (1977): 115–51.

56. Stephen Jay Gould, *The Structure of Evolutionary Theory* (Cambridge, Mass.: Harvard University Press, 2002).
57. Stephen Jay Gould, *Eight Little Piggies* (New York: W. W. Norton, 1993).
58. For a discussion of eugenics, which includes important comments regarding Galton, see Edwin Black, *War Against the Weak: Eugenics and America's Campaign to Create a Master Race* (New York: Four Walls Eight Windows, 2003). For a biography of Galton that includes a presentation of both his "dark visions and bright ideas," see Martin Brookes, *Extreme Measures: The Dark Visions and Bright Ideas of Francis Galton* (New York: Bloomsbury, 2004).
59. Gould, *Eight Little Piggies*, 384–85.
60. Paul Falkowski, Robert J. Scholes, Ed Boyle, Josep Canadell, D. Canfield, J. Elser, N. Gruber, K. Hibbard, P. Högberg, S. Linder, F.T. Mackenzie, B. Moore III, T. Pedersen, Y. Rosenthal, S. Seitzinger, V. Smetacek, and W. Steffen, "The Global Carbon Cycle," *Science* 290 (2000): 291–96; Roldan Muradian, "Ecological Thresholds: A Survey," *Ecological Economics* 38 (2001): 7–24; Marten Scheffer, Steve Carpenter, Jonathan A. Foley, Carl Folke, and Brian Walker, "Catastrophic Shifts in Ecosystems," *Nature* 413 (2001): 591–96; Peter M. Vitousek, "Beyond Global Warming," *Ecology* 75/7 (1994): 1861–76.
61. Richard B. Alley, *The Two-Mile Time Machine* (Princeton: Princeton University Press, 2000); Wallace S. Broecker, "Does the Trigger for Abrupt Climate Change Reside in the Ocean or in the Atmosphere?" *Science* 300 (2003): 1519–22; John Bellamy Foster, *The Ecological Revolution* (New York: Monthly Review Press, 2009); John T. Houghton, Yihui Ding, Dave J. Griggs, Maria Noguer, Paul J. van der Linden, and Xiaosu Dai, *Climate Change 2001: The Scientific Basis* (Cambridge: Cambridge University Press, 2001); National Research Council, *Abrupt Climate Change: Inevitable Surprises* (Washington, D.C.: National Academy Press, 2002).
62. Franz. J. Broswimmer, *Ecocide* (London and Sterling, Va.: Pluto Press, 2002); Niles Eldredge, *The Miner's Canary* (Princeton: Princeton University Press, 1991); Niles Eldredge, *Dominion* (New York: Henry Holt, 1995); D. U. Hooper, F. S. Chapin III, J. J. Ewel, A. Hector, P. Inchausti, S. Lavorel, J. H. Lawton, D. M. Lodge, M. Loreau, S. Naeem, B. Schmid, H. Setälä, A. J. Symstad, J. Vandermerr, and D. A. Wardle, "Effects of Biodiversity on Ecosystem Functioning," *Ecological Monographs* 75/1 (2005): 3–35; Richard Leakey and Roger Lewin, *The Sixth Extinction* (New York: Anchor Books, 1996).

63. Broswimmer, *Ecocide*.
64. Leakey and Lewin, *The Sixth Extinction*, 233.
65. Eldredge, *Dominion*.
66. Ibid., 125–32.
67. Gerardo Ceballos and Paul R. Ehrlich, "Mammal Population Losses and the Extinction Crisis," *Science* 296 (2002): 904–7; Jennifer B. Hughes, Gretchen C. Daily, and Paul R. Ehrlich, "Population Diversity," *Science* 278 (1997): 689–92; Stuart L. Pimm and Peter Raven, "Extinction by Numbers," *Nature* 403 (2000): 843–45; United Nations Environment Programme, *Global Environment Outlook 3* (Sterling, Va.: Earthscan, 2002).
68. L. P. Koh, R. R. Dunn, N. S. Sodhi, R. K. Colwell, H. C. Proctor, and V. S. Smith, "Species Coextinctions and the Biodiversity Crisis," *Science* 305 (2004): 1632–34.
69. Koh et al., "Species Coextinctions."
70. United Nations Environment Programme, *Global Environment Outlook 1* (Sterling, Va.: Earthscan, 1997), chapter 4.
71. Buell, *From Apocalypse to Way of Life*.
72. Lewontin, *The Triple Helix*, 68.

CHAPTER THIRTEEN: MARX'S *GRUNDRISSE* AND THE ECOLOGY OF CAPITALISM

This chapter is an extensively rewritten and abridged version of the following book chapter: John Bellamy Foster, "Marx's *Grundrisse* and the Ecological Contradictions of Capitalism," in *Karl Marx's* Grundrisse*: Foundations of the Critique of Political Economy 150 Years Later*, ed. Marcello Musto (London: Routledge, 2008).

1. Karl Marx and Frederick Engels, *Collected Works* (New York: International Publishers, 1975), vol. 11, 103.
2. Paul Burkett, *Marx and Nature* (New York: St. Martin's Press, 1999); John Bellamy Foster, *Marx's Ecology: Materialism and Nature* (New York: Monthly Review Press, 2000); Peter Dickens, *Society and Nature* (Malden, Mass.: Polity, 2004).
3. Karl Marx, *Grundrisse* (New York: Vintage, 1973), 320.
4. Ibid., 320.
5. Ibid., 85–86.
6. Ibid., 267–68.
7. Marx and Engels, *Collected Works*, vol. 35, 187, 194; translation according to Karl Marx, *Capital*, vol. 1 (New York: Vintage, 1976), 283, 290. The significance of both labor in general and production in general was recognized

by Georg Lukács. In the former, he observed, Marx abstracted "from all the social moments of the labour process, in order to work out clearly those moments . . . *common to all processes of labour*." While an identical logic was evident in the concept of production in general. See Georg Lukács, *A Defence of History and Class Consciousness: Tailism and the Dialectic* (London: Verso, 2000), 98.

8. Marx, *Grundrisse*, 667.
9. Ibid., 489.
10. Marx and Engels, *Collected Works*, vol. 3, 276.
11. G. W. F. Hegel, *The Philosophy of Nature*, vol. 3 (Atlantic Highlands, N.J.: Humanities Press, 1970), 185; John Bellamy Foster and Paul Burkett, "The Dialectic of Organic/Inorganic Relations," *Organization & Environment* 13/4 (2000): 403–25.
12. Marx and Engels, *Collected Works*, vol. 3, 275.
13. Marx, *Grundrisse*, 474, 488. For a systematic analysis of this part of Marx's analysis, see Foster and Burkett, "The Dialectic."
14. Ibid., 460.
15. Ibid., 527. For treatments of Marx's theory of metabolic rift, see John Bellamy Foster, "Marx's Theory of Metabolic Rift," *American Journal of Sociology* 105/2 (1999): 366–405; Foster, *Marx's Ecology*, 155–63; Paul Burkett, *Marxism and Ecological Economics* (Boston: Brill, 2006), 202–7. Paul Burkett discusses how the development of science, e.g., with respect to agriculture, in Marx's conception, gave new insights into production in general, the understanding of which was formed by "the *natural-scientific study* of human production and its natural conditions across different modes of production, and not just capitalism." See Burkett, *Marxism and Ecological Economics*, 89–90.
16. Marx, *Grundrisse*, 881.
17. Ibid., 612.
18. Ibid., 494; see also Michael Lebowitz, *Beyond* Capital (New York: Palgrave Macmillan, 2003), 30–32.
19. See Foster, *Marx's Ecology*.
20. Marx and Engels, *Collected Works*, vol. 37, 799.
21. Marx and Engels, *Collected Works*, vol. 5, 31.
22. Marx, *Grundrisse*, 92.
23. Ibid., 488.
24. Ibid., 488.
25. Ibid., 471.
26. These and the other subheadings in the *Grundrisse* were added by the 1939/1953 editors based on the index he provided for his seven notebooks. See Marx, *Grundrisse*, 66.
27. For a useful discussion of this part of the *Grundrisse*, see Eric J.

Hobsbawm, "Introduction," in Karl Marx, *Pre-Capitalist Economic Formations* (New York: International Publishers, 1964).

28. Marx and Engels, *Collected Works*, vol. 37, 798, emphasis added.
29. Marx, *Grundrisse*, 882.
30. Ibid., 497.
31. Ibid., 498–99.
32. Ibid., 276–77.
33. Ibid., 458.
34. Ibid., 488, emphasis added. Edward Wakefield's theory of colonialism argued that the only way to create a basis for industrial wage labor in the colonies was to first create monopolies in the land to prevent workers from escaping into small subsistence plots. This view was, according to Marx, of "infinite importance" in understanding the presuppositions of capitalism. See ibid., 278.
35. Ibid., 335; see also István Mészáros, *Beyond Capital* (New York: Monthly Review Press, 1995), 568.
36. Marx, *Grundrisse*, 334–35.
37. G. W. F. Hegel, *The Science of Logic* (London: George Allen and Unwin, 1969), 131–37; G. W. F. Hegel, *Hegel's Logic* (Oxford: Oxford University Press, 1975), 136–37.
38. Marx, *Grundrisse*, 539.
39. Ibid., 409–10.
40. Francis Bacon, *Novum Organum* (Chicago: Open Court, 1994), 29, 43. Bacon's complex notion of the domination and subjugation of nature, while frequently expounded in the form of metaphors drawn from the domination within society, was compatible with notions of sustainability insofar as it demanded that society follow "nature's laws." The Baconian ruse was that nature could be mastered through its own laws. But nature's laws *if followed completely* nonetheless put restrictions on production—those necessitated by reproduction and sustainability. For a discussion of the full complexity of the Baconian view in this respect, see William Leiss, *The Domination of Nature* (Boston: Beacon Press, 1974).
41. This paragraph borrows from John Bellamy Foster, "*The Communist Manifesto* and the Environment," in *The Socialist Register, 1998*, ed. Leo Panitch and Colin Leys (New York: Monthly Review Press, 1998), 169–89, quote on 184. Michael Lebowitz has demonstrated that Marx pointed to two kinds of barriers to capital, leading to accumulation of contradictions and crises: general barriers common to production in general, and thus having to do with natural conditions, and more specific historical barriers immanent to capital itself. See Michael Lebowitz, "The General and Specific in Marx's Theory of Crisis," *Studies in Political Economy* 7 (1982): 5–25.

42. Mészáros, *Beyond Capital*, 142.
43. Ibid., 173.
44. Ibid., 759.
45. István Mészáros, *Socialism or Barbarism* (New York: Monthly Review Press, 2001), 61; John Bellamy Foster, *The Ecological Revolution* (New York: Monthly Review Press, 2009), chap. 1.
46. Marx and Engels, *Collected Works*, vol. 37, 799; translation according to Karl Marx, *Capital*, vol. 2 (New York: Vintage, 1981), 949.
47. "*Après moi le déluge!* is the watchword of every capitalist and of every capitalist nation. Hence Capital is reckless of the health or length of life of the labourer, unless under compulsion from society." See Marx and Engels, *Collected Works*, vol. 35, 275.
48. Marx, *Grundrisse*, 409.
49. Marx and Engels, *Collected Works*, vol. 35, 505–8; translation according to Marx, *Capital*, vol. 1, 636–69.

CHAPTER FOURTEEN: THE SOCIOLOGY OF ECOLOGY

This chapter is adapted and revised for this book from the following article: John Bellamy Foster and Brett Clark, "The Sociology of Ecology: Ecological Organicism versus Ecosystem Ecology in the Social Construction of Ecological Science, 1926–1935," *Organization & Environment* 21/3 (2008): 311–52.

1. Roy Bhaskar, *The Possibility of Naturalism* (London: Routledge, 1979).
2. David Bloor, *Knowledge and Social Imagery* (Chicago: University of Chicago Press, 1991), 175.
3. More recent notions of "reflexive modernity" transcend the realist-constructionist divide by in a sense abolishing the distinction. Such approaches are perhaps too quick to accept "the end of nature" (the "end of nature" as independent of human beings) and thus to translate environmental crises into questions of pure "risk" to be fully acceptable to most environmental sociologists, who are disinclined to believe that nature has in any sense "ended" and are concerned rather with the dialectical relations between nature and society. See Anthony Giddens and Christopher Pierson, *Conversations with Anthony Giddens* (London: Polity, 1998), 204–17.
4. In presenting these three principles in this form, we have taken certain liberties in all three of the "quotations" here. The notion that Galileo reached down and touched the earth and said "it still moves" is of course the stuff of legend but can be taken as a fundamental principle of realism. Vico argued that in contradistinction to the natural world, which, since "God made it, he alone knows," the "world of nations" was one that "since men . . . made it, men could come to know" it. This can be seen as the fundamental principle

of historical-humanist constructionism. Marx stated: "Men make their own history, but they do not make it just as they please; they do not make it under circumstances chosen by themselves, but under circumstances directly encountered, given and transmitted from the past." This is the fundamental principle of historical materialism. We have expanded it to encompass natural–environmental conditions as well. All three of these principles are central to what we refer to in this analysis as "realist constructionism." See Michael H. Hart, *The 100: A Ranking of the Most Influential Persons in History* (New York: Hart Publishing, 1978), 102–4; Giambattista Vico, *The New Science* (Ithaca, N.Y.: Cornell University Press, 1984), 96; Karl Marx, *Capital*, vol. 1 (New York: Vintage, 1976), 493; Karl Marx, *The Eighteenth Brumaire of Louis Bonaparte* (New York: International Publishers, 1963), 15; Antonio Gramsci, *Selections from the Prison Notebooks* (New York: International Publishers, 1971), part 3.

5. Imre Lakatos, *The Methodology of Scientific Research Programmes: Philosophical Papers*, vol. 1 (Cambridge: Cambridge University Press, 1978).
6. Bhaskar, *The Possibility of Naturalism*; Roy Bhaskar, *Scientific Realism and Human Emancipation* (London: Verso, 1986); Andrew Sayer, *Method in Social Science* (New York: Routledge, 1992).
7. David Pepper, *Modern Environmentalism* (New York: Routledge, 1996), 242.
8. See Michael G. Barbour, "Ecological Fragmentation in the Fifties," in *Uncommon Ground: Toward Reinventing Nature*, ed. William Cronon (New York: W.W. Norton, 1995), 233–55; Carolyn Merchant, *The Death of Nature: Women, Ecology, and the Scientific Revolution* (New York: Harper and Row, 1980), 252; Carolyn Merchant, *Radical Ecology* (New York: Routledge, 1992), 59; Pepper, *Modern Environmentalism*, 233–34, 242–45; Donald Worster, *Nature's Economy* (Cambridge: Cambridge University Press, 1977), 301–4, 316–23.
9. Frederick H. Buttel, Peter Dickens, Riley E. Dunlap, and August Gijswijt, "Sociological Theory and the Environment," in *Sociological Theory and the Environment*, ed. Riley E. Dunlap, Frederick H. Buttel, Peter Dickens, and August Gijswijt (New York: Rowman and Littlefield, 2002), 3–32, quote on 22. As this quote indicates, "realism" is sometimes presented as an epistemological position and materialism as an ontological one. Bhaskar defines realism as "the theory that the ultimate objects of scientific inquiry exist and act (for the most part) quite independently of scientists and their activity." Realism thus points to a materialist ontology. Consequently, there is no consistency in distinguishing realism and materialism even among critical realists, with Bhaskar using both terms to cover basically the same conceptual range but with somewhat different emphases—so that one can

also refer to a materialist epistemology and a realist ontology. We therefore use the terms somewhat interchangeably in this article, giving preference to "realism" in the specific context of the "realism" versus "constructionist" debate itself and giving preference to "materialism" where the polarity to be stressed is that of "materialism" versus "idealism." See Roy Bhaskar, *Reclaiming Reality* (London: Verso, 1989), 12; Roy Bhaskar, "Materialism," in *A Dictionary of Marxist Thought*, ed. Tom Bottomore (Oxford: Blackwell, 1983), 369-73, and Bhaskar, "Realism," 458-60, in the same volume.

10. Ted Benton, "Biology and Social Theory in the Environmental Debate," in *Social Theory and the Global Environment*, ed. Michael Redclift and Ted Benton (New York: Routledge, 1994), 28-50; Frederick H. Buttel, "Sociology and the Environment," *International Social Science Journal* 109 (1986): 337-56; Frederick H. Buttel, "Environmental and Resources Sociology," *Rural Sociology* 61/1 (1996): 56-76; Riley E. Dunlap and William R. Catton, Jr., "Environmental Sociology," in *Progress in Resource Management and Environmental Planning*, ed. Timothy O'Riordan and Ralph C. D'Arge (New York: John Wiley & Sons, 1979), 57-85; John Bellamy Foster, "Marx's Theory of Metabolic Rift: Classical Foundations for Environmental Sociology," *American Journal of Sociology* 105/2 (1999): 366-405; Michael Goldman and Rachel Schurman, "Closing the 'Great Divide': New Social Theory on Society and Nature," *Annual Review of Sociology* 26 (2000): 563-84; Raymond Murphy, *Rationality and Nature* (Boulder, Colo.: Westview Press, 1994); Raymond Murphy, *Sociology and Nature* (Boulder, Colo.: Westview Press, 1997).
11. William R. Catton, Jr., and Riley E. Dunlap, "Environmental Sociology: A New Paradigm," *American Sociologist* 13 (1978): 41-49; Riley E. Dunlap and William R. Catton, Jr., "Struggling with Human Exemptionalism," *American Sociologist* 25/1 (1994): 5-30.
12. See Rachel Carson, *The Silent Spring* (Boston: Houghton Mifflin, 1962); Barry Commoner, *The Closing Circle* (New York: Knopf, 1971).
13. For a realist discussion, see Benton, "Biology and Social Theory," 44-46; Peter Dickens, *Reconstructing Nature* (London: Routledge, 1996); Luke Martel, *Ecology and Society* (Amherst: University of Massachusetts Press, 1994); John O'Neill, *Ecology, Policy and Politics* (New York: Routledge, 1993), 148-55. For an antirealist discussion, see Keith Tester, *Animals and Society* (New York: Routledge, 1991), 46.
14. See Ian Hacking, "The Participant Irrealist At Large in the Laboratory," *British Journal of the Philosophy of Science* 39 (1988): 277-94; Ian Hacking, *Historical Ontology* (Cambridge, Mass.: Harvard University Press, 2002).
15. Bruno Latour and Steve Woolgar, *Laboratory Life* (Beverly Hills: Sage,

1979); see also Frederick H. Buttel and Peter J. Taylor, "Environmental Sociology and Global Environmental Change," *Society and Natural Resources* 5 (1992): 211–30; David Demeritt, "Science, Constructivism and Nature," in *Remaking Reality*, ed. Bruce Braun and Noel Castree (New York: Routledge, 1998), 173–93; John A. Hannigan, *Environmental Sociology* (New York: Routledge, 1995); Peter J. Taylor and Frederick H. Buttel, "How Do We Know We Have Global Environmental Problems?" *Geoforum* 23/3 (1992): 405–16.

16. Steven Yearly, *The Green Case* (New York: Harper and Row, 1991), 136–37.
17. Dunlap and William R. Catton, Jr., "Struggling with Human Exemptionalism," 22–23; Michael Redclift and Graham Woodgate, "Sustainability and Social Construction," in their *International Handbook of Environmental Sociology* (Northampton, Mass.: Edward Elgar, 1997), 55–70.
18. Richard Rorty, *Philosophy and the Mirror of Nature* (Princeton: Princeton University Press, 1979).
19. Roy Bhaskar, *Plato Etc.* (New York: Verso, 1994), 253; see also Sergio Sismondo, "Some Social Constructions," *Social Studies of Science* 27/3 (1993): 515–53, esp. 535.
20. Harry M. Collins, "Stages in the Empirical Programme of Relativism," *Social Studies of Science* 11 (1981): 3–10, quote on 3.
21. Steve Woolgar, *Science: The Very Idea!* (London: Tavistock, 1988), 55–56, 65.
22. Klaus Eder, *The Social Construction of Nature* (London: Sage Publications, 1996), 20.
23. Phil Macnaghten and John Urry, *Contested Natures* (Thousand Oaks: Sage, 1998), 16.
24. Macnaghten and Urry, *Contested Natures*, 89.
25. Ibid., 217–18.
26. Stephen Jay Gould, *Mismeasure of Man* (New York: W.W. Norton & Company, 1996), 53–54.
27. Robert K. Merton, *Social Theory and Social Structure* (Glencoe, Ill.: Free Press, 1957); Robert K. Merton, *Science, Technology and Society in Seventeenth-Century England* (New York: Harper and Row, 1970).
28. See Stephen Jay Gould, *The Panda's Thumb* (New York: W. W. Norton, 1980), 47–48; Stephen Jay Gould, *An Urchin in the Storm* (New York: W. W. Norton, 1987), 52; Stephen Jay Gould, *Dinosaur in a Haystack* (New York: Harmony Books, 1995), 78–80.
29. Demeritt, "Science, Constructivism and Nature," 176–77.
30. David Bloor, *Knowledge and Social Imagery* (Chicago: University of Chicago Press, 1991), 33–37.

31. See Peter L. Berger and Thomas Luckman, *The Social Construction of Reality* (Garden City, N.Y.: Doubleday, 1966); Donna Jeanne Haraway, *Simians, Cyborgs, and Women: The Reinvention of Nature* (New York: Routledge, 1991); Bruno Latour, *Science in Action* (Cambridge, Mass.: Harvard University Press, 1987).
32. Demeritt, "Science, Constructivism and Nature," 176, 178–79.
33. Benton, "Biology and Social Theory."
34. For a discussion of coevolution, see Richard Norgaard, *Development Betrayed* (New York: Routledge, 1994); for a discussion of co-construction, see William R Freudenburg, Scott Frickel, and Robert Gramling, "Beyond the Nature/Society Divide: Learning to Think about a Mountain," *Sociological Forum* 10/3 (1995): 361–92; Alan Irwin, *Sociology and the Environment* (Cambridge: Polity, 2001).
35. See Haraway, *Simians, Cyborgs, and Women*; Sandra Harding, *Whose Science?, Whose Knowledge?* (Ithaca, N.Y.: Cornell University Press, 1991); Merchant, *Radical Ecology*; Kate Soper, *What Is Nature?* (Oxford: Blackwell, 1995).
36. Soper, *What Is Nature?* 151–52.
37. Ibid., 160.
38. Hacking, *Historical Ontology*, 17; Sismondo, "Some Social Constructions," 537. In the famous "epistemological chicken" debate between Collins and Yearly, and Callon and Latour, the latter two are presented as philosophical radicals who are shifting away from "social realism" (starting from the social nature of truth) toward "natural realism" (starting from natural objects). Latour and Callon, however, claim that their actor-network theory cannot be seen as subservient to traditional "natural realism" of science in that it is aimed at questioning an ontological understanding that even at the level of basic vocabulary is rooted in human-centered terms. Many social-ecological theorists therefore see this kind of radical questioning as liberatory. See Harry M. Collins and Steven Yearly, "Epistemological Chicken," in *Science as Practice and Culture*, ed. Andrew Pickering (Chicago: University of Chicago Press, 1992), 301–26; and Michael Callon and Bruno Latour, "Don't Throw the Baby Out with the Bath School!: A Reply to Collins and Yearly," 343–68, in the same volume.
39. Bruno Latour, *We Have Never Been Modern* (Cambridge, Mass.: Harvard University Press, 1993), 8.
40. See Bruce Braun and Noel Castree, *Remaking Reality* (New York: Routledge, 1998); Noel Castree, "Marxism and the Production of Nature," *Capital & Class* 72 (2000): 5–36; Brian J. Gareau, "We Have Never Been Human: Agential Nature, ANT, and Marxist Political Ecology," *Capitalism, Nature, Socialism* 16/4 (2005): 127–40.
41. Roy Bhaskar, *A Realist Theory of Science* (London: Verso, 1975); Roy

Bhaskar, *The Possibility of Naturalism*; Sayer, *Method in Social Science*; Ted Benton and I. Craib, *Philosophy of Social Science* (London: Palgrave, 2001); Dickens, *Reconstructing Nature*; John Bellamy Foster, *Marx's Ecology* (New York: Monthly Review Press, 2000); Soper, *What Is Nature?*

42. Sayer, *Method in Social Science*, 34.
43. Phil Macnaghten and John Urry, "Towards a Sociology of Nature," *Sociology* 29/2 (1995): 203–20; Macnaghten and Urry, *Contested Natures*, 15–19.
44. Macnaghten and Urry, "Towards a Sociology of Nature," 210.
45. Jack Morrell, "Externalism," in *Dictionary of the History of Science*, ed. W. F. Bynum, E. J. Browne, and Roy Porter (Princeton: Princeton University Press, 1981), 145–46; Steven Shapin, "Hessen Thesis," 185–86, and Steven Shapin, "Merton Thesis," 262, in the same volume.
46. J. G. Crowther, *The Social Relations of Science* (London: Cresset Press, 1967), 432; also see I. Bernard Cohen, "Introduction: The Impact of the Merton Thesis," in *Puritanism and the Rise of Modern Science*, ed. I. Bernard Cohen (New Brunswick: Rutgers University Press, 1990), 1–111, esp. 55. The impact that Hessen had on sociologists, historians, and philosophers of science is dramatized by the fact that the personal library of one of us (John Bellamy Foster) includes a copy of the 1971 edition of *Science at the Cross Roads* containing Hessen's paper. The book was previously part of the library of sociologist Edward Shils (1910–1995), a major contributor to the sociology of science. It was sent to Shils by the distinguished chemist and philosopher of science Michael Polanyi (1891–1976). This copy of *Science at the Cross Roads* still contains a printed card that says "with the compliments of Michael Polanyi." Significantly, the card was found on the opening page of Hessen's essay.
47. Boris Hessen, "The Social and Economic Roots of Newton's 'Principia,'" in *Science at the Cross Roads*, ed. N. I. Bukharin et al. (London: Frank Cass, 1971), 147–212; see also Loren R. Graham, "The Socio-Political Roots of Boris Hessen," *Social Studies of Science* 15/4 (1985): 703–22; David Joravsky, *Soviet Marxism and Nature Science 1917–1932* (New York: Columbia University Press, 1961); Neal Wood, *Communism and British Intellectuals* (Cambridge: Cambridge University Press. 1959), 145.
48. As stated in a letter from Engels to W. Borgius, January 25, 1894. See Karl Marx and Frederick Engels, *Selected Correspondence 1844–1895* (Moscow: Progress, 1975), 441.
49. Edgar Zilsel, "The Sociological Roots of Science," *American Journal of Sociology* 47/4 (1942): 544–62, esp. 558; Edgar Zilsel, *The Social Origins of Modern Science* (Boston: Kluwer Academic, 2000).
50. Steven Shapin, "Discipline and Bounding," *History of Science* 30 (1992): 333–69; Hessen, "The Social and Economic Roots"; J. D. Bernal, *The Social Function of Science* (New York: Macmillan, 1939); Benjamin

Farrington, *Francis Bacon* (New York: Schuman, 1949); Joseph Needham, *Science and Civilization in China* (Cambridge: Cambridge University Press, 1954).

51. See Clifford D. Conner, *A People's History of Science* (New York: Nation Books, 2005), 275–82; Pamela H. Smith, *The Body of the Artisan* (Chicago: University of Chicago Press, 2004), 151, 239.
52. Merton, *Social Theory and Social Structure*; Merton, *Science, Technology and Society in Seventeenth-Century England.*
53. Shapin, "Discipline and Bounding," 342.
54. Jan Golinski, *Making Natural Knowledge* (Chicago: University of Chicago Press, 2005), 48–50.
55. Stephen Cole, *Making Science: Between Nature and Society* (Cambridge, Mass.: Harvard University Press, 1992), 4; Merton, *Social Theory and Social Structure*, 554.
56. Robert K. Merton, *On the Shoulders of Giants* (New York: Harcourt Brace Jovanovich, 1985), 134.
57. Piotr Sztompka, *Robert K. Merton: An Intellectual Profile* (New York: St. Martin's Press, 1986), 35.
58. Robert K. Merton, *The Sociology of Science* (Chicago: University of Chicago Press, 1973), 217.
59. See Shapin, "Discipline and Bounding," 337.
60. Merton's relation to Hessen was complex. Although distancing himself from Hessen's Marxian views and his strongly materialist externalism in the sociology of science, Merton nonetheless openly defended Hessen against those who sought to discard his insights completely. See Robert K. Merton, "Science and the Economy of Seventeenth Century England," *Science & Society* 3/1 (1939): 3–27. What was to be the most influential part of Merton's classic *Science, Technology and Society in Seventeenth-Century England* addressed Protestantism and its influence on science and thus has sometimes been thought of as somewhat Weberian in emphasis. But the second part of this work focused on the materialist-technological conditions (what Zuckerman has called "The Other Merton Thesis"). See Harriet Zuckerman, "The Other Merton Thesis," *Science in Context* 3 (1989): 239–67. This part of the analysis, as Merton himself was at great pains to point out, was concerned with the economic and materialistic interpretation of history, inspired by Hessen's, if somewhat "crude," Marxian analysis. See Merton, *Science, Technology and Society in Seventeenth-Century England*, 142–43, 185–87, 201–6.
61. Shapin, "Discipline and Bounding," 338.
62. Sztompka, *Robert K. Merton*, 78–79.
63. Nancy C. M. Hartsock, *Money, Sex, and Power: Toward a Feminist Historical Materialism* (New York: Longman, 1983); Dorothy E. Smith,

The Everyday World as Problematic: A Feminist Sociology (Boston: Northeastern University Press, 1987); Harding, *Whose Science? Whose Knowledge?*; Sandra Harding, *Is Science Multicultural?* (Bloomington: Indiana University Press, 1998); Haraway, *Simians, Cyborgs, and Women*; Soper, *What Is Nature?*

64. Harding, *Whose Science? Whose Knowledge?*, 120.
65. Fredric Jameson, "History and Class Consciousness as an 'Unfinished Project,'" *Rethinking Marxism* 1/1 (1988): 49–72, esp. 64; Georg Lukács, *History and Class Consciousness* (London: Merlin Press, 1971).
66. Gould, *Mismeasure of Man.*
67. Harding, *Whose Science? Whose Knowledge?*, 119.
68. Ibid., 134–35.
69. Ibid., 148–49.
70. Ibid., 163.
71. Ibid., 36; William Leiss, *The Domination of Nature* (Montreal: McGill–Queen's University Press, 1974); also see Merchant, *The Death of Nature.*
72. Harding, *Whose Science? Whose Knowledge?*, 149.
73. Lakatos, *The Methodology of Scientific Research Programmes*; Thomas S. Kuhn, *The Structure of Scientific Revolutions* (Chicago: University of Chicago Press, 1962); Thomas S. Kuhn, *The Road Since Structure* (Chicago: University of Chicago Press, 2000).
74. Lakatos, *The Methodology of Scientific Research Programmes*, 134; Brendan Larvor, *Lakatos: An Introduction* (New York: Routledge, 1998), 50.
75. Although Lakatos normally refers to research programs as "progressive" or "degenerating," we often follow Kuhn in his treatment of Lakatos in referring to the latter form as "degenerative." See Kuhn, *The Road since Structure*, 132.
76. Lakatos, *The Methodology of Scientific Research Programmes*, 112.
77. Ibid., 114.
78. Michael Burawoy, "Marxism as Science," *American Sociological Review* 55 (1990): 775–93, esp. 777.
79. Kuhn, *The Structure of Scientific Revolutions.*
80. Lakatos, *The Methodology of Scientific Research Programmes*, 69.
81. Despite that Kuhn thought of himself primarily as an internalist, concerned with the logic and historical development of scientific rationality in its own terms, he was well acquainted with classical externalist contributions by sociologists, having carefully studied Hessen, Merton, and Zilsel. See Kuhn, *The Road Since Structure*, 131, 287.
82. Raymond Williams, *Problems in Materialism and Culture* (London: New York, 1980), 67.

83. Karl Marx and Friedrich Engels, *Collected Works* (New York: International Publishers, 1975), vol. 45, 106–8.
84. Ibid.
85. Marx and Engels, *Selected Correspondence 1844–1895*, 120.
86. Marx and Engels, *Collected Works*, vol. 25, 584; Marx and Engels, *Selected Correspondence 1844–1895*, 284; see also, Michael Bell, *An Invitation to Environmental Sociology* (Thousand Oaks, Calif.: Pine Forge Press, 2004), 184.
87. Richard Hofstadter, *Social Darwinism in American Thought* (Boston: Beacon, 1955).
88. Marx and Engels, *Collected Works*, vol. 25, 586–87; capitalization in the original; Paul Burkett and John Bellamy Foster, "The Podolinsky Myth," *Historical Materialism* 16 (2008): 115–61, see esp. 133–34.
89. Richard Dawkins, *The Selfish Gene* (New York: Oxford University Press, 1976).
90. Bell, *An Invitation to Environmental Sociology*, 184.
91. Marx and Engels, *Collected Works*, vol. 45, 106–8.
92. Marx and Engels, *Collected Works*, vol. 5, 475–76, 479–81.
93. Karl Marx, *Early Writings* (New York: Vintage, 1974), 174.
94. G. W. F. Hegel, *Philosophy of Nature*, vol. 1 (New York: Humanities Press, 1970), 212.
95. Auguste Cornu, *The Origins of Marxist Thought* (Springfiled, Ill.: Charles C. Thomas, 1957), 437–40.
96. David R. Keller and Frank B. Golley, *The Philosophy of Ecology* (Athens: University of Georgia Press, 2000), 7–9; Robert P. McIntosh, *The Background of Ecology* (Cambridge: Cambridge University Press, 1985), 22, 29.
97. Frederic Clements and Ralph Chaney, *Environment and Life in the Great Plains* (Washington, D.C.: Carnegie Institution, 1937), 51.
98. Frederic Clements, "Preface to *Plant Succession*," in *The Philosophy of Ecology*, ed. David R. Keller and Frank B. Golley (Athens: University of Georgia Press, 2000), 35–41, quote on 36; Keller and Golley, *The Philosophy of Ecology*; Richard Levins and Richard Lewontin, *The Dialectical Biologist* (Cambridge, Mass.: Harvard University Press, 1985); Worster, *Nature's Economy*.
99. Clements and Chaney, *Environment and Life*, 47–51.
100. Ronald C. Tobey, *Saving the Prairies: The Life Cycle of the Founding School of American Plant Ecology, 1895–1955* (Berkeley: University of California Press, 1981), 182.
101. Levins and Richard Lewontin, *The Dialectical Biologist*, 135.
102. Tobey, *Saving the Prairies*, 213–21.

103. Heinz L. Ansbacher, "On the Origin of Holism," *Individual Psychology* 50/4 (1994): 486–92.
104. John William Bews, *Plant Forms and Their Evolution in South Africa* (London: Longmans, Green and Co., 1925).
105. Jan Christian Smuts, *Holism and Evolution* (New York: Macmillan Company, 1926).
106. Robert Harvey, *The Fall of Apartheid: The Insider Story from Smuts to Mbeki* (Hampshire: Palgrave, 2001), 36–38.
107. Jan Christian Smuts, *Greater South Africa: Plans for a Better World* (Johannesburg: The Truth Legion, 1940), 2–3.
108. See Peder Anker, *Imperial Ecology: Environmental Order in the British Empire, 1895–1945* (Cambridge, Mass.: Harvard University Press, 2001), 45–47; T. R. H. Davenport and Christopher Saunders, *South Africa: A Modern History* (New York: St. Martin's Press, 2000), 244–45; W. K. Hancock, *Smuts: The Sanguine Years 1870–1919* (Cambridge: Cambridge University Press, 1962), 325–47. A crucially important source for our analysis of the debate between Smuts and Tansley and the divisions in ecology in this period is Peder Anker's landmark work *Imperial Ecology: Environmental Order in the British Empire, 1895–1945*. Anker's analysis of the political divisions associated with the ecological debate represented by Smuts and Tansley is far superior to earlier accounts. His research is concerned with two different models of ecological and social management in the empire. Ours is focused on the materialist versus idealist origins of ecological science in this period and its implications for a sociology of ecology (and for the field of environmental sociology). Hence we draw on different materials and arrive at different (but not necessarily divergent) conclusions.
109. Anker, *Imperial Ecology*, 46–51; Davenport and Christopher Saunders, *South Africa*, 292–93; Sven Lindqvist, *A History of Bombing* (London: Granta, 2000), sec. 107.
110. Roy Campbell, *Adamastor* (London: Faber and Faber, 1930), 103.
111. Smuts, *Holism and Evolution*, 21.
112. Ibid., 340.
113. Ibid., 80–82.
114. Ibid., 99.
115. Ibid., 213.
116. Anker, *Imperial Ecology*, 72, 191–92.
117. Jan Christian Smuts, "The Scientific World-Picture of Today," Presidential Address to the British Association for the Advancement of Science, in *Report of The Centenary Meeting* (London: British Association for the Advancement of Science, 1932), 1–18, quote on 17–18.
118. Alfred Adler, *Social Interest* (New York: Capricorn Books, 1964), 68; see also Ansbacher, "On the Origin of Holism," 491.

119. Smuts, *Holism and Evolution*, 297–313.
120. As quoted in Anker, *Imperial Ecology*, 191.
121. W. E. B. Du Bois, *The World and Africa* (New York: Viking Press, 1947), 43.
122. Jan Christian Smuts, *Africa and Some World Problems* (Oxford: Clarendon Press, 1930), 30–32, 43.
123. Smuts, *Africa and Some World Problems*, 75–76.
124. In *Holism and Evolution*, Smuts incorporated the "ontogeny recapitulates phylogeny" theory in support of his theory of holism. See 74, 115.
125. Stephen Jay Gould, *Ontogeny and Phylogeny* (Cambridge, Mass.: Harvard University Press, 1977), 126; see also Gould, *Mismeasure of Man*, 142–51.
126. Smuts, *Africa and Some World Problems*, 92–93.
127. Ibid., 76; see also Jan Christian Smuts, "Climate and Man in Africa," *South African Journal of Science* 29 (1932): 98–131, esp. 127–30.
128. Smuts, *Africa and Some World Problems*, 33, 75–78.
129. See Marx and Engels, *Collected Works*, vol. 45, 50; Smuts, "The Scientific World-Picture of Today," 10; John Tyndall, "Address Delivered before the British Association Assembled at Belfast, with Additions," in *Nineteenth Century Science: An Anthology*, ed. A. S. Weber (Peterborough, Ont.: Broadview Press, 2000), 359–85.
130. Smuts, "The Scientific World-Picture of Today," 12–13, emphasis added.
131. Anker, *Imperial Ecology*, 137–43; Tobey, *Saving the Prairies*. John Scott Haldane was J. B. S. Haldane's father. The latter was a staunch materialist and one of the Baconian Marxists.
132. Karl Popper, *The Open Society and Its Enemies*, vol. 2 (New York: Harper and Row, 1962), 29, 304–5.
133. John Phillips, "A Tribute to Frederic E. Clements and His Concepts in Ecology," *Ecology* 35/2 (1954): 114–15.
134. John Phillips, "Man at the Cross-Roads," in *Our Changing World-View*, ed. Jan Christian Smuts (Johannesburg: University of Witwatersrand Press, 1932).
135. Anker, *Imperial Ecology*, 192.
136. Ibid., 148; Phillips, "Man at the Cross-Roads," 51–52.
137. John William Bews, *Human Ecology* (London: Oxford University Press, 1935).
138. Anker, *Imperial Ecology*, 167; John William Bews, "The Ecological Viewpoint," *South African Journal of Science* 28 (1931): 1–15, esp. 4; Bews, *Human Ecology*, 18–20, 54, 155, 256.
139. Anker, *Imperial Ecology*, 171–75.
140. Ibid., 122.
141. Lancelot Hogben, *The Nature of Living Matter* (New York: Alfred A. Knopf, 1931), 224.

142. Hyman Levy, *The Universe of Science* (London: Century Co., 1933).
143. Hyman Levy, *A Philosophy for a Modern Man* (New York: Alfred A. Knopf, 1938).
144. Like his contemporaries H. G. Wells and Julian Huxley, Tansley was a moderate socialist or social democrat and an adamant materialist. He believed in what he once called a "semi-socialist society." His views thus overlapped in certain critical areas with those of noted British radical scientists of his day, such as Bernal, Haldane, Hogben, Levy, and Needham. However, he was strongly critical of the Soviet Union and what he perceived as its "totalitarian" manner of organizing scientific research. During the famous "social function of science" debate of the 1940s, Tansley was one of the two founders of the Society for Freedom in Science, which opposed the proposals of Bernal and others on the social organization of science. See Anker, *Imperial Ecology*, 22, 224; Gary Werskey, *The Visible College* (New York: Holt, Rinehart and Winston, 1978), 281–82.
145. Stephen Jay Gould, *I Have Landed* (New York: Three Rivers Press, 2003), 113–29; Joseph Lester, *E. Ray Lankester and the Making of Modern British Biology* (Oxford: British Society for the History of Science, 1995).
146. John Bellamy Foster, "E. Ray Lankester, Ecological Materialist: An Introduction to Lankester's 'The Effacement of Nature by Man,'" *Organization & Environment* 13/2 (2000): 233–35; E. Ray Lankester, *Science from an Easy Chair: A Second Series* (New York: Henry Holt, 1913), 365–72.
147. E. Ray Lankester, "Preface," in Hugh S. R. Elliot, *Modern Science and the Illusions of Professor Bergson* (New York: Longmans, Green, and Co., 1912), vii–xvii.
148. Arthur G. Tansley, *Practical Plant Ecology* (London: George Allen and Unwin, 1926), 21–25.
149. Arthur G. Tansley, *The New Psychology and Its Relation to Life* (London: G. Allen & Unwin, 1920).
150. Arthur G. Tansley, *Mind and Life* (London: George Allen and Unwin, 1952).
151. Keller and Frank B. Golley, *The Philosophy of Ecology*, 12.
152. As quoted in Anker, *Imperial Ecology*, 141–42.
153 Arthur G. Tansley, "The Use and Abuse of Vegetational Concepts and Terms," *Ecology* 16/3 (1935): 284–307.
154. John Phillips, "The Complex Organism," part 3 of "Succession, Development, the Climax, and the Complex Organism," *Journal of Ecology* 23/2 (1935): 488–502.
155. Tansley, "The Use and Abuse," 299.
156. Anker, *Imperial Ecology*, 153.
157. Tansley, "The Use and Abuse," 300.

158. Ibid., 299.
159. For a discussion of the crucial role of abstraction (alongside the concept of internal relations) in dialectical thinking, see Bertell Ollman, *Dance of the Dialectic* (Urbana: University of Illinois Press, 2003), 59–112. It is noteworthy that Ollman emphasizes the impact of Levy on his own thinking. See Bertell Ollman, *Alienation* (Cambridge: Cambridge University Press, 1976), 286.
160. Tansley, "The Use and Abuse," 300.
161. Tobey, *Saving the Prairies*, 177–78.
162. Tansley, "The Use and Abuse," 297.
163. Ibid., 303.
164. Ibid., 289.
165. Ibid., 304.
166. Arthur G. Tansley, *The British Islands and Their Vegetation* (Cambridge: University Press, 1939), vi.
167. Ibid., 194–95.
168. Ibid., 149–64.
169. Ibid., 128.
170. Ibid., 128.
171. Anker, *Imperial Ecology*, 299.
172. Tansley, "The Use and Abuse," 285.
173. Ibid., 287–88, 304–5; see also Tansley, *The British Islands*.
174. H. G. Wells, Julian S. Huxley, and G. P. Wells, *The Science of Life* (New York: Literary Guild, 1934).
175. Anker, *Imperial Ecology*, 248–50.
176. H. G. Wells, *The Fate of Man* (New York: Longmans, Green and Co., 1939), 191–92.
177. Tansley, "The Use and Abuse," 305.
178. McIntosh, *The Background of Ecology*, 299; Arthur G. Tansley, *Our Heritage of Wild Nature: A Plea for Organized Nature Conservation* (Cambridge: At the University Press, 1945).
179. Wells, *The Fate of Man*, 191–92.
180. Daniel Simberloff, "A Succession of Paradigms in Ecology," in *The Philosophy of Ecology*, ed. David R. Keller and Frank B. Golley (Athens: University of Georgia Press, 2000), 71–80, quote on 77.
181. Keller and Golley, *The Philosophy of Ecology*, 176.
182. Frederic Clements, "Nature and Structure of the Climax," *Journal of Ecology* 24/1 (1936): 252–84.
183. Phillips, "The Complex Organism."
184. Ibid., 505.
185. Ibid., 489.
186. John Steinbeck, *The Grapes of Wrath* (New York: Viking, 1939); Keller and Golley, *The Philosophy of Ecology*, 28; Tobey, *Saving the Prairies*.

187. Tobey, *Saving the Prairies*, 213–21.
188. Leiss, *The Domination of Nature*.
189. Simberloff, "A Succession of Paradigms in Ecology."
190. Raymond L. Lindeman, "The Trophic-Dynamic Aspect of Ecology," *Ecology* 23 (1942): 399–418; McIntosh, *The Background of Ecology*, 196.
191. Frank Benjamin Golley, *A History of the Ecosystem Concept in Ecology* (New Haven: Yale University Press, 1993); Joel B. Hagen, *An Entangled Bank: The Origins of Ecosystem Ecology* (New Brunswick, N.J.: Rutgers University Press, 1992), 100–107.
192. Peter Dickens, *Society and Nature* (Cambridge: Polity Press, 2004); Foster, "Marx's Theory of Metabolic Rift"; Foster, *Marx's Ecology*.
193. Rachel Carson, *Lost Woods* (Boston: Beacon Press, 1998); Commoner, *The Closing Circle*; see also John Bellamy Foster and Brett Clark, "Rachel Carson's Ecological Critique," *Monthly Review* 59/9 (2008): 1–17. In a study of international environmental treaties from 1870 to 1990, David John Frank empirically demonstrated "the world-level rise and consolidation of the scientific ecosystem model of nature in the post–World War II period." See David John Frank, "Science, Nature, and the Globalization of the Environment, 1870–1990," *Social Forces* 76/2 (1997): 409–37, quote on 428.
194. Wendell Berry, *Life Is a Miracle* (Washington, D.C.: Counterpoint, 2000); Levins and Lewontin, *The Dialectical Biologist*; Lewontin and Levins, *Biology Under the Influence* (New York: Monthly Review Press, 2007).
195. Ironically, some ecological theorists and historians have suggested that the organicism and holism represented by Clements and Smuts persists in the work of ecological systems theorists such as Eugene Odum. Yet Odum's systems ecology descends from the materialist ecosystems analysis initiated by Tansley rather than the philosophical idealism and conception of climax communities propounded by Clements and Smuts. See Barbour, "Ecological Fragmentation in the Fifties"; Robert P. McIntosh, "The Background and Some Current Problems of Theoretical Ecology," *Synthese* 43 (1980): 195–255, esp. 204, 243; Simberloff, "A Succession of Paradigms in Ecology," 77. The teleological view represented by Smuts can also be seen in the work of such figures as Fritjof Capra. Such analysis tries dialectically to transcend the idealist–materialist divide in the interest of a broader ecological holism. See Fritjof Capra, *The Web of Life* (New York: Anchor Books, 1996).
196. Paul E. Stepansky, *In Freud's Shadow: Adler in Context* (Hillsdale, N.J.: Analytic Press, 1983), 254.
197. Adler's journal published a translation of Smuts's 1931 presidential address to the BAAS, which had defended scientific idealism. See Ansbacher, "On the Origin of Holism," 490.

198. Anker, *Imperial Ecology*, 180–81; Phyllis Bottome, *Alfred Adler* (London: Faber and Faber, 1957), 83–84.
199. Pepper, *Modern Environmentalism*, 233–34, 242–45; Worster, *Nature's Economy*, 301–4, 316–23; see also Barbour, "Ecological Fragmentation in the Fifties."
200. In his later *Wealth of Nature*, Worster writes much more positively of Tansley, and Smuts is no longer posed as an alternative. See Donald Worster, *The Wealth of Nature* (Oxford: Oxford University Press, 1993), 175.
201. Merchant, *The Death of Nature*, 292–93.
202. Ibid.
203. Ibid., 252.
204. See chapter 12 in this book; see also Werskey, *The Visible College*.
205. Levins and Lewontin, *The Dialectical Biologist*.
206. Roy Bhaskar, *Dialectic: The Pulse of Freedom* (London: Verso, 1993).
207. Harding, *Whose Science? Whose Knowledge?*
208. For example, see Merchant, *The Death of Nature*; Worster, *Nature's Economy*.
209. See Gould, *Mismeasure of Man*.
210. Sayer, *Method in Social Science*, 40.
211. Ibid.
212. Taylor and Buttel, "How Do We Know We Have Global Environmental Problems?"; Yearly, *The Green Case*, 136–37.
213. Spencer R. Weart, *The Discovery of Global Warming* (Cambridge, Mass.: Harvard University Press, 2003), 3–4.
214. See chapter 5 in this book.
215. Soper, *What Is Nature?*, 249.
216. Ibid.

CHAPTER FIFTEEN: IMPERIALISM AND ECOLOGICAL METABOLISM

This chapter is adapted and revised for this book from the following article: Brett Clark and John Bellamy Foster, "Ecological Imperialism and the Global Metabolic Rift: Unequal Exchange and the Guano/Nitrates Trade," *International Journal of Comparative Sociology* 50/3–4 (2009): 311–34.

1. Alfred W. Crosby, *Ecological Imperialism* (Cambridge: Cambridge University Press, 1986).
2. Ecological degradation and environmental problems are not limited to the capitalist economic system. Ecological contradictions are present in all societies, as noted in numerous environmental histories. See Jared

Diamond, *Collapse* (New York: Viking, 2005); John Bellamy Foster, *The Vulnerable Planet* (New York: Monthly Review Press, 1994); Clive Ponting, *A Green History of the World* (New York: Penguin, 1993).

3. Andre Gunder Frank, *Capitalism and Underdevelopment in Latin America* (New York: Monthly Review Press, 1967); Immanuel Wallerstein, *The Modern World-System I* (New York: Academic Press, 1974).
4. Stephen Bunker and Paul Ciccantell, *Globalization and the Race for Resources* (Baltimore: Johns Hopkins University Press, 2005).
5. Stephen Bunker, "Modes of Extraction, Unequal Exchange, and the Progressive Underdevelopment of an Extreme Periphery," *American Journal of Sociology* 89 (1984): 1017–64; Paul Burkett, *Marx and Nature* (New York: St. Martin's Press, 1999); Alf Hornborg, "Cornucopia or Zero-Sum Game? The Epistemology of Sustainability," *Journal of World-Systems Research* 9/2 (2003): 205–16. See Burkett for a detailed analysis of the relationship between material ecological flows (usually expressed in terms of use values) and value flows in Marx's analysis.
6. Stephen Bunker, *Underdeveloping the Amazon* (Urbana: University of Illinois Press, 1985).
7. See Arghiri Emmanuel, *Unequal Exchange* (New York: Monthly Review Press, 1972); R. Scott Frey, "The International Traffic in Hazardous Wastes," *Journal of Environmental Systems* 23 (1994): 165–77; Hornborg, "Cornucopia or Zero-Sum Game?"; James Rice, "Ecological Unequal Exchange," *International Journal of Comparative Sociology* 48 (2007): 43–72.
8. Alf Hornborg, "Towards an Ecological Theory of Unequal Exchange," *Ecological Economics* 25 (1998):127–36; Alf Hornborg, *The Power of the Machine* (Walnut Creek, Calif.: AltaMira Press, 2001); Andrew K. Jorgenson, "Unequal Ecological Exchange and Environmental Degradation," *Rural Sociology* 71 (2006): 685–712.
9. Mark Elvin, *The Retreat of the Elephants: An Environmental History of China* (New Haven: Yale University Press, 2004), 470.
10. Karl Marx, *Capital*, vol. 1 (New York: Vintage, 1976), 283, 637–38.
11. John Bellamy Foster, *Marx's Ecology* (New York: Monthly Review Press, 2000).
12. István Mészáros, *Beyond Capital* (New York: Monthly Review Press, 1995).
13. Ibid., 44, 170–71.
14. Ibid., 41–45.
15. Jason W. Moore, "Environmental Crises and the Metabolic Rift in World-Historical Perspective," *Organization & Environment* 13/2 (2000): 123–57, quote on 124.
16. Marx, *Capital*, vol. 1, 915.
17. Karl Marx, *Grundrisse* (New York: Penguin Books, 1993), 409–10.

18. John Chalmers Morton, "On the Forces Used in Agriculture," *Journal of the Society of Arts* (December 9, 1859): 53–68.
19. Morton's *Cyclopedia of Agriculture, Practical and Scientific*, with which Marx was closely familiar, contained detailed scientific articles on "Guano," "Manure," "Sewage Manure," as well as articles on "Labour" and agricultural technology. The article on guano dealt with Peru's guano islands. See John Chalmers Morton, ed., *A Cyclopedia of Agriculture, Practical and Scientific* (London: Blackie and Son, 1855).
20. Karl Marx, *Capital*, vol. 3 (New York: Penguin Books, 1991), 950.
21. Marx, *Capital*, vol. 1, 637.
22. Justus von Liebig, *Letters on Modern Agriculture* (London: Walton & Maberly, 1859), 130–31.
23. As quoted in Erland Mårald, "Everything Circulates," *Environment and History* 8 (2002): 65–84, quote on 74.
24. Marx, *Capital*, vol. 1, 860.
25. Ibid., 579–80.
26. Marx, *Capital*, vol. 3, 753.
27. Hal Draper, *The Marx-Engels Glossary*, vol. 3 (New York: Schocken Books, 1986), 15.
28. August Bebel, *Woman in the Past, Present and Future* (London: Zwan Publications, 1988), 208.
29. Elvin, *The Retreat of the Elephants*, 470.
30. Father Joseph de Acosta, *The Natural and Moral History of the Indies*, vol. 1 (New York: Burt Franklin, 1880 [1604]).
31. Jimmy M. Skaggs, *The Great Guano Rush* (New York: St. Martin's Griffin, 1994).
32. Ibid.
33. Liebig, *Letters on Modern Agriculture*.
34. Justus von Liebig, *Familiar Letters*, 3rd ed. (London: Taylor, Walton, and Maberley, 1851), 473.
35. Marx, *Capital*, vol. 3, 949; John Bellamy Foster, "Marx's Theory of Metabolic Rift," *American Journal of Sociology* 105/2 (1999): 366–405.
36. C. R. Fay, "The Movement Towards Free Trade, 1820–1853," in *The Cambridge History of the British Empire*, ed. J. Holland Rose, A.P. Newton, and E.A. Benians (Cambridge: Cambridge University Press, 1940), 388–414, quote on 395.
37. William Jefferson Dennis, *Tacna and Arica* (New Haven: Yale University Press, 1931); Bruce W. Farcau, *The Ten Cents War* (Westport, Conn.: Praeger, 2000).
38. Eduardo Galeano, *Open Veins of Latin America* (New York: Monthly Review Press, 1973), 72–73.
39. George W. Peck, *Melbourne and the Chincha Islands* (New York: Charles Scribner, 1854).

40. W. M. Mathew, "Foreign Contractors and the Peruvian Government at the Outset of the Guano Trade," *Hispanic American Historical Review* 52/4 (1972): 598–620; W. M. Mathew, "A Primitive Export Sector," *Journal of Latin American Studies* 9/1 (1977): 35–57; W. M. Mathew, *The House of Gibbs and the Peruvian Guano Monopoly* (London: Royal Historical Society, 1981).
41. Anthony Gibbs and Sons, Ltd., *Guano* (London: William Clowes & Sons, 1843).
42. J. H. Sheppard, *A Practical Treatise on the Use of Peruvian and Ichaboe African Guano* (London: Simpkin, Marshall & Co., 1844); Joseph A. Smith, *Productive Farming* (London: William Tiat, 1843); Edward Solly, *Rural Chemistry* (London: Office of "The Gardeners' Chronicle," 1843); Joshua Trimmer, *Science with Practice* (London: John Thomas Norris, 1843).
43. Marx, *Capital*, vol. 1, 348.
44. Marx, *Grundrisse*, 527.
45. Eugene D. Genovese, *The Political Economy of Slavery* (Vintage: New York, 1967).
46. Skaggs, *The Great Guano Rush*.
47. Ibid.
48. W. M. Mathew, "The Imperialism of Free Trade," *Economic History Review* 21/3 (1968): 562–79.
49. C. Alexander G. de Secada, "Arms, Guano, and Shipping," *Business History Review* 59/4 (1985): 597–621.
50. Dennis, *Tacna and Arica*; Farcau, *The Ten Cents War*.
51. Alanson Nash, "Peruvian Guano," *Plough, the Loom and the Anvil*, August 1857; "Guano Trade," *New York Observer and Chronicle*, July 24, 1856.
52. Robert Craig, "The African Guano Trade," *The Mariner's Mirror* 50/1 (1964): 25–55, see quote on 35–37.
53. Heraclio Bonilla, "Peru and Bolivia," in *Spanish America After Independence c. 1820– c. 1870*, ed. Leslie Bethell (Cambridge: Cambridge University Press, 1987), 239–82.
54. Shane Hunt, *Growth and Guano in Nineteenth Century Peru*, Discussion Paper no. 34 for the Research Program in Economic Development, Princeton University, unpublished, 1973.
55. A. J. Duffield, *Peru in the Guano Age* (London: Richard Bentley and Son, 1877).
56. Peter Blanchard, "The 'Transitional Man' in Nineteenth-Century Latin America," *Bulletin of Latin American Research* 15/2 (1996): 157–76; Stephen M. Gorman, "The State, Elite, and Export in Nineteenth Century Peru," *Journal of Interamerican Studies and World Affairs* 21/3 (1979): 395–418.

57. Paul Gootenberg, *Imagining Development* (Berkeley: University of California Press, 1993).
58. Duffield, *Peru in the Guano Age*, 89.
59. J. B. Boussingault, *Rural Economy* (New York: D. Appleton & Co., 1845), 290.
60. Robert Cushman Murphy, *Bird Islands of Peru* (New York: G. P. Putnam's Sons, 1925).
61. Michael J. Gonzales, "Chinese Plantation Workers and Social Conflict in Peru in the Late Nineteenth Century," *Journal of Latin American Studies* 21 (1955): 385–424, quote on 390–91.
62. Evelyn Hu-DeHart, "Coolies, Shopkeepers, Pioneers," *Amerasia Journal* 15/2 (1989): 91–116; Evelyn Hu-DeHart, "*Huagong* and *Huashang*," *Amerasia Journal* 28/2 (2002): 64–90.
63. Lawrence A. Clayton, "Chinese Indentured Labor in Peru," *History Today* 30/6 (1980): 19–23; Hu-DeHart, "Coolies, Shopkeepers, Pioneers."
64. Charles Wingfield, *The China Coolie Traffic from Macao to Peru and Cuba* (London: British and Foreign Anti-Slavery Society, 1873), 4.
65. Karl Marx, *The Poverty of Philosophy* (New York: International Publishers, 1963), 112; Karl Marx and Frederick Engels, *On Colonialism* (New York: International Publishers, 1972), 123.
66. Gonzales, "Chinese Plantation Workers."
67. *Friends' Intelligencer*, "Guano Trade," August 4, 1855; Mathew, "A Primitive Export Sector"; Alanson Nash, "Peruvian Guano."
68. Peck, *Melbourne and the Chincha Islands*, 207.
69. Duffield, *Peru in the Guano Age*, 77–78.
70. Clayton, "Chinese Indentured Labor"; "Chincha Islands," *Friends' Intelligencer*, February 11, 1854; Hu-DeHart, "Coolies, Shopkeepers, Pioneers."
71. Nash, "Peruvian Guano."
72. "Chincha Islands, "*Friends' Intelligencer*.
73. Basil Lubbock, *Coolie Ships and Oil Sailers* (Glasgow: Brown, Son & Ferguson, 1955), 35.
74. "Chincha Islands," *Nautical Magazine and Naval Chronicle*, April 1856.
75. Mathew, "A Primitive Export Sector," 44.
76. "Chinese Coolie Trade," *The Christian Review*, April 1862.
77. Wingfield, *The China Coolie Traffic*, 5.
78. Gonzales, "Chinese Plantation Workers," 391.
79. William S. Coker, "The War of the Ten *Centavos*," *Southern Quarterly* 7/2 (1969): 113–29, quote on 118; de Secada, "Arms, Guano, and Shipping."
80. Farcau, *The Ten Cents War*.
81. William Columbus Davis, *The Last Conquistadors* (Atlanta: University of Georgia Press, 1950).

82. Galeano, *Open Veins of Latin America*, 155.
83. de Secada, "Arms, Guano, and Shipping," 609; Farcau, *The Ten Cents War*.
84. Rafael Larco Herrera, *Tacna–Arica* (New York: n.p., 1924), Appendix 2.
85. Dr. I. Alzamora, *Peru and Chile* (Pamphlet: n.p., n.d.); Harold Blakemore, *British Nitrates and Chilean Politics, 1886–1896* (London: University of London, 1974); Heraclio Bonilla, "The War of the Pacific and the National and Colonial Problem in Peru," *Past & Present* 81 (1978): 92–118; Dennis, *Tacna and Arica*; Henry Clay Evans, *Chile and Its Relations with the United States* (Durham, NC: Duke University Press, 1927); Farcau, *The Ten Cents War*; Michael Montéon, *Chile in the Nitrate Era* (Madison: University of Wisconsin Press, 1982); William F. Sater, *Chile and the War of the Pacific* (Lincoln: University of Nebraska Press, 1986).
86. Luis Ortega, "Nitrates, Chilean Entrepreneurs and the Origins of the War of the Pacific," *Journal of Latin American Studies* 16/2 (1984): 337–80.
87. Galeano, *Open Veins of Latin America*, 156–57.
88. John Mayo, *British Merchants and Chilean Development, 1851–1886* (Boulder, Colo.: Westview Press, 1987).
89. U.S. House of Representatives, 47th Congress, 1st Session, House Reports, Report no. 1790, *Chili–Peru*. Washington, D.C., 1882, 217–18; Perry Belmont, *An American Democrat* (New York: Columbia University Press, 1941), 255–62.
90. Dennis, *Tacna and Arica*.
91. Edward P. Crapol, *James G. Blaine* (Wilmington, Del.: Scholarly Resources Inc., 2000).
92. José Carlos Mariátegui, *Seven Interpretive Essays on Peruvian Reality* (Austin: University of Texas Press, 1971), 9–13.
93. Farcau, *The Ten Cents War*, 14.
94. Mariátegui, *Seven Interpretive Essays*, 12.
95. Ibid., 13.
96. Rory Miller, "The Making of the Grace Contract," *Journal of Latin American Studies* 8 (1976): 73–100; Ortega, "Nitrates, Chilean Entrepreneurs."
97. J. R. Brown, "The Frustration of Chile's Nitrate Imperialism," *Pacific Historical Review* 32/4 (1963): 383–96; Irving Stone, "British Long-Term Investment in Latin America, 1865–1913," *Business History Review* 42/3 (1968): 311–39.
98. Andre Gunder Frank, *The Development of Underdevelopment in Latin America* (New York: Monthly Review Press, 1969); Galeano, *Open Veins of Latin America*; J. R. McNeill, *Something New Under the Sun* (New York: W.W. Norton, 2000). During the events leading up to the civil war in Chile, U.S. foreign policy, headed by Blaine, who was again secretary of state, was sympathetic toward Balmaceda, whose nationalism was seen as a curb on British power.

99. August Bebel, *Society of the Future* (Moscow: Progress, 1971), 79.
100. Frank, *Capitalism and Underdevelopment in Latin America*; Mariátegui, *Seven Interpretive Essays*.
101. Simon Romero, "Peru Guards Its Guano as Demand Soars Again," *New York Times*, May 30, 2008.

CHAPTER SIXTEEN: THE ECOLOGY OF CONSUMPTION

1. William Morris, *News from Nowhere and Selected Writings and Designs* (London: Penguin, 1962), 121–22
2. See Richard York, Eugene A. Rosa, and Thomas Dietz, "Footprints on the Earth: The Environmental Consequences of Modernity," *American Sociological Review* 68/2 (2003): 279–300 and "A Rift in Modernity?: Assessing the Anthropogenic Sources of Global Climate Change with the STIRPAT Model," *International Journal of Sociology and Social Policy* 23/10 (2003): 31–51; Mathis Wackernagel and William Rees, *Our Ecological Footprint* (Philadelphia: New Society Publishers, 1996).
3. The term "economic Malthusianism" is applied in this way to explain the shift in the Malthusian argument from overpopulation to overconsumption in Democratic Socialist Perspective, "Symptoms and Causes of the Environmental Crisis," http://www.dsp.org.au/node/87, accessed June 15, 2010.
4. Worldwatch Institute, *The State of the World, 2010* (New York: W. W. Norton, 2010), back cover.
5. Thomas Robert Malthus, *Pamphlets* (New York: Augustus M. Kelley, 1970), 18; Thomas Robert Malthus, *An Essay on the Principle of Population and a Summary View of the Principle of Population* (London: Penguin, 1970), 177.
6. Thomas Robert Malthus, *An Essay on the Principle of Population; or a View of its Past and Present Effects on Human Happiness; With an Inquiry into Our Prospects Respecting the Future Removal or Mitigation of the Evils which it Occasions* (Cambridge: Cambridge University Press, 1989), vol. 2, 127–28. For a systematic critique of Malthus from an environmental perspective see John Bellamy Foster, *Marx's Ecology* (New York: Monthly Review Press, 2000), 81–104.
7. Thomas Robert Malthus, *Principles of Political Economy* (Cambridge: Cambridge University Press, 1989), vol. 1, 463–90. See also David Ricardo, *Notes on Malthus* (Baltimore: Johns Hopkins Press, 1928), 232–46.
8. Lester Brown's emphasis in the late 1990s on China's food needs had shifted a decade later to the threat of China in terms of the consumption of oil and its ownership of cars and the effect on the world environment.

Compare Lester Brown, *Who Will Feed China?* (New York: W.W. Norton, 1995) to Lester Brown, *Plan B 2.0* (New York: W. W. Norton, 2006).

9. Erik Assadourian, "The Rise and Fall of Consumer Cultures," in Worldwatch Institute, *State of the World, 2010* (New York: W. W. Norton, 2010), 7.
10. Thomas Princen, Michael Maniates, and Ken Conca, "Confronting Consumption," and Thomas Princen, "Consumption and Its Externalities," in *Confronting Consumption*, ed. Princen, Maniates, and Conca (Cambridge, Mass.: MIT Press, 2002), 16–17, 30; Annie Leonard, *The Story of Stuff* (New York: Free Press, 2010), xxix. Princen, Maniates, and Conca recognize that the perspective of environmental consumption means that it is not just final consumers in the economic sense that are at fault. But they make little or no use of this insight.
11. Karl Marx, *Capital*, vol. 2 (London: Penguin, 1978), 468–602. On the relation between Marx's departments and the Keynesian aggregates see Shigeto Tsuru, "Keynes versus Marx: The Methodology of Aggregates," in *Marx and Modern Economics*, ed. David Horowitz (New York: Monthly Review Press, 1968), 168–202, and Michal Kalecki, "The Marxian Equations of Reproduction and Modern Economics," in *The Faltering Economy: The Problem of Accumulation under Monopoly Capitalism*, ed. John Bellamy Foster and Henryk Szlajfer (New York: Monthly Review Press, 1984), 159–66.
12. See John Bellamy Foster, "Marxian Economics and the State," in Foster and Szlajfer, *The Faltering Economy*, 340–41.
13. It is worth noting that some segments of the anti-consumerism movement advocate that people should *earn* less and spend less, and thus don't necessarily save more.
14. Leonard, *The Story of Stuff*, 185–87; Joel Makower, "Calculating the Gross National Trash," March 20, 2009, http://www.greenbiz.com.
15. Derrick Jensen and Aric McBay, *What We Leave Behind* (New York: Seven Stories, 2009), 290–91.
16. See the discussion in Michael Dawson, *The Consumer Trap* (Urbana: University of Illinois Press, 2003), 11–14.
17. Karl Marx, *The Poverty of Philosophy* (New York: International Publishers, 1963), 41–42.
18. John Kenneth Galbraith, *The Affluent Society* (New York: New American Library, 1984), 126.
19. John Kenneth Galbraith, *The Economics of Peace and Laughter* (New York: New American Library, 1971), 75–77. See also Thomas Princen, Michael Maniates, and Ken Conca, "Conclusion: To Confront Consumption," in Princen, Maniates, and Conca, *Confronting Consumption*, 321–26.
20. Mohan Munasinghe, "Can Sustainable Consumers and Producers Save the

Planet?," *Journal of Industrial Ecology* 14/1 (2010), 4–6; Randall Krantz, "A New Vision of Sustainable Consumption: The Business Challenge," *Journal of Industrial Ecology* 14/1 (2010), 7–9.

21. Marx, *Capital*, vol. 1 (London: Penguin, 1976), 165.
22. Alan Durning, *How Much Is Enough? The Consumer Society and the Future of the Earth* (New York: W. W. Norton, 1992), 11.
23. Raymond Williams, *Marxism and Literature* (New York: Oxford University Press, 1977), 5.
24. Assadourian, "Rise and Fall of Consumer Cultures," 3, 7. Raymond Williams has noted in *Keywords* (New York: Oxford University Press, 1983) that "in archaeology and in cultural anthropology their reference to culture or a culture is primarily to material production, while in history and cultural studies their reference is primarily to signifying and symbolic systems" (91). Assadourian uses the latter, more culturalist approach, emphasizing symbolic systems, while his focus is on the economic—the role of the consumer. This has the apologetic advantage of separating the issue of consumption from material production, thereby avoiding the realities of capitalism.
25. Assadourian, "Rise and Fall of Consumer Cultures," 3.
26. What is being referred to in our argument here as "economic Malthusianism," because of its emphasis on the average economic consumer as the source of environmental problems, depends heavily, since the role of production is downplayed, on the reification of consumption, such that it becomes *its own presupposition*. The economy itself (or the market) thus becomes simply a product of consumer culture.
27. Assadourian, "Rise and Fall of Consumer Cultures," 11.
28. This thinness is also to be found in the one source Assadourian cites here: Peter N. Stearns, *Consumerism in World History* (London: Routledge, 2001), where a very few isolated examples of upper middle-class consumption and early advertisements are taken as standing for a whole process of economic development.
29. Assadourian, "Rise and Fall of Consumer Cultures," 4, 18–20.
30. Despite side discussions of "voluntary simplicity," applicable mainly to the middle class, Worldwatch's economic Malthusian argument, as represented by Assadourian, does not generally adopt the common approach of seeing consumers as capable of solving environmental problems through a kind of moral or ecological restraint, as popularized by the Environmental Defense Fund in its annual calendars depicting ten point programs—whereby individuals, simply by recycling, conserving household energy, etc., can "save the earth." (See the critical analysis in Michael Maniates, "Individuation," in Princen, Maniates, and Conca, *Confronting Consumption*, 43–66.) Rather, Assadourian, like economic Malthusianism

in general, adopts the even more elitist view that consumers must be "driven" by green business entrepreneurs (cultural pioneers) to the right decisions. Nevertheless, the "consumer society" argument feeds in various ways on the notion that it is consumers who are culpable, and the source of the original economic sin.

31. On the connection between big business and advertising/marketing see Robert W. McChesney, John Bellamy Foster, Inger L. Stole, and Hannah Holleman, "The Sales Effort and Monopoly Capital," *Monthly Review* 60/11 (April 2009): 1–23.
32. Assadourian, "Rise and Fall of Consumer Cultures," 19, "Business and Economy: Management Priorities," in Worldwatch, *State of the World, 2010*, 18, 83–84.
33. Ray Anderson, Mona Amodeo, and Ida Kubiszewski, "Changing Business Cultures from Within," Worldwatch, *State of the World, 2010*, 99–101; "Editing Out Unsustainable Behavior," Worldwatch, *State of the World, 2010*, 125–26. Worldwatch's admiration for Wal-Mart is long-standing. See the celebration of Wal-Mart in L. Hunter Lovins, "Rethinking Production," Worldwatch, *State of the World, 2008* (New York: W. W. Norton, 2008), 35–36; and Ben Block, "Wal-Mart Scrutinizes Supply-Chain Sustainability," www.worldwatch.org, July 20, 2009.
34. Stacy Mitchell, "Keep Your Eyes on the Size: The Impossibility of a Green Wal-Mart," www.grist.org, March 28, 2007; Sarah Anderson, "Wal-Mart's New Greenwashing Report," www.alternet.org, November 20, 2007; Wes Jackson quoted in Heather Rogers, *Green Gone Wrong* (New York: Scribner, 2010), 191.
35. Assadourian, "Rise and Fall of Consumer Cultures," 6–7.
36. Durning, *How Much Is Enough?*, 26–29.
37. Bureau of Labor Statistics, U.S. Department of Labor, "Consumer Expenditures in 2008," March 2010, Table 1; Michael Dawson, "Transportation Inequality in America," March 2, 2010, http://www.deathbycar.info; Hugh Mackenzie, Hans Messinger, and Rick Smith, *Size Matters: Canada's Ecological Footprint*, Canadian Centre for Policy Alternatives, www.GrowingGap.ca, June 2008.
38. Matthew Miller and Duncan Greenberg, ed., "The Richest People in America" (2009), *Forbes*, http://forbes.com; Arthur B. Kennickell, "Ponds and Streams: Wealth and Income in the U.S, 1989 to 2007," Board of Governors of the Federal Reserve System (United States), Finance and Economics Discussion Series, Working Paper, Number 2009-13 (2009), 55, 63.
39. Ajay Kapur, Niall Macleod, and Narendra Singh, "Plutonomy: Buying Luxury, Explaining Global Imbalances," Citigroup Research, October 16, 2005, http://www.scribd.com, and "Revisiting Plutonomy: The Rich Get

Richer," Citigroup Research, March 5, 2006; http://www.scribd.com; Robert Frank, "Plutonomics," *Wall Street Journal*, January 8, 2007.

40. Gert Spaargaren, "Sustainable Consumption," in *The Ecological Modernisation Reader*, ed. Arthur P.J. Mol, David Sonnenfeld, and Gert Spaargaren (London: Routledge, 2009), 322.
41. Marx, *Capital*, vol. 1, 125.
42. Marx, *Grundrisse* (London: Penguin, 1973), 90–94, 99–100.
43. The concept of "consumer," as a way of designating the main economic role of individuals in society, and of the end-point of the economic process, is a distortion, as Raymond Williams and Michael Dawson have emphasized, imparting the ideal role that individuals are supposed to serve from the standpoint of a monopoly capitalist society, where the realization of surplus value (profit) is always a problem because of the lack of effective demand. It neither describes how people think of themselves, or adequately captures the process of the satisfaction of material needs. Although consumption itself is a term that needs to be subjected to critique, the idealization of the consumer constitutes a deeper fetishism. See Raymond Williams, *Problems in Materialism and Culture* (London: Verso, 1980), 187–91; Dawson, *The Consumer Trap*, 5–6.
44. Fredric Jameson, *Valences of the Dialectic* (London: Verso, 2009), 266.
45. Alfred North Whitehead, *Science and the Modern World* (New York: Free Press, 1925), 51.
46. Karl Marx and Frederick Engels, *Collected Works* (New York: International Publishers, 1975), vol. 5, 518.
47. Williams, *Problems in Materialism and Culture*, 185.
48. Thorstein Veblen, *Absentee Ownership and Business Ownership in Modern Times* (New York: Augustus M. Kelley, 1964), 284–325; Paul A. Baran and Paul M. Sweezy, *Monopoly Capital* (New York: Monthly Review Press, 1966), 131–38. See also Allan Schnaiberg, *The Environment: From Surplus to Scarcity* (New York: Oxford University Press, 1980), 157–204.
49. McChesney et. al., "The Sales Effort and Monopoly Capital," 6.
50. Juliet B. Schor, *Plenitude* (New York: Penguin, 2010), 41.
51. Epicurus, *The Epicurus Reader* (Indianapolis: Hackett Publishing, 1994), 33. Translation follows David Konstan, *A Life Worthy of the Gods: The Materialist Psychology of Epicurus* (Las Vegas: Parmenides, 2008), 47. In the latter work, Konstan explains that in Epicurean psychology the search for "endless accumulation" is motivated by the fear of death and the search for immortality in a world where poverty appears to be "death's anteroom" (xii–xiv, 44–47).
52. For Hegel, a "bad infinity" was the notion of a straight line, extending infinitely in both directions, and thus without limits—the epitome of what is now seen as an unecological view. A true infinity, in contrast, was a circle,

and thus more compatible with what we now see as an ecological worldview. See Michael Inwood, *A Hegel Dictionary* (Oxford: Blackwell, 1992), 141.

53. Morris, *News from Nowhere and Selected Writings and Designs*, 121–22; John Bellamy Foster, "William Morris's Letters on Epping Forest: An Introduction," *Organization & Environment* 11/1 (March 1998): 90–92.
54. Dawson, *The Consumer Trap*, 132–54. As Dawson indicated on the opening page of his book, the United States spent over $1 trillion on marketing in 1992 at a time when the U.S. economy was much smaller than today.
55. István Mészáros, *Beyond Capital* (New York: Monthly Review Press, 1995), 739–70.
56. Aldo Leopold, *The Sand County Almanac* (New York: Oxford University Press, 1949), viii.
57. On this conception of sustainable development, see Paul Burkett, "Marx's Concept of Sustainable Human Development," *Monthly Review* 57/5 (October 2005): 34–62; Michael A. Lebowitz, *Beyond Capital* (New York: St. Martin's Press, 1992).
58. Harry Magdoff and Fred Magdoff, "Approaching Socialism," *Monthly Review* 57/3 (July–August 2005): 59–60.
59. The truth, as Herman E. Daly argues, is that "further growth in GNP" (economic growth as currently measured) does not necessarily "make us richer. It may make us poorer." This is because the public costs of the expansion of private riches are seldom included in the calculation since to do this would contradict the nature of capitalism itself. Herman Daly, *Steady-State Economics* (Washington, D.C.: Island Press, 1991), 100. Indeed, much of what is counted as economic growth is actually the wasteful and unsustainable use of human and natural resources (some of which are irreplaceable). It does not contribute therefore to real wealth or human welfare. See Herman E. Daly and John B. Cobb, Jr., *For the Common Good* (Boston: Beacon Press, 1994), 443–92.
60. An economic slowdown has already occurred in the advanced capitalist economies in the form of a deepening tendency toward economic stagnation from the 1970s to the present, reflected in the current deep economic malaise. In some ways this is currently taking pressure off the environment. Yet this needs to be seen as what it is: a failed capitalist growth economy, which is enormously destructive of both human beings and the environment. It signals that capitalism has moved from creative destruction, as Schumpeter called it, to an era of unqualified destructiveness. It therefore makes even more imperative the shift to a new, rational social order.
61. Epicurus, *The Epicurus Reader*, 39. On the way in which capitalism serves to promote private riches by destroying public wealth (the Lauderdale Paradox) see chapter 1 of this book.

62. Schor, *Plenitude*, 4–12, 99–134. Schor avoids any direct structural treatment of capitalism (equating it simply with the market). She thus depicts the economy in orthodox terms as consisting simply of households and firms. She then claims that growth is not an imperative in the contemporary economy. A no-growth economy thus simply becomes, as in green theory in general, a case of individual choice (169–74). In this sense, Schor's analysis fits with the voluntaristic emphasis of the "de-growth movement," which emphasizes "a voluntary transition toward a just, participatory and ecologically sustainable economy." See Marko Ulvila and Jarna Pasanen, *Sustainable Futures* (Finland: Ministry of Foreign Affairs, 2009), 45.
63. István Mészáros, "The Challenge of Sustainable Development and the Culture of Substantive Equality," *Monthly Review* 53/7 (December 2001): 10–19.
64. Antonio Gramsci, *Selections from the Prison Notebooks* (New York: International Publishers, 1971), 137, 366.
65. Frederick Engels, *The Condition of the Working Class in England* (Chicago: Academy Chicago Publishers, 1984); Brett Clark and John Bellamy Foster, "The Environmental Conditions of the Working Class: An Introduction to Selections from Frederick Engels's *Condition of the Working Class in England in 1844*," *Organization & Environment* 19/3 (September 2006): 375–88. On the concept of the environmental proletariat, see chapter 18 of this book.
66. María Fernanda Espinosa, "Climate Crisis: A Symptom of the Development Model of the World Capitalist System," speech delivered to the Panel on Structural Causes of Climate Change, World People's Conference on Climate Change and the Rights of Mother Earth, Cochabamba, Bolivia, April 20, 2010, http://mrzine.monthlyreview.org, translation by Fred Magdoff and Victor Wallis.

CHAPTER SEVENTEEN: THE METABOLISM OF TWENTY-FIRST CENTURY SOCIALISM

This chapter is adapted and revised for this book from the following article: Brett Clark and John Bellamy Foster, "The Dialectic of Social and Ecological Metabolism: Marx, Mészáros, and the Absolute Limits of Capital," *Socialism and Democracy* 24/2 (2010): 124–38.

1. Karl Marx, *Capital*, vol. 1 (New York: Vintage, 1976), 283.
2. Georg Lukács, *A Defense of "History and Class Consciousness": Tailism and the Dialectic* (London: Verso, 2000), 96. See also John Bellamy Foster, "The Dialectics of Nature and Marxist Ecology," in *Dialectics for the New Century*, ed. Bertell Ollman and Tony Smith (New York: Palgrave Macmillan, 2008), 50–82.

3. See the following for important discussions and applications of metabolic analysis: Paul Burkett, *Marx and Nature* (New York: St. Martin's Press, 1999); Brett Clark and Richard York, "Carbon Metabolism: Global Capitalism, Climate Change, and the Biospheric Rift," *Theory and Society* 34 (2005): 391–428; Rebecca Clausen and Brett Clark, "The Metabolic Rift and Marine Ecology: An Analysis of the Oceanic Crisis within Capitalist Production," *Organization & Environment* 18 (2005): 422–44; John Bellamy Foster, "Marx's Theory of Metabolic Rift: Classical Foundations for Environmental Sociology," *American Journal of Sociology* 105 (1999): 366–405; John Bellamy Foster, *Marx's Ecology* (New York: Monthly Review Press, 2000); Philip Mancus, "Nitrogen Fertilizer Dependency and Its Contradictions: A Theoretical Exploration of Social-Ecological Metabolism," *Rural Sociology* 272 (2007): 269–88; Jason W. Moore, "Environmental Crises and the Metabolic Rift in World-Historical Perspective," *Organization & Environment* 13 (2000): 123–57.
4. István Mészáros, *Beyond Capital* (New York: Monthly Review Press, 1995).
5. Alfred Schmidt argued that Marx was influenced by Jakob Moleschott's conception of metabolism. Although it is true that Marx was aware of Moleschott's usage and attended lectures by him, the term metabolism was already established in the literature before Moleschott used it. For years, Marx studied Liebig's agricultural work closely and in *Capital* he employed the concept in a similar fashion. See Foster, *Marx's Ecology*, 161–62; Foster, "Marx's Theory of Metabolic Rift," 381; Alfred Schmidt, *The Concept of Nature in Marx* (London: New Left Books, 1971).
6. Karl Marx, *Texts on Method* (Oxford: Basil Blackwell, 1975), 209; Marx, *Capital*, vol. 1, 133, 283, 290, 637–38.
7. Erwin Schrödinger, *What Is Life?* (Cambridge: Cambridge University Press, 1945), 70–71.
8. On the role of thermodynamics in Marx's critique of political economy see Paul Burkett and John Bellamy Foster, "Metabolism, Energy, and Entropy in Marx's Critique of Political Economy," *Theory and Society* 35/1 (February 2006): 109–56.
9. Karl Marx, *A Contribution to the Critique of Political Economy* (New York: International Publishers, 1972), 51–52.
10. Marx, *Capital*, vol. 1, 637.
11. Karl Marx, *Grundrisse* (New York: Penguin Books, 1993), 409–10; Karl Marx, *Capital*, vol. 3 (New York: Penguin Books, 1991), 950; Marx, *Capital*, vol. 1, 638.
12. See chapter 15 in this book.
13. Marx, *Capital*, vol. 1, 637–38; Marx, *Capital*, vol. 3, 959.
14. Marx, *Capital*, vol. 3, 911.

15. István Mészáros, *Marx's Theory of Alienation* (London: Merlin Press, 1970), 104–08.
16. István Mészáros, *The Necessity of Social Control* (London: Merlin Press, 1971), later published as a chapter in *Beyond Capital*. Also see, Donella Meadows, Dennis H. Meadows, Jørgen Randers, and William W. Behrens III, *The Limits to Growth: A Report for the Club of Rome's Project on the Predicament of Mankind* (New York: Universe Books, 1972). For Mészáros's critique of the *Limits to Growth* view, see István Mészáros, *The Challenge and Burden of Historical Time* (New York: Monthly Review Press, 2008), 275–78.
17. Mészáros, *Beyond Capital*, 874–75.
18. Chávez first called Mészáros "Pathfinder" (*Señalador de caminos*)—referring to his role in illuminating the transition to socialism—in an inscription that he wrote in a copy of Simón Rodríguez's *Collected Works*, which he gave to Mészáros at a dinner in the Miraflores Palace on September 10, 2001. On the same occasion they discussed Mészáros's *Beyond Capital*. Chávez exhibited the copious notes he had made in his copy.
19. István Mészáros, "The Challenge of Sustainable Development and the Culture of Substantive Equality," *Monthly Review* 53/7 (December 2001): 10–19.
20. Mészáros, *Beyond Capital*, 41.
21. Ibid., 44, 170–71; Marx, *Capital*, vol. 1, 342.
22. Mészáros, *Beyond Capital*, 45–46; Mészáros, *The Challenge and Burden of Historical Time*; Mészáros, *The Structural Crisis of Capital* (New York: Monthly Review Press, 2009). See also Michael A. Lebowitz, *Build It Now* (New York: Monthly Review Press, 2006); as well as chapter 1 in this book.
23. On the barriers/boundaries dialectic, which Marx inherited from Hegel and applied to capital, and which Mészáros developed further, see chapter 13 in this book.
24. Mészáros, *The Challenge and Burden of Historical Time*, 99.
25. Mészáros, *Beyond Capital*, 893–94 (italics in all quotations are in the original).
26. Mészáros, *Challenge and Burden of Historical Time*, 98–100; Mészáros, *Beyond Capital*, 566–79; Joseph A. Schumpeter, *Capitalism, Socialism and Democracy* (New York: Harper and Row, 1942), 82–83.
27. Mészáros, *Beyond Capital*, 876–77, 881.
28. Ibid., 142.
29. World Wildlife Fund, *Living Planet Report 2008*, http://assets. panda.org, 1–4; Benjamin S. Halpern et al., "A Global Map of Human Impact on Marine Ecosystems," *Science* 319 (2008): 948–52; Gerardo Ceballos and Paul R. Ehrlich, "Mammal Population Losses and the Extinction Crisis," *Science* 296 (2002): 904–7; United Nations Environment Programme,

Global Environment Outlook 1 (Sterling, Va.: Earthscan, 1997); James Hansen, "Tipping Point," in *State of the Wild 2008–2009*, ed. Eva Fearn (Washington, D.C.: Island Press, 2008), 6–15; Intergovernmental Panel on Climate Change, *Climate Change 2007: The Physical Science Basis* (Cambridge: Cambridge University Press, 2007); Marah J. Hardt and Carl Safina, "Threatening Ocean Life from the Inside Out," *Scientific American* 303/2 (August 2010): 66–73.

30. Mészáros, *Beyond Capital*, 47, 173–74.
31. Ibid., 44.
32. See chapter 3 in this book.
33. Mészáros, *Challenge and Burden*, 149, 252; Mészáros, *Socialism or Barbarism* (New York: Monthly Review Press, 2001), 97–107.
34. Mészáros, *Beyond Capital*, 149, 187–92.
35. Ariel Salleh, "From Metabolic Rift to 'Metabolic Value': Reflections on Environmental Sociology and the Alternative Globalization Movement," *Organization & Environment* 23/2 (2010): 205–19; Ariel Salleh, ed., *Eco-Sufficiency and Global Justice* (New York: Pluto Press, 2009). See also Rebecca Clausen, "Time to Pay the Piper," *Monthly Review* 62/2 (June 2010): 51–55.
36. Michael A. Lebowitz, *The Socialist Alternative* (New York: Monthly Review Press, 2010), 23–25, 80–81; Mészáros, *Beyond Capital*, 823.
37. Mészáros, *The Structural Crisis of Capital*, 140.
38. Mészáros, *Beyond Capital*; Gretchen C. Daly, ed., *Nature's Services* (Washington D.C.: Island Press, 1997); Marx, *Capital*, vol. 3, 283, 949, 959.
39. Bolívar's conception of equality as the "law of laws" is the defining principle of the Bolivarian Revolution as articulated by Mészáros in particular. On this see John Bellamy Foster, "Foreword" to Marta Harnecker, "Latin America and Twenty-First Century Socialism," *Monthly Review* 62/3 (July–August 2010): ix–xiv; Simón Bolívar, *Selected Works* (New York: Colonial Press, 1951), vol. 2, 603.
40. Michael Lebowitz, "The Path to Human Development," *Monthly Review* 60/9 (February 2009): 41–63.
41. See John Bellamy Foster, *The Ecological Revolution* (New York: Monthly Review Press, 2009), 32–35.
42. Marx, *Capital*, vol. 3, 911; see also Paul Burkett, "Marx's Vision of Sustainable Human Development," *Monthly Review* 57/5 (2005): 34–62.
43. Michael Lebowitz, "New Wings for Socialism," *Monthly Review* 58/11 (2007): 34–41; Michael Lebowitz, "An Alternative Worth Struggling For," *Monthly Review* 60/5 (2008): 20–21; Lebowitz, *Build It Now*; Prensa Latina, "Chávez Stresses the Importance of Getting Rid of the Oil Rentier Model in Venezuela," MRzine, http://mrzine.org (January 11, 2010).

44. Esmé McAvoy, "The World's First Really Green Oil Deal," *Independent* (UK), August 8, 2010; María Fernanda Espinosa, "Climate Crisis: A Symptom of the Development Model of the World Capitalist System," speech delivered to the Panel on Structural Causes of Climate Change, World People's Conference on Climate Change and the Rights of Mother Earth, Cochabamba, Bolivia, April 20, 2010, http://mrzine.monthlyreview.org, translation by Fred Magdoff and Victor Wallis; Alberto Acosta et. al., "Leave Crude Oil in the Ground or the Search for Paradise Lost," *Revista Polis* 23 (2009), http://www.revistapolis.cl/polis%20final/english/23e/acosta.html.
45. Christina Schiavoni and William Camacaro, "The Venezuelan Effort to Build a New Food and Agriculture System," *Monthly Review* 61/3 (2009): 129–41.
46. http://pwccc.wordpress.com/2010/04/28/peoples-agreement. See also "Notes from the Editors," *Monthly Review*, June 2010.
47. Mészáros, *Challenge and Burden*, 292.
48. Mészáros, *Beyond Capital*, 147.
49. Mészáros, *Challenge and Burden*, 278.

CHAPTER EIGHTEEN: WHY ECOLOGICAL REVOLUTION?

This chapter is adapted and revised for this book from the following article: John Bellamy Foster, "Why Ecological Revolution?" *Monthly Review* 61/8 (2010): 1–18.

1. On the long-term aspects of the current financial-economic crisis, see John Bellamy Foster and Fred Magdoff, *The Great Financial Crisis* (New York: Monthly Review Press, 2009).
2. James E. Hansen, "Strategies to Address Global Warming" (July 13, 2009), http//www.columbia.edu.
3. Ibid.; "Seas Grow Less Effective at Absorbing Emissions," *New York Times*, November 19, 2009; S. Khatiwala. F. Primeau and T. Hall, "Reconstruction of the History of Anthropogenic CO_2 Concentrations in the Ocean," *Nature* 462/9 (November 2009), 346–50.
4. Agence France Presse (AFP), "UN Warns of 70 Percent Desertification by 2025," October 4, 2005.
5. Ulka Kelkar and Suruchi Badwal, *South Asian Regional Study on Climate Change Impacts and Adaptation*, UN Human Development Report 2007/2008: Occasional Paper, http://www.undp.org.
6. "Arctic Seas Turn to Acid, Putting Vital Food Chain at Risk," October 4, 2009, http://www. guardian.com.uk.
7. Hansen, "Strategies to Address Global Warming"; AFP, "Top UN Climate Scientist Backs Ambitious CO_2 Cuts," August 25, 2009.

8. See chapter 7 in this book for an extended discussion of the Jevons Paradox.
9. Bill McKibben, "Response," in Tim Flannery, *Now or Never* (New York: Atlantic Monthly Press, 2009), 116; Al Gore, *Our Choice: A Plan to Solve the Climate Crisis* (Emmaus, PA: Rodale, 2009), 327.
10. Friends of the Earth, "Subprime Carbon?" (March 2009), http://www.foe.org, and *A Dangerous Obsession* (November 2009), http://www.foe.co.uk; James E. Hansen, "Worshipping the Temple of Doom" (May 5, 2009), http://www.columbia.edu.
11. Brian Tokar, "Toward Climate Justice: Can We Turn Back from the Abyss?" *Z Magazine* 22/9 (September 2009), http://www.zmag. org; Hansen, "Strategies to Address Global Warming"; Greenpeace, *Business as Usual* (October 20, 2009), http://www.greenpeace.org; "Climate Bill: Senate Democrats Abandon Comprehensive Energy Bill," *Huffington Post*, July, 22, 2010; "Low Targets Goals Dropped: Copenhagen Ends in Failure," *Guardian*, December 19, 2009.
12. James Lovelock, *The Revenge of Gaia* (New York: Basic Books, 2006), and *The Vanishing Face of Gaia* (New York: Basic Books, 2009), 139–58; Gore, *Our Choice*, 314–15. Hansen, it should be noted, also places hope in the development of fourth generation nuclear power as part of the solution. See James Hansen, *Storms of My Grandchildren* (New York: Bloomsbury USA, 2009), 194–204.
13. Gore, *Our Choice*, 303, 320, 327, 330–32, 346.
14. Karl Marx, *Capital*, vol. 1 (London: Penguin 1976), 247–80. On how Marx's M–C–M′ formula serves to define the "regime of capital," see Robert Heilbroner, *The Nature and Logic of Capitalism* (New York: W. W. Norton, 1985), 33–77.
15. Karl Marx, *Grundrisse* (London: Penguin, 1973), 334–35, 409–10, and *Capital*, 1:742. See also chapter 13 in this book.
16. Marx, *Capital*, vol. 1, 552–53.
17. For an extended discussion of the Lauderdale Paradox, see chapter 1 in this book.
18. Karl Marx, *Capital*, vol. 3 (London: Penguin, 1981), 949, *Critique of the Gotha Programme* (New York: International Publishers, 1938), 3, 15.
19. John Maynard Keynes, "National Self-Sufficiency," in *Collected Writings* (London: Macmillan/Cambridge University Press, 1982), vol. 21, 241–42.
20. Karl Marx, *Capital*, vol. 1, 636–39, *Capital*, vol. 3, 948–50, and *Capital*, vol. 2 (London: Penguin 1978), 322; Foster, *The Ecological Revolution*, 161–200.
21. See chapter 13 in this book.
22. Marx, *Capital*, vol. 1, 914–26.
23. Ibid., 1:381.

24. Ibid., 1:496; Karl Marx and Frederick Engels, *Collected Works* (New York: International Publishers, 1975), vol. 33, 400; Gore, *Our Choice*, 365.
25. James Hansen, Makiko Sato, Pushker Kharecha, David Beerling, Valerie Masson-Delmotte, Mark Pagani, Maureen Raymo, Dana L. Royer, and James C. Zachos, "Target Atmospheric CO_2: Where Should Humanity Aim?" *Open Atmospheric Science Journal* 2 (2008): 217–31; James E. Hansen, "Response to Dr. Martin Parkinson, Secretary of the Australian Department of Climate Change" (May 4, 2009), http://www.columbia.edu; Hansen, "Strategies to Address Global Warming" and "Worshipping the Temple of Doom"; Frank Ackerman, Elizabeth A. Stanton, Stephen J. DeCanio, Eban Goodstein, Richard B. Howarth, Richard B. Norgaard, Catherine S. Norman, and Kristen A. Sheeran, "The Economics of 350," October 2009, 3–4 (document available at http://www.e3network.org).
26. James E. Hansen, "The Sword of Damocles" (February 15, 2009), "Coal River Mountain Action" (June 25, 2009), and "I Just Had a Baby, at Age 68" (November 6, 2009), http://www.columbia.edu; Ken Ward, "The Night I Slept with Jim Hansen" (November 11, 2009), http://www.grist.org.
27. Tom Athanasiou and Paul Baer, *Dead Heat* (New York: Seven Stories Press, 2002).
28. John Bellamy Foster, Hannah Holleman, and Robert W. McChesney, "The U.S. Imperial Triangle and Military Spending," *Monthly Review* 60/5 (October 2008), 9–13. The Bamako Appeal can be found in Samir Amin, *The World We Wish to See* (New York: Monthly Review Press, 2008), 107–34.
29. An important source in understanding Cuban developments is the film *The Power of Community: How Cuba Survived Peak Oil*, http://www.powerof-community.org/cm/index.php. On Venezuela see Christina Schiavoni and William Camacaro, "The Venezuelan Effort to Build a New Food and Agriculture System," *Monthly Review* 61/3 (July–August 2009): 129–41. On Ecuador see Elisa Dennis, "Keep It in the Ground," *Dollars & Sense*, July–August 2010, http://dollarsandsense.org.
30. Karl Marx and Frederick Engels, *The Holy Family* (Moscow: Foreign Languages Publishing House, 1956), 52. The translation follows Paul M. Sweezy, *Modern Capitalism and Other Essays* (New York: Monthly Review Press, 1972), 149.
31. Sweezy, *Modern Capitalism*, 164.
32. John Bellamy Foster, "*The Vulnerable Planet* Fifteen Years Later," *Monthly Review* 54/7 (December 2009): 17–19.
33. On the climate justice movement see Tokar, "Toward Climate Justice."
34. Marx, *Capital*, vol. 3, 959.
35. On the elementary triangles of socialism and ecology see Foster, *The*

Ecological Revolution, 32–35. The failure of Soviet-type societies to conform to these elementary triangles goes a long way toward explaining their decline and fall, despite their socialist pretensions. See John Bellamy Foster, *The Vulnerable Planet* (New York: Monthly Review Press, 1999), 96–101.

36. Lewis Mumford, *The Condition of Man* (New York: Harcourt Brace Jovanovich, 1973), 411; Marx, *Critique of the Gotha Programme*, 10: John Stuart Mill, *Principles of Political Economy* (New York: Longmans, Green and Co., 1904), 453–55.
37. Marx, *Capital*, vol. 3, 911, 959.
38. Paul Burkett, "Marx's Vision of Sustainable Human Development," *Monthly Review* 57/5 (October 2005): 34–62.

INDEX